Biofluid Mechanics

Principles and Applications

Biofluid Mechanics

Principles and Applications

Ali Ostadfar, PhD

Simon Fraser University

ELSEVIER

AMSTERDAM • BOSTON • HEIDELBERG • LONDON
NEW YORK • OXFORD • PARIS • SAN DIEGO
SAN FRANCISCO • SINGAPORE • SYDNEY • TOKYO

Academic Press is an imprint of Elsevier

AP

Academic Press is an imprint of Elsevier
125 London Wall, London EC2Y 5AS, UK
525 B Street, Suite 1800, San Diego, CA 92101-4495, USA
50 Hampshire Street, 5th Floor, Cambridge, MA 02139, USA
The Boulevard, Langford Lane, Kidlington, Oxford OX5 1GB, UK

Notices

Knowledge and best practice in this field are constantly changing. As new research and experience broaden our understanding, changes in research methods, professional practices, or medical treatment may become necessary.

Practitioners and researchers must always rely on their own experience and knowledge in evaluating and using any information, methods, compounds, or experiments described herein. In using such information or methods they should be mindful of their own safety and the safety of others, including parties for whom they have a professional responsibility.

To the fullest extent of the law, neither the Publisher nor the authors, contributors, or editors, assume any liability for any injury and/or damage to persons or property as a matter of products liability, negligence or otherwise, or from any use or operation of any methods, products, instructions, or ideas contained in the material herein.

British Library Cataloguing-in-Publication Data
A catalogue record for this book is available from the British Library.

Library of Congress Cataloging-in-Publication Data
A catalog record for this book is available from the Library of Congress.

ISBN: 978-0-12-802408-9

For Information on all Academic Press publications
visit our website at https://www.elsevier.com/

Publisher: Joe Hayton
Acquisition Editor: Fiona Geraghty
Editorial Project Manager: Lindsay Lawrence
Production Project Manager: Susan Li
Designer: Matthew Limbert

Typeset by MPS Limited, Chennai, India

To my wonderful family

my Parents, Reza and Parvin

my wife, Mojgan

and

my son, Artin

Contents

Preface

Biomedical engineering is a multidisciplinary and a very young field among other engineering fields. Biofluid mechanics is a subsystem of biomedical engineering that is very significant in teaching, research and clinical practice.

The aim of this textbook is to introduce biofluid mechanics to the students, researchers, clinicians and other users with engineering, medical, or clinical backgrounds. Several informative books have been published regarding biofluid mechanics. However, this book is the first one in this field that covers most of the human organs with an emphasis on the micro blood circulation, macro blood circulation, and other biofluid flows inside the human organs, and also contains comprehensive information about blood flow and measurements in related medical devices. With a consideration of novel technologies, the final chapter explains tissue engineering and its role and application in biofluid mechanics.

The textbook begins with the principles of fluid and biofluid mechanics and it is followed by micro- and macrocirculation systems, biofluid dynamics in human organs and artificial, assistive and implantable devices, blood damage, hemolysis and blood storage and transportation, blood reaction and the immune system, biocompatibility and biomaterial, realtime measurement techniques for blood flows and ends with a quick review of the application of tissue engineering in the cardiovascular system. These topics make this book a novel and unique reference in the field of biomedical engineering.

The topics and mathematical level of this book have been designed in a way that all levels of students in related fields will find it to be a valuable source of knowledge and learning. The readers may also benefit from the learning objectives and summaries that have been provided in each chapter; these are extremely helpful for an efficient and quick review both before and after the tutorial.

Despite great precautions, I expect that some mistakes and errors will appear in this book. I welcome suggestions from all readers and colleagues to improve subsequent editions.

Ali Ostadfar PhD

Chapter 1

Fluid Mechanics and Biofluids Principles

LEARNING OBJECTIVES

- History of body fluid and biofluid mechanics.
- Review of biofluid mechanics applications.
- Definition of dimensions and units in engineering.
- Role of dimensional analysis in engineering equations.
- Characterization of basic formulas in fluid mechanics, including density, velocity, pressure, volumetric flow rate and mass flow rate.

1.1 INTRODUCTION

1.1.1 History of Body Fluids

Blood has been known to have been used by humans as an applied fluid for thousands of years. Stone Age painters used several different combinations of materials to make their art. Blood was one of these materials. The cave paintings (40,000–8,000 BC) in Asia, Europe, Africa and Australia prove the use of blood as a paint material, and probably before that they realized the importance of blood for the body during hunting and fighting.

In Chinese culture, Yellow Emperor, Huang Ti (2697–2597 BC) wrote a manuscript about circulation. Huang Ti is famous for the writing of "*internal classics*," fundamental theories of Chinese medicine. Although most Chinese researchers believe that the book was written in warring periods (475–221 BC) by anonymous authors [1,2].

Egyptians began their blood practices around 3000 years ago [3]. They used bloodletting as a healing method, and then other nations (eg, Greeks, Romans, Persians, Arabs and Asians) continued and modified the Egyptian's method.

Ancient physicians in Persian empires (550 BC– AD 637) knew of blood circulation. According to *Bundahishn* book (meaning primal creation),

Biofluid Mechanics. DOI: http://dx.doi.org/10.1016/B978-0-12-802408-9.00001-6

blood circulation is similar to water and rivers; after food digestion, blood feeds the body organs. The book presented a preliminary theory of blood circulation among the pulmonary system, heart, liver and other organs [4].

In western culture, many ideas were documented. Hippocrates (~400 BC) believed that four basic humors in humans are blood, phlegm, black bile and yellow bile and each humor was centered in a particular organ—brain, lung, spleen and gall bladder [3]. After Hippocrates, Aristotle (384–322 BC) wrote that the heart was the center of the blood vessels. Praxagoras, a Greek physician, recognized the difference between arteries (air carriers, as he thought) and veins (blood carriers) [1].

Andreas Vasalius (1514–64), the father of modern dissection, published a reference textbook "*On the Fabric of the Human Body*" in human anatomy [5]. To explain the cardiovascular system, he drew several charts of the anatomy of vessels and heart.

Without any doubt, William Harvey (1578–1657) was a pioneer in the biofluid field in the western world. He was an English physician who earned his degree in medicine from Padua in Italy. He published a book about blood circulation and explained the role of the heart as a blood pump in the circulation system. He proved that blood is recirculated and not consumed. He proved, by calculation, that the blood pumped from the heart in just a few minutes exceeded the complete volume of blood contained in the body [1].

Poiseuille (1797–1869) was a French physiologist and physician. He studied physics and mathematics at Ecole Polythechnique in Paris. Later, he became interested in the blood flow in human narrow vessels (tubes). In 1838, he derived Poiseuille's law which describes the relationship between nonturbulent flow and pressure gradient in tubes with constant cross-section.

In 1900, Karl Landsteiner (1868–1943), an Austrian physician and biologist, classified the blood to several main groups. And in 1930, he received the Nobel Prize in physiology or medicine for differentiating the blood groups. The medical world recognizes him as the father of transfusion medicine and Karl was awarded a Lasker Award in 1946 posthumously. Also, Dr Landsteiner and his two colleagues; Constantin Levaditi and Erwin Popper discovered the Polio virus.

Oswald Hope Robertson (1886–1966) was an English-American medical scientist who pioneered the idea of the blood bank in 1917 during the First World War in France. During the war, he was working for the US army medical corps and using needles he collected blood into glass bottles containing citrate and dextrose solution. He stored these bottles for up to 26 days and transported it to medical camps for wounded soldiers and his idea was proved by this storage and transfusion method. He published his technique in the British Medical Journal in 1918 and because of his significant contribution in the war the British Government awarded him the Distinguished Service Order [6].

From the beginning of the 20th century all sciences, especially biofluids science, made significant advances and it happened because of pioneering researchers who worked to extend the borders of knowledge for human life. Researchers such as:

- John R. Womersley (Pulsatile flow),
- Fahraeus and Lindqvist (Blood flow aspect in narrow vessels),
- Willen Einthoven (Electrocardiograph-ECG),
- Werner Frossman (Catheterization),
- George Haas, Willem Kolff and Nils Alwall (Hemodialysis machine),
- Leland Clark (Biosensors),
- Charles Hufnagel (Artificial heart valves),
- John Heysham Gibbon (Heart-lung machine),
- Andreas Gruntzig (Balloon dilation for stenosed artery),
- Robert Jarvic and Willem Kloff (Total artificial heart)
- and other pioneering scientists at universities and institutions around the world.

1.1.2 Scope of Biofluid Mechanics

Fluid mechanics is a branch of mechanics that studies fluids and the forces on them. Fluid mechanics examines fluids in two subsystems: static and dynamic. Fluids, and especially air and water, have a major role in the life of creatures and ~65% of our body is composed of water. Blood transports fuel and energy to tissues to maintain their usual activities. An understanding of biofluid mechanics and its aspects helps us to find or predict the solutions for disorders or diseases. To realize the role of biofluids research, let us imagine a hospital without equipment and tools related to biofluids, such as syringes, serums, surgery equipment, hemodialysis machines, blood pressure meters, ECG, stents, angiography and balloon therapy equipment, artificial lung and oxygenation systems, defibrillators, biosensors and pulse oximeters, ultrasound blood flow detectors, blood test equipment, etc. Without these instruments and tools, our hospitals would be a building with several rooms with minimum health care and a huge rate of death just like hospitals in the middle ages.

This imagining is the best explanation for the advantage of biofluid mechanics, goals and scope.

The World Health Organization (WHO) estimated around 17 million people die of cardiovascular diseases, mostly heart attacks and strokes, every year [7]. Also, the number of people with end stage of renal disease who receive renal replacement therapy is more than 1.4 million with an increasing rate of 8% per year [8]. This enormous rate of death and therapy for only two diseases demonstrates the importance of biofluid mechanics research.

Micro electro mechanical system and nano electro mechanical systems open new horizons in biofluid mechanics especially in in vivo conditions. These techniques help researchers to make micro–nano sensors or actuators as movable or implantable assistive devices which are popular in research among famous companies and universities.

To create new devices in the biofluids field, we need to understand biofluids phenomena. Although the variety of biofluids is limited to a few fluids (~33 types) in humans, they are more complicated than other fluids in nature. Chapter 1.3 clearly will discuss the aspects of body fluids and blood rheology.

1.1.3 Dimensions and Units

In engineering fields and especially fluid mechanics, dimensions and units are a fundamental tool to explain the characterization of phenomena. A *dimension* is a measure of a physical variable by fundamental quantities without numerical value, such as distance, time, mass, temperature, etc., but a *unit* is a particular way to assign a measurement with numerical value to the dimension.

Examples:

- Length (L) is a *dimension*, but it is measured in *units* of meters (m) or feet (ft).
- Time (T) is a *dimension*, but it is measured in *units* of seconds (s).
- Mass (M) is a *dimension*, but it is measured in *units* of kilograms (kg) or pounds (lb).

In biofluid mechanics, the main dimensions are length, mass, time and temperature. Two units are in use today, the international or metric (SI) system and the imperial (British) system. The SI is very common and most of the countries in the world use the SI system. Unfortunately the difference between SI and imperial leads to calculation errors for both practicing engineers and students due to missing or improper conversion factors. The simple example for difference is in temperature units. In the SI system, the unit of temperature is Kelvin (K) (or centigrade (°C)) and in the imperial system it is Fahrenheit (°F).

Examples:

- Freezing point of water in the SI system (metric) is 0°C (or 273.15 K) but in the imperial system is 32°F.
- Boiling point of water in the SI system (metric) is 100°C (or 373.15 K) but in the imperial system is 212°F.

This book uses the international system (SI) of measurement for units and equations. Appendix A explains more about conversion factors between measurement and unit groups.

1.1.3.1 ***Dimensional Analysis***

Dimensional analysis is an analytical method between different physical quantities and units of measure to understand their relationship by using fundamental dimensions, such as Mass (M), Time (T), Length (L) and Temperature (θ).

In simple words, any physical equation must have the same dimension on both sides (left and right). This application is called dimensional analysis. If you want to derive an equation, the best method to check any correction of the equation is dimension analysis. The following examples clarify this definition.

Examples:

- The unit of speed or velocity is meters per second (m/s) and its dimensional analysis is

$$\{V\} = \{\text{Length } (L)/\text{Time } (T)\} = L/T = LT^{-1}$$

  So dimension analysis of velocity is LT^{-1}
- The unit of acceleration is meters per second squared (m/s^2) and its dimensional analysis is

$$\{a\} = \{\text{velocity } (V)/\text{Time } (T)\} = \frac{LT^{-1}}{T} = LT^{-2}$$

- The unit of force is Newton (in SI) and it is equal to mass (M) multiplied by acceleration

$$\{F\} = \{\text{mass} \times \text{acceleration}\} = M \cdot LT^{-2} = MLT^{-2}$$

- Imagine, you derived an equation for force and you correlated the following formula in the end:

$$F = (@ \times \#)/\sqrt{\%}$$

The dimension analysis of right side of the equation *must be* MLT^{-2}. Otherwise, your formula correlation is not correct.

1.1.4 Basic Concepts and Definitions of Fluid Mechanics

From the viewpoint of fluid mechanics, all substances consist of only two forms, solid and fluid. A fluid is a matter that frequently deforms under a shear stress. Any amount of shear stress (even very small) applied to a fluid will result in a motion of the fluid. Two major groups are defined for the fluids: liquids and gases (in a generic state, plasma is a third group of fluids). A liquidis formed from relatively compressed molecules with powerful cohesive forces. A liquid has a tendency to retain its volume and will shape a free surface under a gravitational field [9]. Gases are the opposite to liquids. Gas elements are generally separated from one another and gases have no defined volume or free surface (see Fig. 1.1).

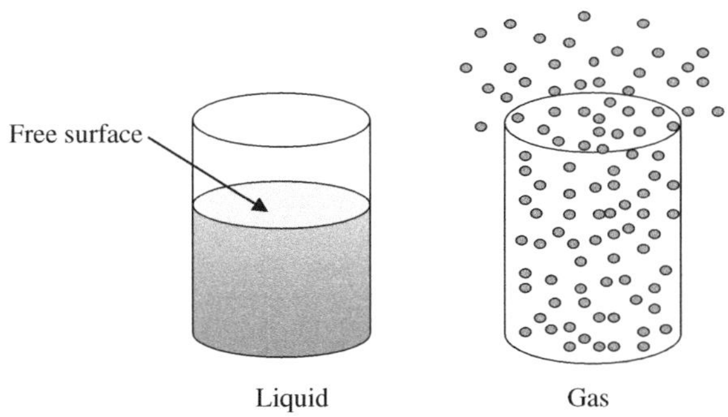

■ **FIGURE 1.1** Liquid container walls control shear stress and the liquid retains its volume. Regarding molecule bonding, Gas expands out of the container and filling available space.

In the following, we explain the basic definitions and equations in fluid mechanics.

- *Density* or volumetric mass density (symbol: ρ—SI unit is kg/m^3) is mass of fluid matter per unit volume,

$$\rho = m/V \tag{1.1}$$

  where m is mass and V is fluid volume.
- *Velocity* (symbol: v—SI unit: m/s) is time rate (t) of position changing (Δx) for a movable object, mathematically velocity is:

$$v = \Delta x/t \tag{1.2}$$

- *Pressure* (symbol: p or P—SI unit: Pascal or Pa) is the ratio of perpendicular force (F) to the unit area (A) of a subject,

$$P = F/A \tag{1.3}$$

Eq. (1.3) is the origin form of pressure. Liquid pressure depends on the liquid density; to derive Eq. (1.3) for hydrostatic pressure, we can replace force with liquid weight instead and rewrite Eq. (1.3) to

$$P = \rho g h \tag{1.4}$$

where ρ is density, g is gravity acceleration and h is height (or depth) of liquid (see Fig. 1.2).

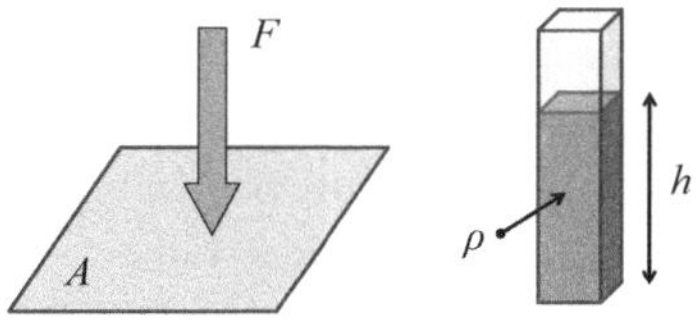

FIGURE 1.2 Left: Schematic of pressure definition by normal force on area unit so $P=F/A$. Right: a liquid tank with density (ρ), height (h) under gravity acceleration (g); $P = (\rho)gh$.

- *Flow rate* or volumetric flow rate (symbol: Q—SI unit: m^3/s) is the fluid volume (flow) which passes through a cross-sectional area of channel per unit of time. Some medicine and biology textbooks use the term of "flow" instead of flow rate, which is popular in mechanical engineering text books. Mathematically the flow rate is

$$Q = vA \tag{1.5}$$

where v is velocity of flow (m/s) and A is cross-section area (m^2) of fluid channel (see Fig. 1.3).

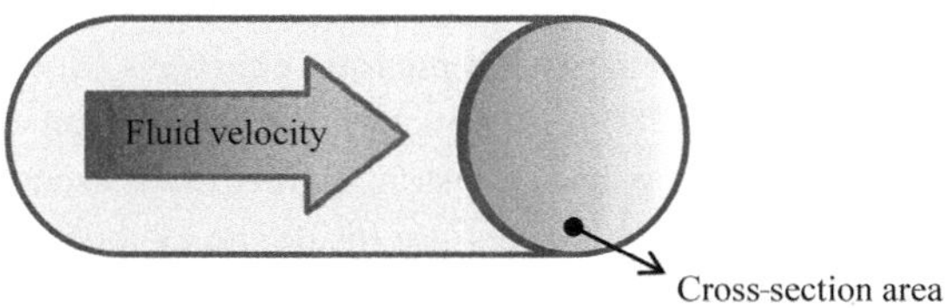

FIGURE 1.3 Schematic of a pipe, flow velocity and cross section area.

- *Mass flow rate* (symbol: $\dot{m}$—SI unit: kg/s) is the fluid mass which passes through cross-sectional area of channel per unit of time. Mathematically the mass flow rate is

$$\dot{m} = \rho Q \tag{1.6}$$

where ρ is density, Q is volumetric flow rate of fluidic channel.

CHAPTER SUMMARY

- Body fluid has been used by humans as an applied fluid for thousands of years.
- Biofluid mechanics is focused on the interaction of physiology and biology matters with fluids mechanics (liquids or gases).
- Biofluid mechanics has a main role in the invention and creation of medical devices, especially in cardiovascular fields.
- A *dimension* is a measure of a physical variable by fundamental quantities without numerical value, but *unit* is a particular way to assign a measurement with numerical value to the dimension.
- Two unit systems are used for measurement, International System (SI) and imperial (British) units.
- Dimensional analysis is a useful tool for determining the units of variables in equations and also is a tool to check equations are correct.
- Fluid mechanics has four essential primary dimension units, mass (M), time (T), length (L) and temperature (θ).
- The basic fluidic terms are:

    Density: $\rho = m/V$

    Velocity: $v = \Delta x/t$

    Pressure: $P = F/A$ and hydrostatic pressure $P = \rho g h$

    Volumetric flow rate: $Q = vA$

    Mass flow rate: $\dot{m} = \rho Q$

PROBLEMS

1. Write out dimensional analysis for density (ρ), Pressure (P) (Eq. (1.3)), volumetric flow rate (Q) and mass flow rate ($\dot{m}$)?
2. Compare dimensional analysis of pressure equations (Eqs. (1.3) and (1.4)) and prove Eq. (1.4) is a correctly derived equation of Eq. (1.3)?
3. A normal force 200 N pushes a piston, cylinder cross-section area is 0.2 m^2, Determine the water pressure inside the cylinder?
4. Determine the pressure at a depth of 115 m of sea?
5. 0.04 m^3/s water flows through a 400 mm tube, later the tube diameter reduces to 200 mm, calculate the average velocities for each tube?

6. By dimensional analysis prove that the Reynolds number (Re) is a dimensionless number,

$$Re = \frac{\rho v d}{\mu},$$

where ρ is density, v is velocity, d is pipe diameter and μ is viscosity $\left(\frac{\text{kg}}{\text{m}\cdot\text{s}}\right)$.

7. Determine the pressure (Pa) at point A, mercury $SG = 13.6$ (see Fig. 1.4);

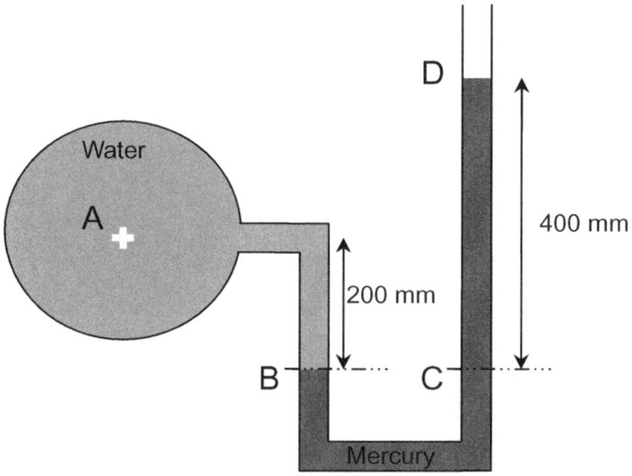

■ **FIGURE 1.4** Pressure meter.

1.2 FUNDAMENTALS OF FLUID MECHANICS

Learning Objectives

- Fluid kinematics and its aspects such as Lagrangian and Eulerian methods, velocity and acceleration fields, flow visualization or flow lines.
- Definition of viscosity.
- Behavior of Newtonian and non-Newtonian fluids and their principles.
- Dimensionless numbers and their role in Biofluid Mechanics.
- Description of steady, unsteady, laminar and turbulent flows.
- Explanation of boundary conditions and no slip boundary conditions.
- Definition of compressible and incompressible flows.
- Stress tensors in fluid element.
- Categorizing of viscoelasticity and viscoplasticity.
- Definition and application of basic equations in fluid mechanics.

- Formulation and explanation of conservation of mass, momentum and energy.
- Formulation of Navier–Stokes, Bernoulli and Hagen–Poiseuille Equations.
- Definition of steady flow along tube.
- Comparison of pulsatile flow in rigid and elastic tubes.
- Definition of resistance, compliance and inertance in fluid mechanics.
- Demonstration of two-phase flows.

1.2.1 Fluid Kinematics

Fluid kinematics collaborates with the motion of fluids without considering the nature of forces and momentums. The geometries of motion are defined by fluid kinematics, which means that the fluid characters such as velocity are a function of position (location) and time in the flow field. In the following we will discuss two important definitions in fluid kinematic: Lagrangian and Eulerian. Both methods are applicable in fluid mechanics but the Eulerian method is more realistic for fluids.

In the Lagrangian description, which is more appropriate with solid mechanics, properties (velocity, pressure, etc.) of each particle are tracked during their motion. As the particle moves in the fluid field the properties of the particle, such as velocity or pressure, etc., may change with time. We can represent these properties as $v_p^{\rightarrow}(t)$, $P_p(t)$, etc. The method of describing the fluid flow by the reporting of particles' data is the Lagrangian description.

To understand the Lagrangian method, we can imagine a snapshot of a particle among particles. At the snapshot time of (t), the particle (p) has defined property of $v^{\rightarrow}$, or $v_p^{\rightarrow}(t)$ (see Fig. 1.5).

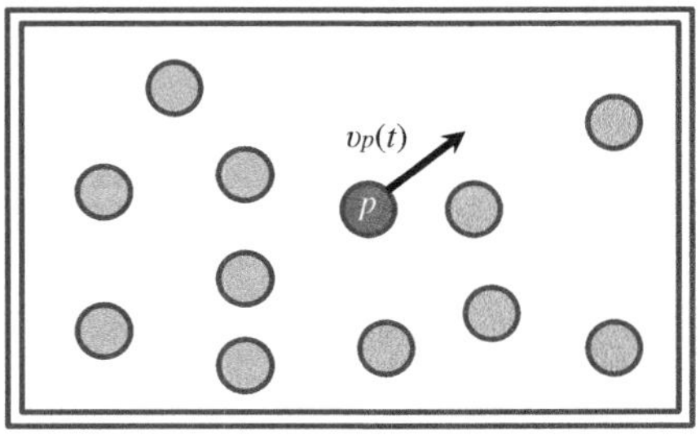

■ **FIGURE 1.5** Lagrangian method description as a snapshot of particle velocity at the time of *t* or *v-p*(*t*).

In the Eulerian description, which is more appropriate for fluids, rather than following all fluid particles we can characterize the flow properties (velocity, pressure, etc.) at every point in a four-dimensional coordinates system (3D space plus time). It means that the flow properties, such as velocity, pressure, can be changed by location and time varying. We can compute the velocity field $v(x, y, z, t)$ or pressure $P(x, y, z, t)$ of the flow model by this field method and its result is not similar to $v_{\vec{p}}(t)$ or $P_p(t)$. One of the features in the Eulerian description of fluid flow is the control volume. The control volume is a finite volume which the fluid flows in or out of. In fluid mechanics, the control volume can be rigid, deformable, fixed (in Eulerian method) or moving depending on the fluidic model (see Fig. 1.6).

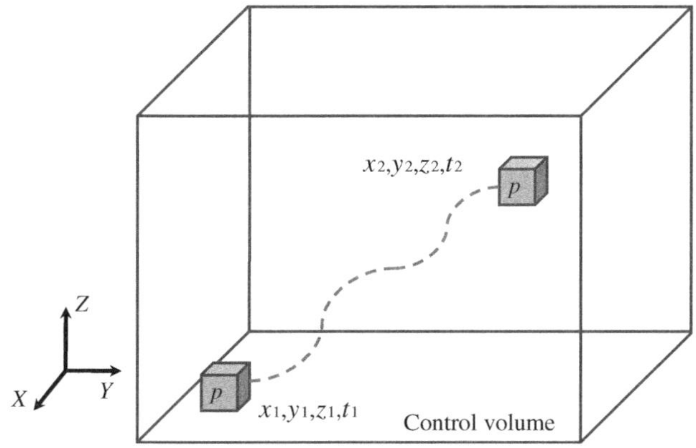

■ **FIGURE 1.6** Eulerian description, trajectory of particle (p) in a control volume.

Velocity field: In the Eulerian method, the velocity is a function of space and time. In Cartesian coordinates, the velocity field is,

$$V = u(x, y, z, t)\,\vec{i} + v(x, y, z, t)\,\vec{j} + w(x, y, z, t)\,\vec{k} \tag{1.7}$$

where (u, v, w) are velocity elements in (x, y, z) coordinates ($V_x = u$, $V_y = v$, $V_z = w$) and time of "t" is a function of these coordinates or $f(t) = (x, y, z)$.

Acceleration field: Regarding acceleration definition $\left(a = \frac{dV}{dt}\right)$, acceleration field in Eulerian method is,

$$a = \frac{dV}{dt} = \frac{du}{dt}\vec{i} + \frac{dv}{dt}\vec{j} + \frac{dw}{dt}\vec{k} = a_x\vec{i} + a_y\vec{j} + a_z\vec{k}$$

$$V_x = \frac{\partial x}{\partial t} = u,\ V_y = \frac{\partial y}{\partial t} = v,\ V_z = \frac{\partial z}{\partial t} = w$$

$$a_x = \frac{\partial u}{\partial t} + u\frac{\partial u}{\partial x} + v\frac{\partial u}{\partial y} + w\frac{\partial u}{\partial z} \tag{1.8}$$

$$a_y = \frac{\partial v}{\partial t} + u\frac{\partial v}{\partial x} + v\frac{\partial v}{\partial y} + w\frac{\partial v}{\partial z}$$

$$a_z = \frac{\partial w}{\partial t} + u\frac{\partial w}{\partial x} + v\frac{\partial w}{\partial y} + w\frac{\partial w}{\partial z}$$

Regarding vector calculus, acceleration is,

$$a = \frac{\partial V}{\partial t} + V\cdot\nabla V \tag{1.9}$$

The first term on the right hand side is *local acceleration* and the second term on right is called *convective acceleration*.

From calculus definitions, we know that the gradient operator (∇, Del or Nabla) is

$$\nabla = \frac{\partial}{\partial x}\vec{i} + \frac{\partial}{\partial y}\vec{j} + \frac{\partial}{\partial z}\vec{k} \tag{1.10}$$

Flow visualization or flow lines: Flow visualization is one of the best methods to learn a flow field when the study of fluid aspects requires advanced mathematic. There are four flow lines for the examination of a flow field: streamlines, pathlines, streaklines and timelines.

Streamlines: A streamline is a curve that is tangential to the velocity vector of fluid flow. They show instantaneous movement of fluid elements through the flow field (see Fig. 1.7).

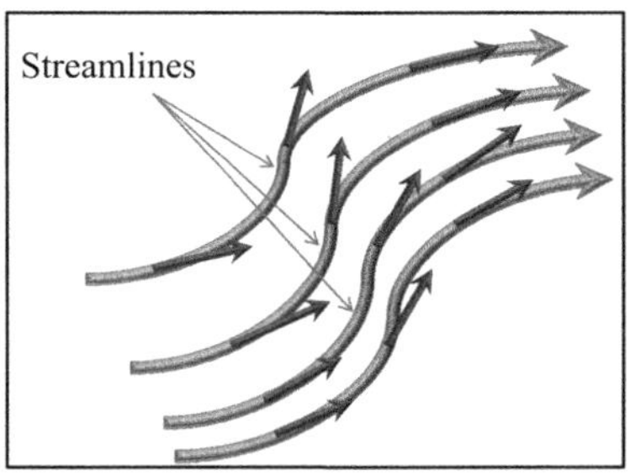

■ **FIGURE 1.7** Schematic of the streamlines in a fluid field.

Pathlines: A pathline is the real path moved by an individual fluid as a time function (see Fig. 1.8).

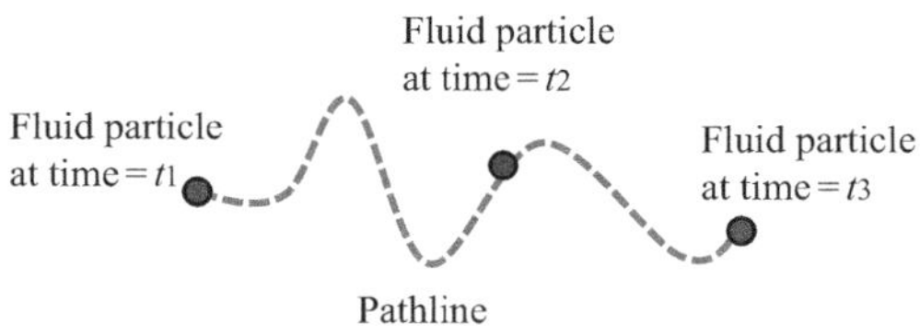

■ **FIGURE 1.8** Schematic of the pathlines for a particle from t_1 to t_3.

Streaklines: A streakline is the locus of fluid particles that have earlier passed sequentially through a spatial (station) point in the flow (see Fig. 1.9).

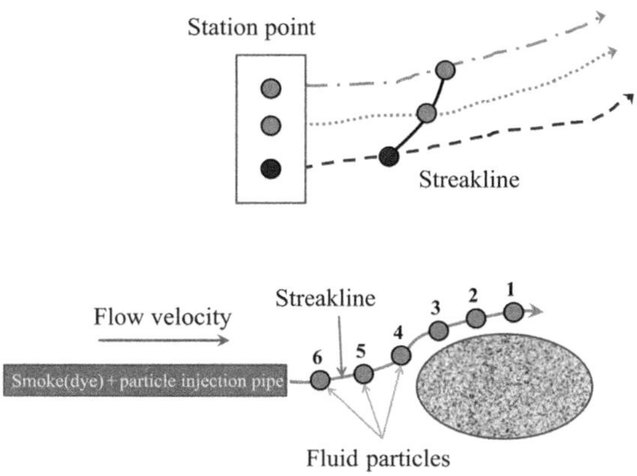

■ **FIGURE 1.9** Schematic of the streaklines for injected particles to the fluid field.

Timelines: A timeline is a set of adjacent fluid particles which were marked at a time instant and a line is formed on these particles and it relocates as the particles move (see Fig. 1.10).

1.2.2 **Viscosity**

In the previous section, we described some of the basic definitions in fluid mechanics, such as pressure, density, flow rate, etc. There are also other essential variables in fluid mechanics. Viscosity is the most important of these variables to characterize the mechanical behavior of fluids.

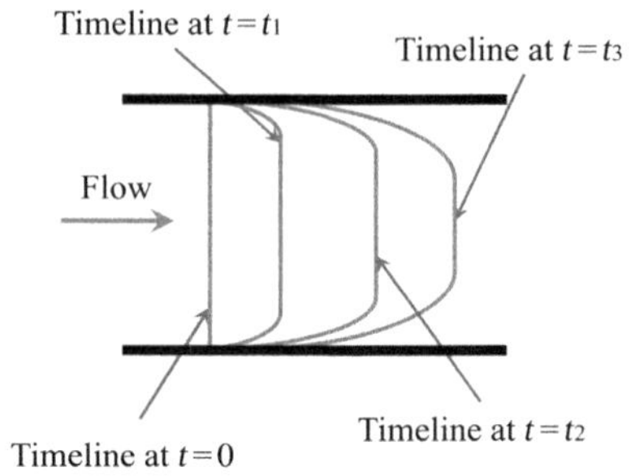

■ **FIGURE 1.10** Schematic of the timelines between $t = 0$ to $t = t_2$.

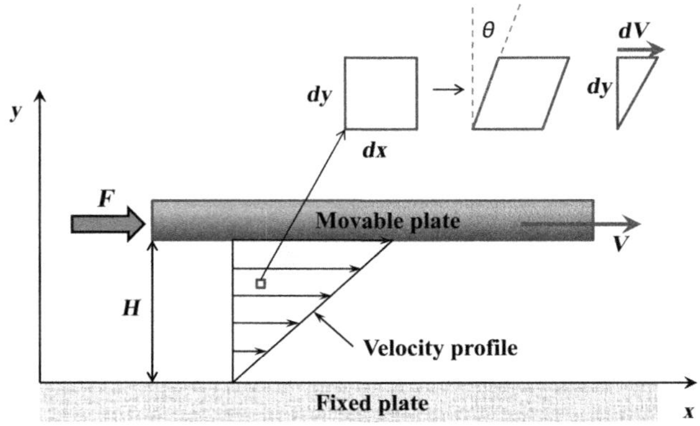

■ **FIGURE 1.11** Shear stress causes a deformation on fluid element.

The simple definition of viscosity is resistance of a fluid to deformation by shear or tensile stress. In other words, viscosity is the internal particles friction of a moving fluid. The ideal fluid (super fluidity) is a fluid with no resistance to shear stress.

Considering Fig. 1.11, we have a liquid droplet between one fixed plate and one movable plate. The movable plate moves by force "F" and constant velocity of "V." The velocity gradient (γ or shear rate) is

$$\text{Velocity gradient } (\text{s}^{-1}) = \gamma = \frac{dV}{dy} = \frac{V}{H} \tag{1.11}$$

The shear stress on a fluid element is

$$\tau = \frac{F}{A} \tag{1.12}$$

There is a relation between shear stress and shear strain rate of fluid element (θ = shear strain angle),

$$\tau \propto \frac{\partial \theta}{\partial t} \tag{1.13}$$

And the shear strain rate has relation to velocity gradient,

$$\frac{d\theta}{dt} = \frac{dV}{dy} \tag{1.14}$$

For linear fluids, a relative constant is defined by the rate of shear stress to shear rate and it is called the viscosity coefficient (μ),

$$\tau = \mu \frac{d\theta}{dt} = \mu \frac{dV}{dy} \tag{1.15}$$

or

$$\mu = \frac{\tau}{\gamma} \tag{1.16}$$

Fig. 1.12 illustrates the general graph of viscosity.

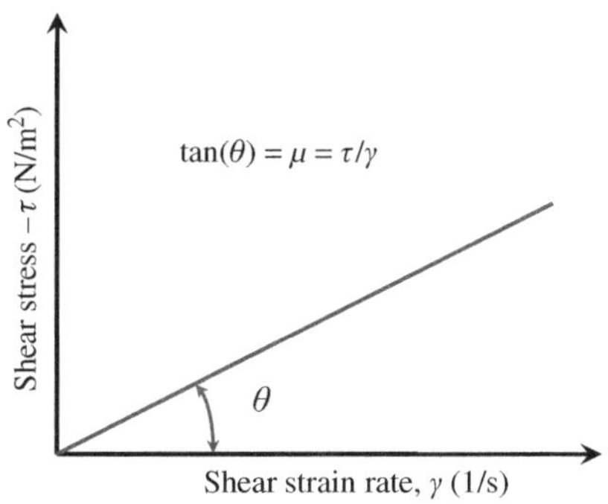

■ **FIGURE 1.12** General graph of viscosity; shear stress (τ) versus rate of shearing strain (γ).

Three categorizations are defined for fluids: the ideal fluid or nonviscous (μ = zero), Newtonian fluids, (μ = constant) and non-Newtonian fluids (μ is changeable).

1.2.3 **Newtonian and Non-Newtonian Fluids**

Sir Isaac Newton proposed the first resistance law and the relation between shear stress and shear strain rate in 1687. The linear fluid which follows Eqs. (1.15) or (1.16) with zero amounts for deviatoric normal

Table 1.1 Examples for Non-Newtonian Substances

Biofluids	Foods	Industry
■ Blood (regarding shear rate) ■ Semen ■ Synovial ■ Saliva	■ Egg white ■ Chocolates ■ Dairy products (Butter, cheese, yogurts) ■ Salad dressings (Ketchup, mayonnaise, etc.) ■ Ice-cream	■ Paints ■ Lubrication oils and greases ■ Soups ■ Pharmaceutical products (Such as: Foams, creams, suspensions) ■ Shampoos and toothpaste ■ Nail polish

stress elements ($\tau_{xx} = \tau_{yy} = \tau_{zz} = 0$) is called a Newtonian fluid (we will explain about stress tensors later). As Fig. 1.12 shows, the Newtonian fluid has a straight line of slope (μ) passing through the origin of $\tau - \gamma$. The viscosity of Newtonian fluids has thermodynamic characteristics and varies with fluidic aspects, such as temperature and pressure [9]. The sensitivity of viscosity to temperature is greater than pressure. For instance, viscosity of air at 20°C is 1.85 E^{-5} kg/ms and at 60°C it is 2.00 E^{-5} kg/ms, but an increase of pressure from 1 to 50 atmospheres will increase the air viscosity only by 10%. Simple examples for Newtonian fluids are water or blood plasma.

In non-Newtonian fluids, a constant viscosity coefficient cannot be defined but this is not the only condition. Some fluids, such as Boger fluids, have constant viscosity but they have been categorized as non-Newtonian fluids because they do not follow zero amounts for deviatoric normal stress elements (in boger fluids $\tau_{ii} \neq 0$) [10].

Table 1.1 illustrates some fluids and materials with non-Newtonian behavior. In fluid mechanics, three fluidic groups are categorized based on non-Newtonian fluids behavior:

1. *Time-independent fluids*: As the name explains, these fluids are not depend on time. It means that the shear rate is a function of stress at that point and at that time. Examples of this behavior are Bingham, Casson, Pseudoplastic and Dilatent fluids.
2. *Time-dependent fluids*: They are complicated fluids and the viscosity depends on the duration of shearing. Examples of this behavior are yogurt, synovial fluid and printer ink.
3. *Viscoelastic fluids*: They are mainly viscous but they have some elastic behaviors, especially after deformation. An example of this fluidic behavior is some lubricants.

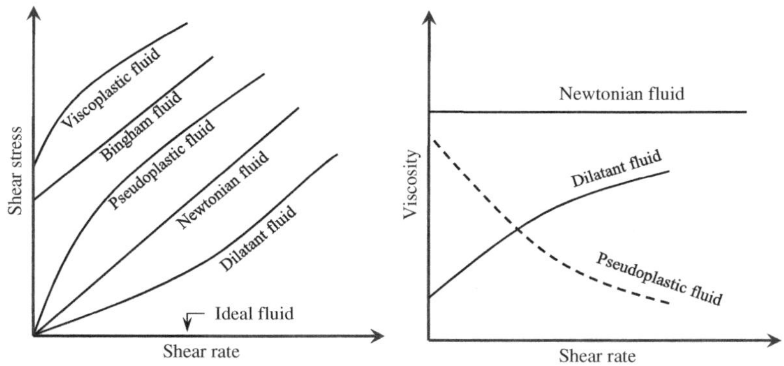

■ **FIGURE 1.13** Left: Types of time-independent fluids, Right: Relations of viscosity-shear rate between Newtonian, Dilatant and Pseudoplastic fluids.

Fig. 1.13 illustrates the types of time-independent non-Newtonian fluids;

- *Dilatant* (or shear thickening), is described by an increase in viscosity with increasing shear rate. An example is the suspension of cornstarch in water.
- *Pseuduplastic* (or shear thinning), is described by a decrease in viscosity with increasing shear rate. Toothpaste is a good example of a shear thinning fluid.
- *Viscoplastic*, is described by the existence of a yield stress which has to be exceeded for the fluid to flow or deform. An example is mayonnaise sauce.
- *Bingham (viscoplastic family member)*, has linear viscosity but a certain amount of yield stress must be applied to the fluid to begin flow or deformation. For instance, tomato ketchup is a bingham fluid.

1.2.4 **Dimensionless Numbers of Biofluid Mechanics**

Dimensionless (pure) numbers represent a property of a physical phenomenon, without any physical units such as distance, time or mass. These numbers are a useful tool to characterize the mechanical behavior of fluids in dynamic conditions. Dimensionless numbers help us to understand correlation between some parameters or forces in the phenomenon and also rescaling of a fluidic experience. Mechanical engineering has more than 40 dimensionless numbers. Some of these numbers are very essential in biofluid mechanics, such as:

- Reynolds number (Re)
- Womersley number (α)
- Strouhal number (St)
- Dean number (De)
- Stokes number (Stk)

Reynolds number (Re), is defined as the ratio of inertial forces to viscose forces when a fluid flows in a channel. Reynolds number is also an assessment number to evaluate flow regimes within a similar fluid. The Reynolds number for a channel (or pipe) is defined as:

$$Re = \frac{\rho V D_h}{\mu} \tag{1.17}$$

where D_h is hydraulic diameter (m), ρ is density (kg/m^3), V is fluid velocity (m/s) and μ is dynamic viscosity (Pa $\cdot$ s).

Womersley number (α), is defined as the ratio of oscillatory inertia to viscose effects for a fluid in pulsatile flow. The Womersley number in the aorta and arteries is large and in arterioles it becomes very small (less than 1). The Womersley number for a channel (or pipe) is defined as:

$$\alpha = \frac{D_h}{2}\sqrt{\frac{\rho(2\pi f)}{\mu}} \tag{1.18}$$

where D_h is hydraulic diameter (m), ρ is density (kg/m^3), f is fluid frequency (Hz) and μ is dynamic viscosity (Pa $\cdot$ s). In the human body, heart beat rate can be considered as frequency.

Strouhal number (St), is a dimensionless number to explain oscillatory flow mechanisms. This number is the ratio of inertial forces as a result of oscillatory flow conditions to inertial forces because of velocity changing between two points of a flow field. The Strouhal number for a channel (or pipe) is defined as:

$$St = \frac{fL}{V} \tag{1.19}$$

where L is characteristic length (for instance hydraulic diameter D_h) (m), f is fluid frequency (Hz) and V is velocity of the fluid.

There is a relation among Womersley, Reynolds and Strouhal numbers [11]:

$$\alpha = \sqrt{\frac{Re\ St}{2}} \tag{1.20}$$

Dean number (De), is the ratio of viscous force on a fluid flowing in a curved pipe to the centrifugal force. The Dean number for a channel (or pipe) is defined as:

$$De = Re \cdot \sqrt{\frac{D_h}{2R}} \tag{1.21}$$

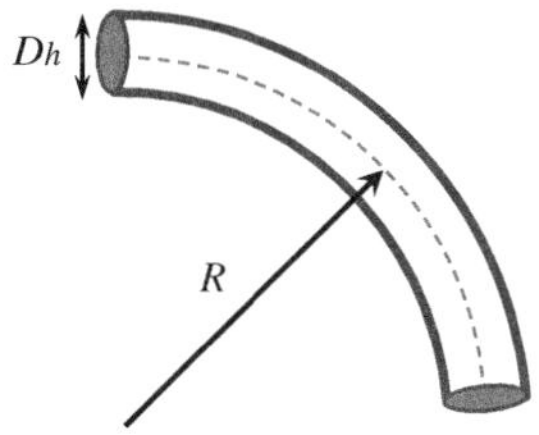

■ **FIGURE 1.14** Geometry of a curved pipe.

where *Re* is Reynolds number, D_h is hydraulic diameter (m) and R is the radius of curvature of the channel (m). A schematic of a curved pipe (vessel) is shown in Fig. 1.14.

Stokes number (Stk), is defined as the ratio of the characteristic reaction time of a suspended particle in fluid flow to a characteristic time of the flow. The reaction time is the time that the particle takes to respond to velocity changes in flow. The Stokes number is defined as:

$$Stk = \frac{\rho_p d_p^2 V}{18\mu D} \tag{1.22}$$

where ρ_p is particle density, d_p is particle diameter, V is fluid velocity, μ is dynamic viscosity of fluid and D is diameter of the obstacle (typically for channels is inner diameter).

1.2.5 Steady versus Unsteady Flow and Laminar Versus Turbulent Flow

Flow is characterized by the fluid properties vectors field (velocity, pressure, etc.). If this properties field is time-independent $\left(\text{or } \frac{d}{dt} = 0\right)$, the flow is steady and it is unsteady if the velocity field is time-dependant. Laminar and turbulent flows are fluidic aspects of viscous flows. In laminar flow, the streamlines are parallel and in the case of a circular pipe they are concentric cylinders around the flow centerline. Transition is a connecting situation from laminar to turbulent flow and it occurs because of fluid velocity or geometry changes. In turbulent flow, the flow fluctuates continually and it has chaotic behaviors. In turbulent flow, the velocity field cannot be predicted with absolute accuracy and the velocity profile will become much blunter at the flow centerline. Because of fluctuations in turbulent flow, a large increase will be present for shear stress among fluid layers.

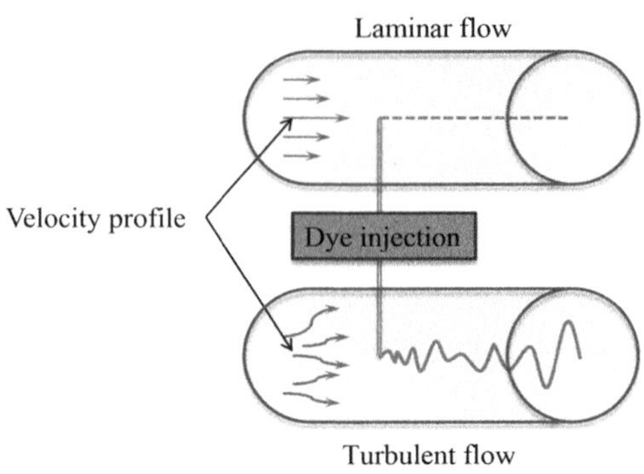

■ **FIGURE 1.15** Dye releasing into a stream of Newtonian fluid to demonstrate laminar (up) and turbulent flow (down).

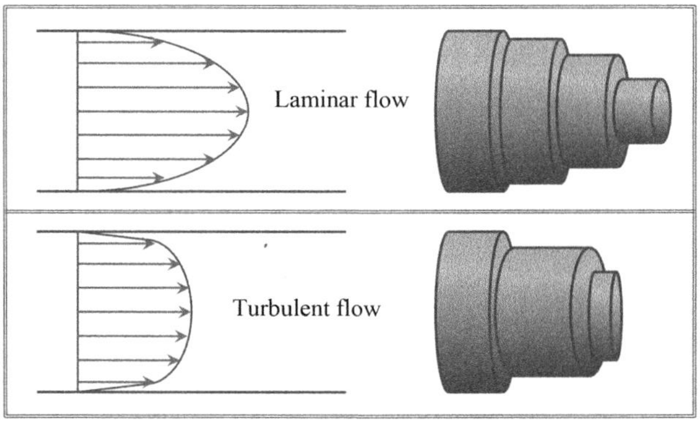

■ **FIGURE 1.16** Schematic of velocity profiles for laminar and turbulent flow in a pipe.

Fig. 1.15 illustrates a schematic of the laminar and turbulent flows.

Velocity profile: Velocity profile is a diagram of the velocity vectors of fluid flow as a function of distance perpendicular to the flow direction. Fig. 1.16 shows the schematic of a velocity profile for laminar and turbulent flow.

For a viscous flow through a circular channel (Fig. 1.17), the axial velocity profiles are given by Eqs. (1.23) and (1.24) [9]. The axial velocity varies from 0 at the wall ($r = R$) up to a maximum at the centerline $r = 0$.

Laminar flow:

$$V_r = V_0\left(1 - \frac{r^2}{R^2}\right) \tag{1.23}$$

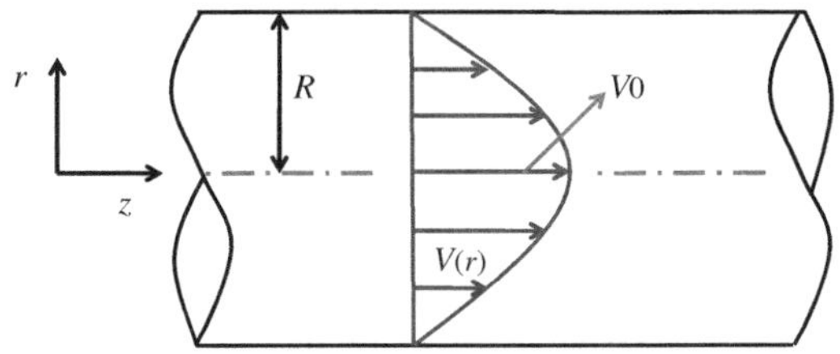

■ **FIGURE 1.17** Velocity profile as a function of radius.

Table 1.2 Approximation of Flow Regimes (Laminar to Turbulent Flow) Based on Reynolds Number [9]

Reynolds Number Range	Flow Regime Conditions
$0 < Re < 1$	Highly viscous laminar "creeping" motion
$1 < Re < 100$	Laminar, strong Reynolds-number dependence
$100 < Re < 10^3$	Laminar, boundary-layer theory useful
$10^3 < Re < 10^4$	Transition to turbulence
$10^4 < Re < 10^6$	Turbulent, moderate Reynolds-number dependence
$10^6 < Re < \infty$	Turbulent, slight Reynolds-number dependence

Turbulent flow:

$$V_r \approx V_0\left(1-\frac{r}{R}\right)^m, \quad \frac{1}{9} \le m \le \frac{1}{5} \tag{1.24}$$

where V_r is axial velocity as a function of radius, V_0 is maximum velocity at the centerline, r is radial distance from the centerline and R is radius of the channel.

As we explained in Section 1.2.4, the Reynolds number is a useful tool to evaluate the flow regimes in fluid mechanics. Table 1.2 represents the range of Reynolds numbers to evaluate flow regimes from laminar to turbulent flow. These ranges differ with surface condition or roughness, stream fluctuations, channel or flow geometry, constrictions and bends.

Table 1.3 represents exact Reynolds numbers which are most practical and approved. This table shows accurate numbers for commercial pipes, although with exceptional care in the manufacturing of rounded inlets, very smooth internal surfaces and steady flow for entrance, Reynolds numbers can be predicted to reach much higher values.

Table 1.3 Accurate Flow Regimes (Laminar to Turbulent Flow) Based on Reynolds Number

Reynolds Number Range	Flow Regime Conditions
$Re < 2100$	Laminar
$2100 < Re < 4200$	Transition
$4200 < Re$	Turbulent

1.2.6 Boundary Conditions and No Slip Boundary Condition

Boundary conditions have a key role in physics applications. The governing equations for physical characteristics, such as velocity and pressure, are defined by partial differential equations. These equations must be applicable for every point of fluid flow. To calculate these equations we need other statements about the physical data, such as velocity field in fluid field. These statements are named as boundary conditions. Several types of boundary conditions are considered for a fluid flow: solid surface (include porous), free liquid surface, liquid–vapor interface, liquid–liquid interface, inlet or exit section [12]. The solid surface boundary conditions cover a wide range in fluid mechanics.

In a viscous fluid, the velocity continues everywhere and the tangential parts of velocity are equal to that of the solid boundary and the velocity is zero for a solid boundary at rest, or *no slip* boundary condition [13]. In plain language, the velocity vector at a fluid–solid boundary is equal to that of the solid boundary; tangential velocity of fluid at interface ($V_{\text{fluid at interface}}$) = $V_{\text{solid}} = 0$ (*no slip*). The "no slip" term is used because of the sticking of fluid particles to solid surfaces.

Fig. 1.18 shows the schematic of this aspect in fluid mechanics.

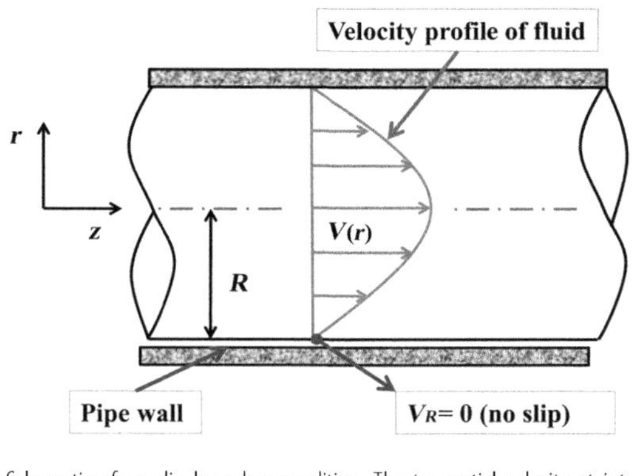

■ **FIGURE 1.18** Schematic of no slip boundary condition. The tangential velocity at interface point between flow and solid wall of pipe is equal to zero ($V_R = 0$ m/s, no slip).

1.2.7 Compressible and Incompressible Flows

Flows can be classified as compressible or incompressible. If the physical (thermodynamic and geometry) considerations cannot change the density (ρ) of fluid, this fluid is incompressible. It is ideal to have fully incompressible fluid but with some assumptions and exceptions we can consider some fluids as incompressible fluids. Water is a simple example of an incompressible fluid. Gases, such as air, will be categorized as compressible fluids.

Mathematically, for an incompressible fluid, the density changing rate in flow is zero,

$$\frac{d\rho}{dt} = 0 \tag{1.25}$$

In the case of compressible fluids, they can have significant changes in density. The continuity equation (it will be explained later in this chapter) explains the behavior of these fluids;

$$\frac{d\rho}{dt} + \nabla\cdot(\rho V) = 0$$

or

$$\frac{d\rho}{dt} = -\nabla\cdot(\rho V) \tag{1.26}$$

where ρ is fluid density, t is time and V is velocity vector field. The divergence of vector field ($\nabla\cdot$) is a scalar function and defines the vector changes in Cartesian coordinates (x, y, z).

For an incompressible fluid, Eq. (1.26) will be changed to,

$$\nabla\cdot V = 0 \tag{1.27}$$

So, the divergence of the velocity field is zero and it means that the fluid and its elements do not have a volume changing rate (incompressible condition).

1.2.8 Stress Tensor

Stress is the result of forces per unit area (Newton per square meter or Pascal, Pa, or dyne per square centimeter). Three stress states are used in mechanical engineering design: normal (axial) stress, shear stress and isotropic stress.

- Normal stress: The force (F) is perpendicular to surface (A) and parallel to the main axis of the object.
- Shear stress: The force is parallel (F) to surface (A) and perpendicular to the axis of the object.

- Isotropic stress: The object body is under equal tension or compression in all directions.

Fig. 1.19 shows a schematic of these stresses.

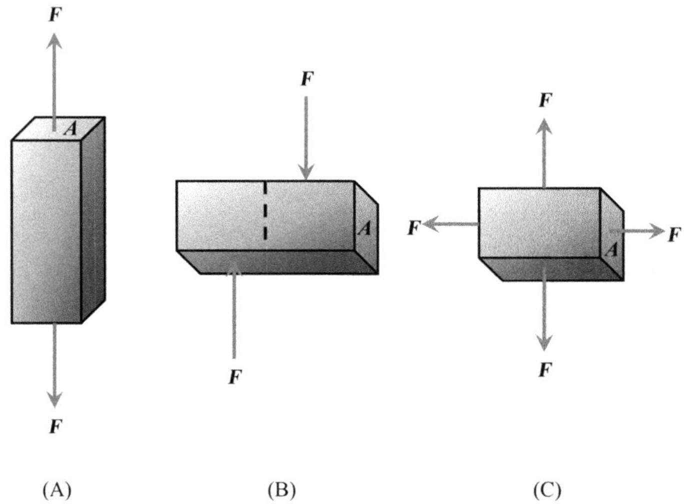

■ **FIGURE 1.19** Schematic of stresses; (A) normal stress (B) shear stress (C) isotropic stress.

The stresses have tensors character, which means that they correlate a vector with a direction. Generally, tensors have a connection role between vector sets in two- or three-dimensional elements. One important tensor in continuum mechanics is the Cauchy stress tensor (known as stress tensor). According to Cauchy's law, the stress tensor $\boldsymbol{\sigma}$ maps normal (perpendicular) to a surface to unit length direction vector performing on that surface,

$$T_i = \sigma_{ij} \cdot n_j \tag{1.28}$$

The stress tensor has nine elements. Fig. 1.20 shows a stress tensor and its elements.

The element σ_{ii} is normal stress on the surface with normal vector (e_i) and the element σ_{ij} $(i \neq j)$ is the shear stress in the direction of e_i on the surface with normal vector (e_j) [14].

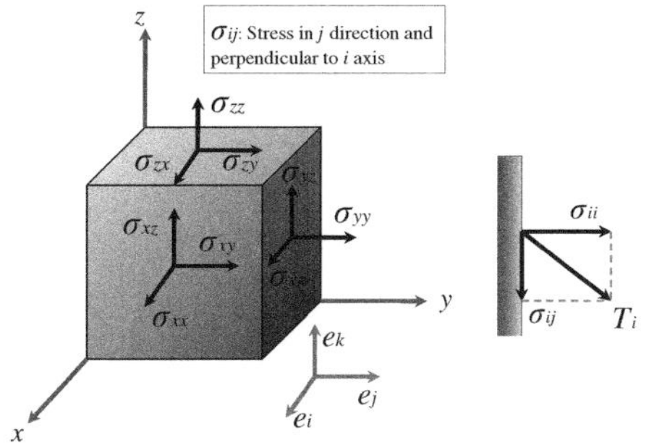

■ **FIGURE 1.20** Left: Elements of stress tensor in a three dimensional fluidic volume, Right: Vectors of normal and shear stress on surface.

The stress tensor can be written as a 3×3 matrix,

$$\sigma_{ij} = \begin{bmatrix} \sigma_{xx} & \sigma_{xy} & \sigma_{xz} \\ \sigma_{yx} & \sigma_{yy} & \sigma_{yz} \\ \sigma_{zx} & \sigma_{zy} & \sigma_{zz} \end{bmatrix}$$

and Eq. (1.27) can be rewritten as T_x

$$\begin{bmatrix} T_x \\ T_y \\ T_z \end{bmatrix} = \begin{bmatrix} \sigma_{xx} & \sigma_{yx} & \sigma_{zx} \\ \sigma_{xy} & \sigma_{yy} & \sigma_{zy} \\ \sigma_{xz} & \sigma_{yz} & \sigma_{zz} \end{bmatrix} \begin{bmatrix} n_x \\ n_y \\ n_z \end{bmatrix} \tag{1.29}$$

1.2.9 Viscoelasticity and Viscoplasticity

Viscoelasticity is the mechanical property of material which demonstrates both viscous and elastic behavior under force deformation. When viscoelastic materials (like honey) are loaded, a deformation will be expected. When the load is removed, the material will be divided into two parts, one portion of the deformation tries to recover immediately and a second portion (major part) recovers with time. Because of this time-dependant behavior we call it viscoelastic material [15]. Fig. 1.21 shows this behavior of viscoelastic materials in a relaxation (stress-time) graph. As the figure shows, the viscoelastic fluid has a longer delay time in comparison to a Newtonian fluid.

Several models are used for viscoelastic models and Maxwell is one of the common models for this field. Maxwell considered a spring (symbol of elasticity) and a damper or dashpot (symbol of viscosity) in series under a stress to model viscoelastic materials, Fig. 1.22 shows a schematic of this model.

The governing equation of the Maxwell model is,

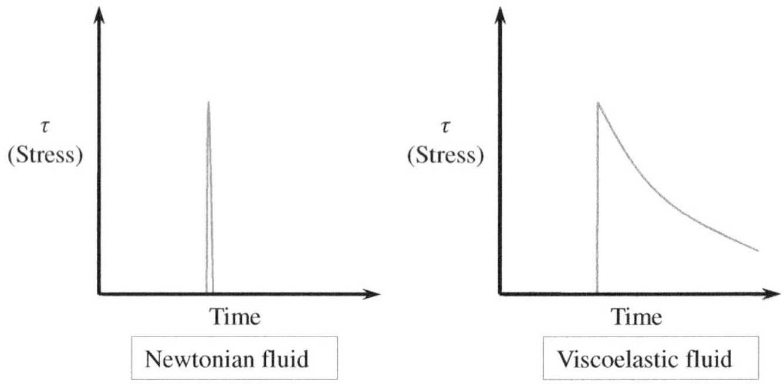

■ **FIGURE 1.21** Schematic of stress relaxation between Newtonian fluid (left) and viscoelastic fluid (right).

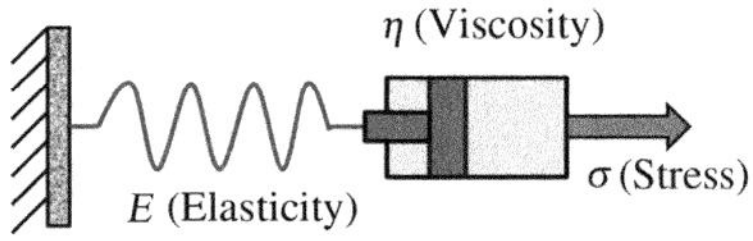

■ **FIGURE 1.22** Maxwell model for viscoelastic material.

$$\frac{d\epsilon}{dt} = \frac{1}{E} \cdot \frac{d\sigma}{dt} + \frac{\sigma}{\eta}$$

or

$$\dot{\epsilon} = \frac{\dot{\sigma}}{E} + \frac{\sigma}{\eta} \tag{1.30}$$

where ϵ is strain or deformation fraction, E is elasticity module of material, σ is stress on material element, η is viscosity and t is time.

Examples of viscoelastic materials are honey, peanut butter, dairy cream, marshmallow cream, and in biological materials, skin, bones and vessel walls are good examples.

Viscoplasticity behavior of a fluid is characterized by a yield (threshold) stress which must be exceeded before the fluid will flow or deform. If the external applied stress be smaller than yield stress, the fluid will deform or flow elastically. The flow curve can be nonlinear or linear if the stress has exceeded the yield stress. Viscoplastic material exhibits an apparent viscosity which decreases with increasing shear rate. Familiar examples for viscoplastic fluid are emulsions, particulate suspensions, blood, drilling muds, etc. [10].

Three common models are used for viscoplastic fluids: Bingham plastic, Herschel–Bulkley and Casson models. The Casson model is applicable for biological materials such as blood. The Casson equation, based on the model is [16],

$$\tau^{1/2} = b^{1/2} + a|\gamma^{1/2}| \tag{1.31}$$

where τ is shear stress, γ is shear rate, a and b are constants (b is yield stress at low shear rate).

Eq. (1.30) is a general form of the Casson model. Recently, more precise equations [10,17] based on the Casson model have been introduced to explain viscoplastic materials such as blood.

1.2.10 **Basic Equations of Fluid Mechanics**

The main purpose in fluid mechanics, especially fluid dynamics, is to describe the main characteristics of fluid flow, such as velocity, pressure and temperature. The best method to characterize these variables is to use related equations. The cornerstones of fluid mechanics are the fundamental equations of

- Conservation of Mass
- Conservation of Momentum
- Conservation of Energy
- Navier–Stokes equation
- Bernoulli equations
- Hagen–Poiseuille equation

These equations speak about the physics aspects and mathematical statements of a fluid field. In the case of modeling, these equations have an essential role in simulating and solving fluidic problems.

1.2.11 **Conservation of Mass, Momentum and Energy**

This famous sentence, "you cannot get something from nothing," is the basis of conservation. It means that the physical properties of an object must be conserved during a process or interaction. There is only a transformation or transportation from one point to another point.

Conservation of Mass: Consider a cube with volume of v (control volume, CV) and a fluid passes through the cube (Fig. 1.23), there is only transportation by the fluid flow, and the conservation of mass for this transportation is [11],

$$\begin{pmatrix}\text{Rate of Mass accumulation} \\ \text{in } CV\end{pmatrix} = \begin{pmatrix}\text{Flow of} \\ \text{mass into } CV\end{pmatrix} - \begin{pmatrix}\text{Flow of} \\ \text{mass from } CV\end{pmatrix}$$

Mathematically, the general form of mass conservation is

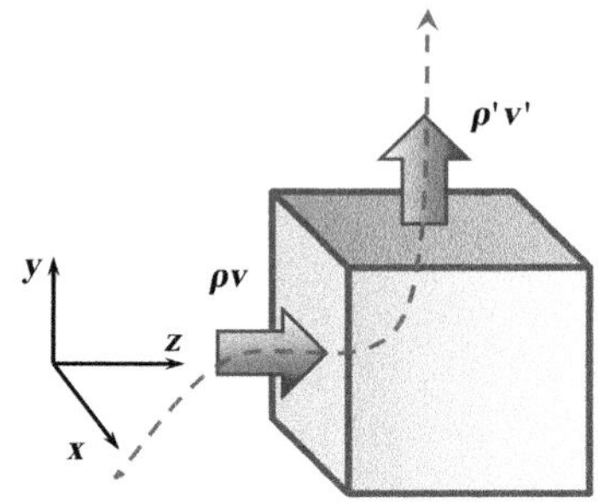

■ **FIGURE 1.23** Schematic of flow through a control volume.

$$\frac{\partial \rho}{\partial t} + \nabla \cdot (\rho \boldsymbol{V}) = 0 \tag{1.32}$$

where ρ is density, t is time and $\boldsymbol{V}$ is fluid velocity. The term $(\rho \boldsymbol{V})$ is the mass flux through the control volume. Eq. (1.32) is called the continuity equation.

In the case of an incompressible fluid ($\rho = \text{constant}$), Eq. (1.32) reduces to

$$\nabla \cdot \boldsymbol{V} = 0 \tag{1.33}$$

Conservation of Momentum: This principle states that the total momentum of a closed system (energy isolated to or from the outside) remains constant through time, it means that the momentum of a system cannot be changed without forces. Conservation of linear momentum is similar to mass conservation, it can be described as [11],

$$\begin{pmatrix}\text{Rate of momentum accumulation}\\ \text{in } CV\end{pmatrix} = \begin{pmatrix}\text{Rate of}\\ \text{momentum flow into } CV\end{pmatrix} - \begin{pmatrix}\text{Rate of}\\ \text{momentum flow from } CV\end{pmatrix} + \sum \text{Forces acting on } CV$$

If we consider a cube control volume such as Fig. 1.23, the conservation of linear momentum for the control volume is

$$\rho \frac{\partial \boldsymbol{V}}{\partial t} + \rho \boldsymbol{V} \cdot \nabla \boldsymbol{V} = -\nabla P + \nabla \cdot \tau + \rho g \tag{1.34}$$

where ρ is density, t is time, $\boldsymbol{V}$ is fluid velocity, P is normal pressure on fluid element surfaces, τ is stress tensor and g is gravity.

Conservation of Energy: This principle states that the total energy of a closed system (energy isolated to or from the outside) remains constant through time and the energy only changes its form, from one kind of energy, such as kinetic energy of fluid flow to electricity in a wind turbine. In other words, it can be [11]

$$\begin{pmatrix}\text{Rate of energy accumulation}\\ \text{in } CV\end{pmatrix}=\begin{pmatrix}\text{Rate of energy}\\ \text{transfer across } CV\end{pmatrix}+\begin{pmatrix}\text{Rate of done work}\\ \text{on } CV\end{pmatrix}+\begin{pmatrix}\text{Rate of energy production}\\ \text{within } CV\end{pmatrix}$$

Mathematically, the energy conservation for "biological energy transportation" is [11]

$$\rho C_p\left(\frac{\partial T}{\partial t}+\boldsymbol{V}\cdot\nabla T\right)=k\nabla^2\mathrm{T}+\dot{Q}_p+\dot{W} \tag{1.35}$$

where ρ is density, C_p is heat capacity at constant pressure, T is temperature, $\boldsymbol{V}$ is velocity, k is thermal conductivity, $\dot{W}$ is rate of other work in system after subtracting the work for fluid flow and $\dot{Q}_p$ is summation of rate of heat production and viscous dissipation.

1.2.12 Navier–Stokes Equations

The equations are named after Claude L. Navier and George G. Stokes because of their contribution in fluid mechanics and present a mathematical model for the motion of fluids. The origin of the equations is Newton's second law with the application of several variables such as velocity (and its components), density, pressure, gravity, temperature and viscous stress tensor to fluid motion. A combination of momentum, mass and energy conservations were applied to derive the equations.

The general form (vector form) of the Navier–Stokes equation is [11]

$$\rho\frac{\partial \boldsymbol{V}}{\partial t}+\rho\boldsymbol{V}\cdot\nabla\boldsymbol{V}=-\nabla P+\mu\nabla^2\boldsymbol{V}+\rho g \tag{1.36}$$

where ρ is density, t is time, $\boldsymbol{V}$ is fluid velocity, P is normal pressure on fluid element surfaces, μ is viscosity and g is gravity.

Table 1.4 shows the Navier–Stokes equation in rectangular and cylindrical coordinates.

Table 1.4 Navier–Stokes Equation for Incompressible Fluids in Rectangular and Cylindrical Coordinates

Rectangular Coordinates:

x direction:

$$\rho\left(\frac{\partial \mathbf{V}_x}{\partial t} + \mathbf{V}_x\frac{\partial \mathbf{V}_x}{\partial x} + \mathbf{V}_y\frac{\partial \mathbf{V}_x}{\partial y} + \mathbf{V}_z\frac{\partial \mathbf{V}_x}{\partial z}\right) = -\frac{\partial P}{\partial x} + \mu\left(\frac{\partial^2 \mathbf{V}_x}{\partial x^2} + \frac{\partial^2 \mathbf{V}_x}{\partial y^2} + \frac{\partial^2 \mathbf{V}_x}{\partial z^2}\right) + \rho g_x \tag{1.37}$$

y direction:

$$\rho\left(\frac{\partial \mathbf{V}_y}{\partial t} + \mathbf{V}_x\frac{\partial \mathbf{V}_y}{\partial x} + \mathbf{V}_y\frac{\partial \mathbf{V}_y}{\partial y} + \mathbf{V}_z\frac{\partial \mathbf{V}_y}{\partial z}\right) = -\frac{\partial P}{\partial y} + \mu\left(\frac{\partial^2 \mathbf{V}_y}{\partial x^2} + \frac{\partial^2 \mathbf{V}_y}{\partial y^2} + \frac{\partial^2 \mathbf{V}_y}{\partial z^2}\right) + \rho g_y \tag{1.38}$$

z direction:

$$\rho\left(\frac{\partial \mathbf{V}_z}{\partial t} + \mathbf{V}_x\frac{\partial \mathbf{V}_z}{\partial x} + \mathbf{V}_y\frac{\partial \mathbf{V}_z}{\partial y} + \mathbf{V}_z\frac{\partial \mathbf{V}_z}{\partial z}\right) = -\frac{\partial P}{\partial z} + \mu\left(\frac{\partial^2 \mathbf{V}_z}{\partial x^2} + \frac{\partial^2 \mathbf{V}_z}{\partial y^2} + \frac{\partial^2 \mathbf{V}_z}{\partial z^2}\right) + \rho g_z \tag{1.39}$$

Cylindrical Coordinates:

r direction:

$$\rho\left(\frac{\partial \mathbf{V}_r}{\partial t} + \mathbf{V}_r\frac{\partial \mathbf{V}_r}{\partial r} + \frac{\mathbf{V}_\theta}{r}\frac{\partial \mathbf{V}_r}{\partial \theta} - \frac{\mathbf{V}_\theta^2}{r} + \mathbf{V}_z\frac{\partial \mathbf{V}_r}{\partial z}\right) = -\frac{\partial P}{\partial r} + \mu\left[\frac{\partial}{\partial r}\left(\frac{1}{r}\frac{\partial(r\mathbf{V}_r)}{\partial r}\right) + \left(\frac{1}{r^2}\frac{\partial^2 \mathbf{V}_r}{\partial \theta^2} - \frac{2}{r^2}\frac{\partial \mathbf{V}_\theta}{\partial \theta} + \frac{\partial^2 \mathbf{V}_r}{\partial z^2}\right)\right] + \rho g_r \tag{1.40}$$

θ direction:

$$\rho\left(\frac{\partial \mathbf{V}_\theta}{\partial t} + \mathbf{V}_r\frac{\partial \mathbf{V}_\theta}{\partial r} + \frac{\mathbf{V}_\theta}{r}\frac{\partial \mathbf{V}_\theta}{\partial \theta} - \frac{\mathbf{V}_r\mathbf{V}_\theta}{r} + \mathbf{V}_z\frac{\partial \mathbf{V}_\theta}{\partial z}\right) = -\frac{1}{r}\frac{\partial P}{\partial \theta} + \mu\left[\frac{\partial}{\partial r}\left(\frac{1}{r}\frac{\partial(r\mathbf{V}_\theta)}{\partial r}\right) + \left(\frac{1}{r^2}\frac{\partial^2 \mathbf{V}_\theta}{\partial \theta^2} - \frac{2}{r^2}\frac{\partial \mathbf{V}_r}{\partial \theta} + \frac{\partial^2 \mathbf{V}_\theta}{\partial z^2}\right)\right] + \rho g_\theta \tag{1.41}$$

z direction:

$$\rho\left(\frac{\partial \mathbf{V}_z}{\partial t} + \mathbf{V}_r\frac{\partial \mathbf{V}_z}{\partial r} + \frac{\mathbf{V}_\theta}{r}\frac{\partial \mathbf{V}_z}{\partial \theta} + \mathbf{V}_z\frac{\partial \mathbf{V}_z}{\partial z}\right) = -\frac{\partial P}{\partial z} + \mu\left[\left(\frac{1}{r}\frac{\partial}{\partial r}\left(r\frac{\partial \mathbf{V}_z}{\partial r}\right) + \frac{1}{r^2}\frac{\partial^2 \mathbf{V}_z}{\partial \theta^2} + \frac{\partial^2 \mathbf{V}_z}{\partial z^2}\right)\right] + \rho g_z \tag{1.42}$$

1.2.13 Bernoulli Equation

The Bernoulli equation is a correlation from the conservation equations to demonstrate a relation between velocity, elevation and pressure in a nonviscous (frictionless) fluid [9]. For frictionless ($\tau = 0$) and steady flow behavior $\left(\frac{\partial V}{\partial t} = 0\right)$, the equation of linear momentum will reduce to

$$\rho \boldsymbol{V} \nabla \cdot \boldsymbol{V} = -\nabla P + \rho g \tag{1.43}$$

In plain language, the Bernoulli equation says that if an incompressible fluid flows through different sizes of pipes, the fluid velocity changes. This change creates an acceleration in the system which creates different forces and pressures on the pipe's cross-section area.

After simplification and integration of Eq. (1.43), the general form of the Bernoulli equation is

$$\frac{1}{2}\rho \boldsymbol{V}^2 + P + \rho g z = \text{constant} \tag{1.44}$$

or between two points along the streamline, the equation is

$$\frac{P_2 - P_1}{\rho} + \frac{1}{2}(\boldsymbol{V}_2^2 - \boldsymbol{V}_1^2) + g(z_2 - z_1) = 0 \tag{1.45}$$

where ρ is density, $\boldsymbol{V}$ is fluid velocity, P is pressure, z is elevation and g is gravity. Fig. 1.24 shows a simple example to clarify these variables in the Bernoulli equation for a fluid flow in a tube.

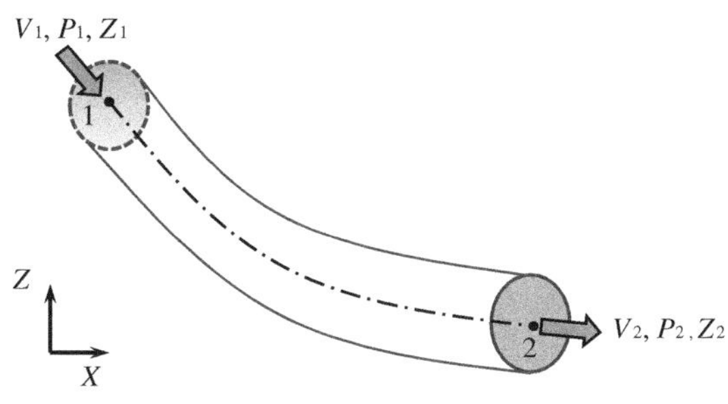

FIGURE 1.24 A fluid flow in a tube between two points with different elevation, velocity and pressure.

1.2.14 Hagen–Poiseuille Equation

The Hagen–Poiseuille Equation (or Poiseuille equation) is a fluidic law to calculate flow pressure drop in a long cylindrical pipe and it was derived separately by Poiseuille and Hagen in 1838 and 1839, respectively.

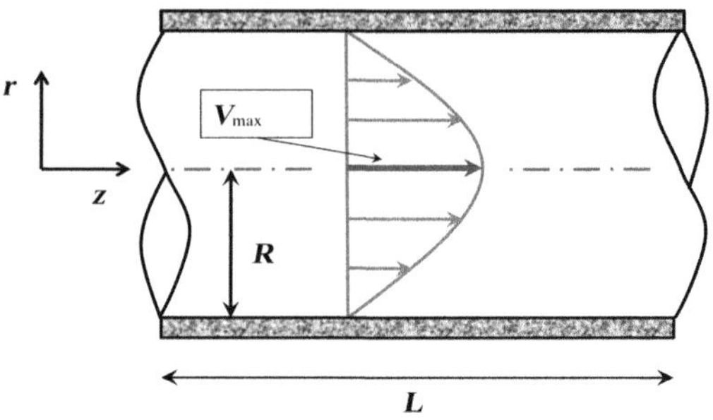

■ **FIGURE 1.25** Longitudinal cutting view of a pipe with a laminar flow and parabolic velocity profile.

Consider a steady flow of an incompressible Newtonian fluid in a long rigid pipe. Using conservation equations and boundary conditions for cylindrical coordinates, the laminar flow has a parabolic profile from zero at the wall to a maximum velocity at the centerline (Fig. 1.25),

$$V_z = \frac{\Delta P\, R^2}{4\mu L}\left(1 - \frac{r^2}{R^2}\right) \qquad 0 \le r \le R \tag{1.46}$$

As a result,

$$V_{\max} = \frac{\Delta P\, R^2}{4\mu L} \tag{1.47}$$

Considering the definition of average velocity in cylindrical coordinates and Eq. (1.46), the average velocity in the pipe is [9]

$$\overline{V} = \frac{1}{2} V_{\max} \tag{1.48}$$

The volumetric flow rate (Q) equals cross-section area multiplied by average velocity ($Q = \overline{V}A$), so

$$\Delta P = \frac{8\mu L Q}{\pi R^4} \tag{1.49}$$

Eq. (1.47) is known as the Poiseuille equation.

1.2.15 Steady Flow Along Tube

As it was explained before, in the steady flow, the fluid properties at a point in the fluidic system do not change over time (time-independent). Consider two tubes with the same length but different diameters in two

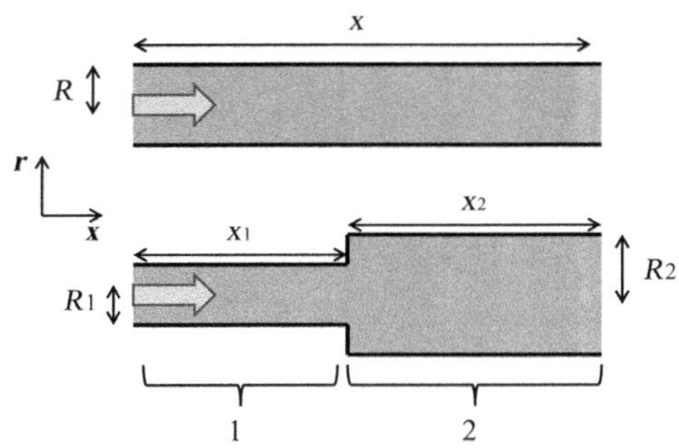

■ **FIGURE 1.26** Geometry of regular tube vs. segmented tube.

segments. Fig. 1.26 shows a schematic of these tubes. Considering the Poiseuille equation, the pressure drop as a function of length is

$$\Delta P = \frac{8\mu qx}{\pi R^4} \tag{1.50}$$

or

$$P_{\text{inlet}} - P_x = \frac{8\mu qx}{\pi R^4}$$

or

$$P_x = P_{\text{inlet}} - \frac{8\mu qx}{\pi R^4} \tag{1.51}$$

where P is pressure, q is flow rate, μ is viscosity, R is radius and x is a point in axis length.

If the tube has several segments, each segment has its pressure drop based on segment geometry. In this case, the tube has two segments in series, to be identified by subscripts 1 and 2 (Fig. 1.26). The amount of flow rate in the tube segments with different diameter and length is similar, the pressure drops in the system are (based on Eq. (1.51))

$$P_{x_1} = P_{\text{inlet1}} - \frac{8\mu qx_1}{\pi R_1^4} \tag{1.52}$$

$$P_{x_2} = P_{\text{inlet2}} - \frac{8\mu qx_2}{\pi R_2^4} \tag{1.53}$$

The general form of pressure drop for a segmented tube with "n" segments is

$$P_{x_n} = P_{\text{inlet}(n)} - \frac{8\mu qx_n}{\pi R_n^4} \tag{1.54}$$

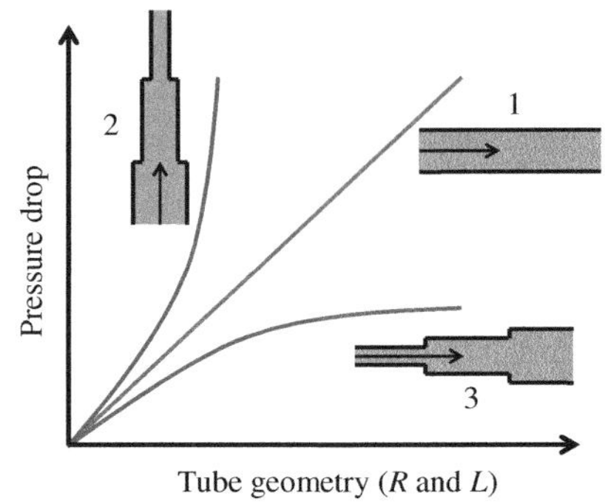

■ **FIGURE 1.27** Schematic of pressure drop curves for (1) straight, (2) converging, and (3) diverging tubes.

And total pressure drop in tube system is

$$P_{\text{total}} = \sum_{1}^{n} P_{x_n} \tag{1.55}$$

Fig. 1.27 illustrates a schematic of pressure drop in straight, diverging and converging tubes for an equal flow rate. Regarding the fourth order of tube radius in the denominator of the Poiseuille equation (see Eq. (1.50)), by decreasing diameter (converging shape) the pressure drop will increase significantly. But the pressure drop for a regular tube is linear even when changing the length due to its order in the numerator of Eq. (1.50).

1.2.16 **Pulsatile Flow in Rigid and Elastic Tubes**

Understanding the physics of pulsatile (oscillating) flow is of primary interest for many medical and industrial applications. As it was explained before, when the flow properties are a function of time, pressure drop within the tube becomes a function of time too (unsteady flow).

In a rigid tube:

If a waveform pressure plays the role of an inlet pressure with nonzero mean in a tube, the mean value will be a constant pressure drop and the remaining part will serve as an oscillatory pressure [18], so

$$\Delta P(t) = \Delta P_S + \Delta P_\varphi(t) \tag{1.56}$$

where ΔP_S is the constant part and ΔP_φ is the oscillatory part of input pressure. Fig. 1.28 shows a schematic of a pressure-time graph to explain the constant and oscillatory parts. As the figure shows, the constant

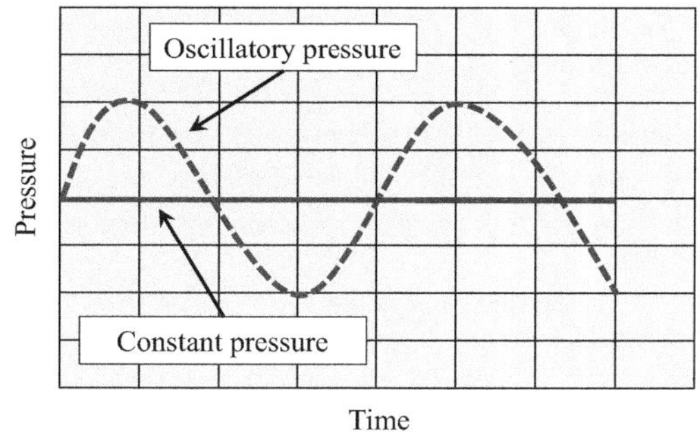

■ **FIGURE 1.28** A pressure-time diagram, Constant pressure (solid line) and oscillatory pressure (dash line).

pressure is not a function of time $\left(\frac{\partial P}{\partial t}=0\right)$, nevertheless the oscillatory pressure will change with time $\left(\frac{\partial P}{\partial t}\neq 0\right)$.

In terms of pressure gradient, the oscillatory pressure gradient is

$$\frac{dP_\varphi(t)}{dx}=P_0\mathrm{e}^{i\omega t} \tag{1.57}$$

or

$$\frac{dP_\varphi(t)}{dx}=P_0(\cos\omega t+i\sin\omega t) \tag{1.58}$$

where $i=\sqrt{-1}$, $\omega=2\pi f$ and f is frequency of flow, t is time and P_0 is wave amplitude of pressure. In this formulation, $P_0(\cos\omega t)$ and $P_0(i\sin\omega t)$ describe the real part and the imaginary part of the driving pressure, respectively [18].

Inside the tube, oscillatory flow velocity can be written as [18]

$$U_\varphi(r,t)=\frac{iP_0R^2}{\mu\Omega^2}\left(1-\frac{J_0(\zeta)}{J_0(\Lambda)}\right)\mathrm{e}^{i\omega t} \tag{1.59}$$

where J_0 (Bessel function), Ω, Λ and ζ are

$$J_0(x)=1-\frac{x^2}{2^2}+\frac{x^4}{2^4 2!}-\frac{x^6}{2^6(3!)^2}+\cdots \tag{1.60}$$

$$\Omega=\sqrt{\frac{\rho\omega}{\mu}}R \tag{1.61}$$

$$\Lambda=\left(\frac{i-1}{\sqrt{2}}\right)\Omega \tag{1.62}$$

$$\zeta(r)=\Lambda\left(\frac{r}{R}\right), \tag{1.63}$$

where R is tube radius, r is radial coordinate, ρ is density, μ is viscosity and ω is angular frequency ($\omega = 2\pi f$).

An easy comparison is made between steady and oscillatory velocity profiles by considering the amplitude of oscillatory pressure gradient to be the same as the steady flow pressure gradient [18,19], so

$$\text{Pressure gradient} = P_{\square} = \frac{dP}{dx} = \frac{\Delta P}{L} \tag{1.64}$$

$$P_o = P_s \tag{1.65}$$

According to the flow rate definition, oscillatory flow rate or $Q_\varphi(t)$ [18] is

$$Q_\varphi(t) = \int_0^R 2\pi r U_\varphi(r,t)dr \tag{1.66}$$

$$= \frac{i\pi P_s R^4}{\mu\Omega^2}\left(1 - \frac{2J_1(\zeta)}{\Lambda J_0(\Lambda)}\right)e^{i\omega t} \tag{1.67}$$

And mathematically, the steady term of flow rate is

$$U_s(r) = \frac{\Delta P}{4\mu L}(R^2 - r^2) \tag{1.68}$$

$$Q_s = \int_0^R 2\pi r U_s(r)dr \tag{1.69}$$

In an elastic tube:

The main difference between incompressible flow in an elastic tube and in a rigid tube is that in a rigid tube any change in the pressure will be sensed immediately all along the tube, but in an elastic tube some part of the pressure change will be absorbed by the elastic wall. This characteristic creates a delay for flow properties in an elastic tube.

In an elastic tube with pulsatile flow, a pressure change will generate a wave movement to flow and the speed of this wave is recognized as the wave speed factor. The Moen–Korteweg equation calculates this wave speed factor for a tube wall which is entirely elastic and very thin (wall thickness $\ll$ tube diameter) [18],

$$c_o = \sqrt{\frac{Eh}{\rho d}} \tag{1.70}$$

where E is elasticity (Young's modulus) of the tube wall, h is the thickness of the tube wall, ρ is the fluid density and d is the tube diameter.

The Moen–Korteweg equation has been derived for inviscid (no viscosity) conditions which means that we cannot use it for regular and common "no slip" conditions. This situation creates a complex issue and to solve it we have to derive a solution for the movement of fluid and tube wall.

A solution has been found for this problem. We know that the frequency parameter "Ω" of pulsatile flow plays a main role in the governing equations of rigid tubes. Similarly in an elastic tube, frequency parameter Ω has a key role too, but it shares its role with another important parameter, wave speed factor "c" [18].

$$c = \sqrt{\frac{2}{(1-\sigma^2)z}}\, c_o \tag{1.71}$$

where σ is Poisson's ratio of the elastic tube wall and z is a solution of the following equation

$$\{(g-1)(\sigma^2-1)\}z^2 + \left\{\frac{\rho_w h}{\rho R}(g-1) + \left(2\sigma - \frac{1}{2}\right)g - 2\right\}z + \frac{2\rho_w h}{\rho R} + g = 0 \tag{1.72}$$

where ρ_w is density of tube wall and g is

$$g = \frac{2J_1(\Lambda)}{\Lambda J_0(\Lambda)} \tag{1.73}$$

The wave speed factor c is complex of real and imaginary parts, which depend on the value of Ω (a frequency parameter). An increase in the value of Ω results in an increase in the real part (to one) and decrease for the imaginary part (to zero).

In comparison to a rigid tube, the flow properties in an elastic tube have a second term for velocity (V_φ), which is known as radial velocity (Fig. 1.29). The pulsatile flow characters in an elastic tube, such as velocities and flow rate, can be written as [18]

$$\frac{U_\varphi(x,r,t)}{\widehat{U_s}} = \frac{4}{i\Omega^2}\left(1 - G\frac{J_0(\zeta)}{J_0(\Lambda)}\right)e^{i\omega(t-x/c)} \tag{1.74}$$

$$\frac{V_\varphi(x,r,t)}{\widehat{U_s}} = \frac{2R\omega}{i\Lambda^2 c}\left(\frac{r}{R} - G\frac{2J_1(\zeta)}{\Lambda J_0(\Lambda)}\right)e^{i\omega(t-x/c)} \tag{1.75}$$

$$\frac{Q_\varphi(x,t)}{Q_s} = \frac{8}{i\Omega^2}(1 - Gg)e^{i\omega(t-x/c)} \tag{1.76}$$

$$\widehat{U_s} = U_s(r=0) = \frac{\Delta P R^2}{4\mu L} \tag{1.77}$$

Elasticity factor or

$$G = \frac{2 + z(2\sigma - 1)}{z(2\sigma - g)} \tag{1.78}$$

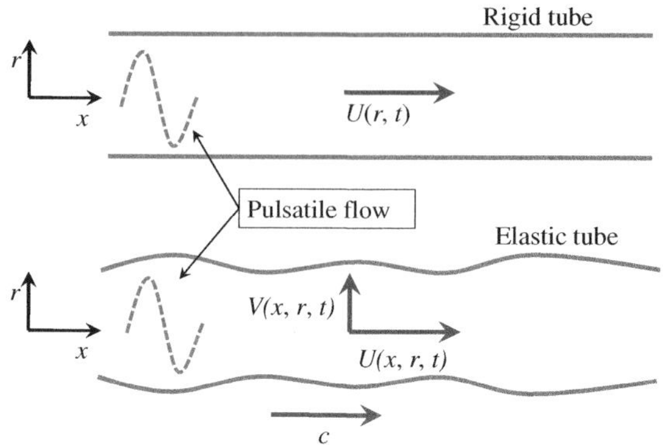

■ **FIGURE 1.29** Schematic of pulsatile flow in a rigid tube (top) and elastic tube (down). Due to elasticity behavior of tube wall, an additional velocity (radial velocity) or $V(x, r, t)$ is represented for elastic tube.

1.2.17 **Resistance, Compliance and Inertance**

Resistance:

Two aspects determine blood flow through a vessel: (1) pressure drop (or difference) between inlet and outlet of vessel and (2) vascular resistance due to the impediment to blood flow through the vessel. Vascular resistance occurs as an effect of friction forces between the blood flow and the intravascular endothelium of the vessel (Fig. 1.30). Ohm's law is a fundamental equation in the cardiovascular system and can be calculated by the following equation [20]:

$$R = \frac{\Delta P}{Q} \tag{1.79}$$

where Q is (volumetric) flow rate, ΔP is pressure drop and R is resistance. Fig. 1.30 illustrates these relationships for a blood vessel in the circulatory system.

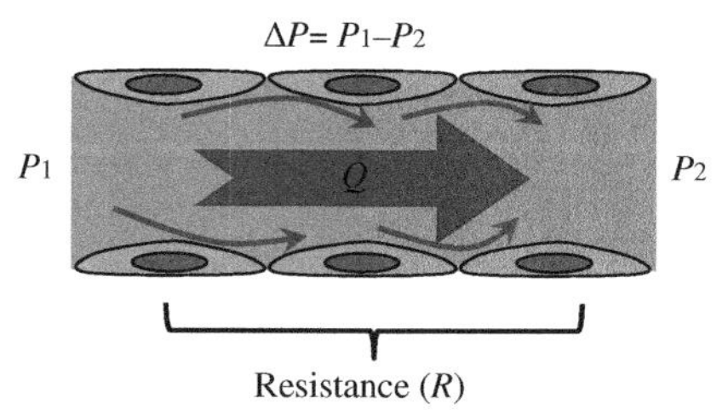

■ **FIGURE 1.30** Cutting view of blood vessel and interrelationships among pressure drop (ΔP), resistance (R), and blood flow rate (Q).

For parallel vessels (mother vessel to daughter vessels)

$$\frac{1}{R_{\text{total}}} = \frac{1}{R_1} + \frac{1}{R_2} + \frac{1}{R_3} + \cdots \tag{1.80}$$

Compliance:

In in vivo (in the living organism or body) conditions, the relation between pressure and volume can be measured. In a closed circuit or limited space, a volume increase results in a pressure increase and vice versa. This aspect does not change the material (in most cases) or organ structure but it may affect the performance of the organ. Hollow organs such as vessels are a good example for this physiologic parameter.

Mathematically, compliance in the cardiovascular system is [20]

$$C = \frac{\Delta V}{\Delta P} \tag{1.81}$$

where C is compliance, ΔV and ΔP are volume and pressure changes, respectively (see Fig. 1.31).

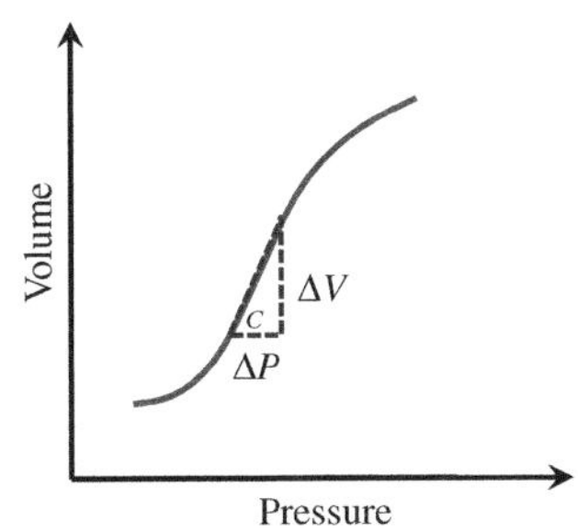

■ **FIGURE 1.31** Volume-pressure curve of compliance "c" and its relation to pressure and volume difference.

Elastance (E) is the inverse of compliance (or $E = \frac{\Delta P}{\Delta V}$) and the organ or biological system wants to recoil toward the original condition or dimension [21].

Several techniques are available to determine volumes, such as Magnetic Resonance Imaging (MRI), Computed Tomography or Ultrasound method. For instance, MRI can measure diameter changes by wall-tracking for large vessels. The compliance or elastance can perform as a useful quantitative method to measure the mechanical properties of an organ due to aging or diseases [21].

Inertance:

Inertance occurs as the effect of momentum in fluid flow and it relates the pressure difference with a change in the volumetric flow rate with time. We consider an incompressible fluid element with length of l, density of ρ and mass of m in a tube with cross-section area of A under pressure difference of ΔP (Fig. 1.32). According to Newton's second law,

$$F = m \cdot a = (\rho \cdot (l \cdot A)) \cdot \frac{dv}{dt} \tag{1.82}$$

$$F = \Delta P \cdot A \tag{1.83}$$

$$dv = \frac{dQ}{A} \tag{1.84}$$

Considering Eqs. (1.82)–(1.84),

$$\Delta P = \frac{\rho \cdot l}{A} \cdot \frac{dQ}{dt} \tag{1.85}$$

Inertance:

$$I = \frac{\rho \cdot l}{A} \tag{1.86}$$

$$\Delta P = I \cdot \frac{dQ}{dt} \tag{1.87}$$

where ΔP is pressure difference, I is inertance, Q is flow rate, v is velocity and t is time.

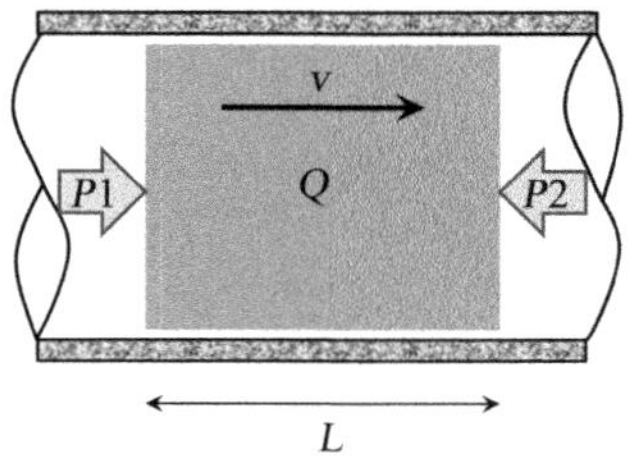

■ **FIGURE 1.32** A fluid element in a tube with variables of inertance.

Inertance has a key role in accelerating and decelerating blood flow in the cardiovascular system [21].

1.2.18 **Two-Phase Flows**

The most common group of multiphase flows is two-phase flow. In two-phase flow, the fluid has two parts which are not chemically or physically (usually) related and each phase has its own flow properties. The common examples for two-phase flows are aerosols, suspensions, and, for a biofluid example, blood.

Blood is one of the most common two-phase (or multiphase) fluids. The first phase is the plasma (it contains water, proteins and electrolytes) and the second phase is created by the blood cells. The blood cells can be considered as a fluid because they are mostly composed of water surrounded by an elastic membrane [22]. Generally, at the macro scale of blood simulation, the blood cells' properties are considered the same as the blood fluid property, but this is not accurate because these cells affect the blood flow.

One of the important items in two-phase flows is to know the exact location of each particle (or particle trajectory). Lagrangian, Eulerian and Boltzman models can be considered as an appropriate approach to solve the two-phase flows [22,23].

There are several corresponding conditions for two-phase problems: thermal, mechanical and chemical boundary conditions. One of the ideas to solve two-phase problems is using basic fluid mechanics equations (mass conservation, momentum conservation, Navier–Stokes, etc.) for each phase individually, but this method needs an extra boundary condition as a balance interface between phases which makes it a complex state [22].

It is necessary to introduce two proper velocity fields in the formulation with the purpose of considering the effect of relative motion (mass diffusion, energy and momentum) between phases [23].

CHAPTER SUMMARY

- In fluid kinematics, *Lagrangian* and *Eulerian* are useful tools but the Eulerian method is more realistic for fluids.
- In the *Lagrangian* method, properties (velocity, pressure, etc.) of each particle are tracked during its motion.
- The *Eulerian* method, which is more appropriate for fluids, characterizes the flow properties (velocity, pressure, etc.) at every point in the four-dimensional coordinates system (4D).
- *Velocity field*, in the Eulerian method, is a function of space and time and for Cartesian coordinates is

$$V = u(x,y,z,t)\,\vec{i} + v(x,y,z,t)\,\vec{j} + w(x,y,z,t)\,\vec{k}$$

- *Acceleration field* in the Eulerian method for Cartesian coordinates is

$$a = \frac{dV}{dt} = \frac{du}{dt}\vec{i} + \frac{dv}{dt}\vec{j} + \frac{dw}{dt}\vec{k} = a_x\vec{i} + a_y\vec{j} + a_z\vec{k}$$

or

$$a = \frac{\partial V}{\partial t} + V \cdot \nabla V$$

- In *flow visualization*, four flow lines examine the flow field: streamlines, pathlines, streaklines and timelines.
- *Viscosity* is the resistance of a fluid to deformation (or flow) by shear or tensile stress. For linear fluids, the viscosity is $\mu = \frac{\tau}{\gamma}$.
- In *Newtonian fluids*, a constant (linear) viscosity coefficient can be defined and the deviatoric normal stress elements must be zero ($\tau_{ii} = 0$).
- *Dimensionless (pure) numbers* are a useful tool to characterize the mechanical behavior of fluids in dynamic conditions:
  Reynolds number: $Re = \frac{\rho V D_h}{\mu}$
  Womersley number: $\alpha = \frac{D_h}{2}\sqrt{\frac{\rho(2\pi f)}{\mu}}$
  Strouhal number: $St = \frac{fL}{V}$
  Dean number: $De = Re \cdot \sqrt{\frac{D_h}{2R}}$
  Stokes number: $Stk = \frac{\rho_p d_p^2 V}{18 \mu D}$
- If fluid properties (velocity) are time-independent $\left(\text{or } \frac{d}{dt} = 0\right)$, the flow is *steady* and it is *unsteady* if the velocity field is time-dependant.
- In *laminar flow*, the streamlines are parallel.
- In *turbulent flow*, the velocity field cannot be predicted with absolute accuracy.

Reynolds Number Range	Flow Regime Conditions
$Re < 2100$	Laminar
$2100 < Re < 4200$	Transition
$4200 < Re$	Turbulent

- In a *no slip* condition, the tangential velocity at the interface point between the flow and the solid wall of a pipe is zero.
- Fluid is *incompressible*, if the physical (thermodynamic and geometry) considerations cannot change the density (ρ) of fluid.
- The *stress tensor* $\boldsymbol{\sigma}$ maps normal (perpendicular) to a surface to a unit length direction vector performing on that surface, $T_i = \sigma_{ij} \cdot n_j$
- *Viscoelasticity* is the mechanical property of material which demonstrates both viscous and elastic behavior under force deformation.
- *Viscoplasticity* behavior of a fluid is characterized by a yield (threshold) stress which must be exceeded before the fluid will flow or deform

- *Mass conservation* is

$$\frac{\partial \rho}{\partial t} + \nabla \cdot (\rho \boldsymbol{V}) = 0$$

- *Conservation of linear momentum* for the control volume is

$$\rho \frac{\partial \boldsymbol{V}}{\partial t} + \rho \boldsymbol{V} \cdot \nabla \boldsymbol{V} = -\nabla P + \nabla \cdot \tau + \rho g$$

- The *energy conservation* for "biological energy transportation" is

$$\rho C_p \left(\frac{\partial T}{\partial t} + \boldsymbol{V} \cdot \nabla T \right) = k \nabla^2 \mathrm{T} + \dot{Q}_p + \dot{W}$$

- The general form (vector form) of the *Navier–Stokes equation* is

$$\rho \frac{\partial \boldsymbol{V}}{\partial t} + \rho \boldsymbol{V} \cdot \nabla \boldsymbol{V} = -\nabla P + \mu \nabla^2 \boldsymbol{V} + \rho g$$

- The general form of the *Bernoulli equation* is

$$\frac{1}{2} \rho \boldsymbol{V}^2 + P + \rho g z = \text{constant}$$

or

$$\frac{1}{2} \rho V_1^2 + P_1 + \rho g z_1 = \frac{1}{2} \rho V_2^2 + P_2 + \rho g z_2$$

- The general form of the *Poiseuille equation* is

$$\Delta P = \frac{8 \mu L Q}{\pi R^4}$$

- *Pressure drop in pulsatile flow* is related to constant pressure and oscillatory pressure

$$\Delta P(t) = \Delta P_s + \Delta P_\varphi(t)$$

- *Pulsatile flow in a rigid tube versus elastic tube.* According to the elasticity behavior of a tube wall, an additional velocity (radial velocity) or *V(x, r, t)* is represented for an elastic tube (Fig. 1.29),
- *Vascular resistance (R)* occurs as an effect of friction forces between the blood flow and the intravascular endothelium of the vessel,

$$R = \frac{\Delta P}{Q}$$

- *Compliance* in the cardiovascular system is the rate of volume change to pressure change in a closed circuit or limited space,

$$C = \frac{\Delta V}{\Delta P}$$

- *Inertance* relates the pressure difference to a change in volumetric flow rate with time.

PROBLEMS

1. Discuss the fluid aspects for the curves that have been drawn in Fig. 1.13.
2. The velocity elements for a steady incompressible flow are:

$$u = (3x + 2y)t, \quad v = (4x - 3y)t, \quad w = 0$$

   Is the continuity equation satisfied?
3. Are the following values possible for steady incompressible 2D flow?
   a. $u = (4x^2y + y^2), \quad v = (-8xy + 3x)$
   b. $u = (2xy + y^2), \quad v = (4xy - 3x^2)$
4. A nozzle discharges water into the atmosphere, inlet and outlet areas are 400 and 300 mm^2, respectively. Determine the flow rate when the inlet pressure is 500 Pa (No energy loss)? (see Fig. 1.33)

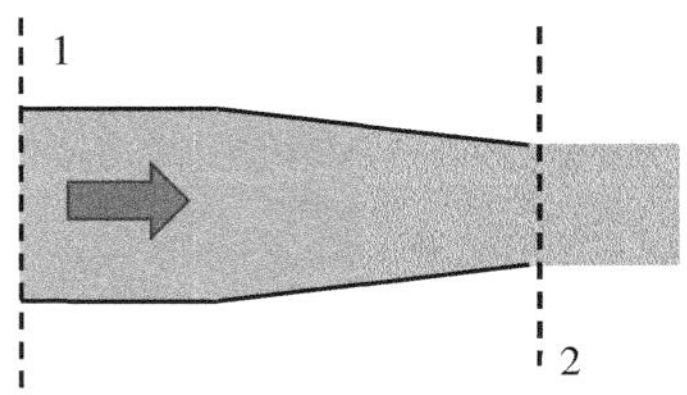

FIGURE 1.33 A discharging nozzle.

5. A fluid flows in a tube with 40 mm length and 1.5 mm diameter, the required head to produce 10 mm^3/s is 35 mm. Determine the viscosity of the fluid, the fluid density is 800 kg/m^3?
6. Determine the flow rate that causes a wall shear stress of 1 N/m^2 in a 7 mm diameter vessel? Would the flow be turbulent or laminar?
7. A syringe with inside diameter of 15 mm horizontally ejects liquid with density of 1040 kg/m, through a needle with inside diameter of 1 mm at a velocity of 0.5 m/s. Determine the pressure in the syringe?
8. Use the Navier–Stokes equations in the Cartesian coordinate system to determine the velocity profile for the blood flow in an artery with 0.001 m diameter and length of 0.04 m, inlet ($P1$) and outlet ($P2$)

pressure is 85 and 75 mmHg respectively, (Assume the $v_y = v_z = 0$, incompressible steady flow) (Fig. 1.34)

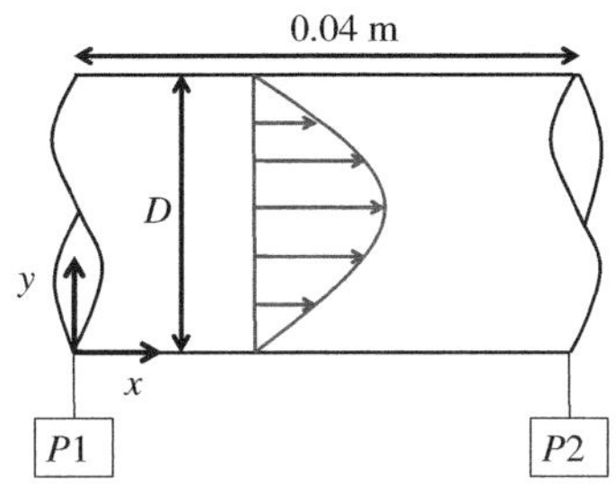

■ **FIGURE 1.34** Blood flow in an artery for problem 8 .

1.3 HEMATOLOGY AND BLOOD RHEOLOGY

Learning Objectives

- Explanation of bodily fluid and blood components.
- Categorization of blood types.
- Definition of pH and bodily fluids pH.
- Categorization of blood cells and definition of erythrocytes, leukocytes and thrombocytes.
- Definition of plasma, electrolytes, proteins of blood
- Blood viscosity aspects
- Formulation and explanation of rheological models of blood
- Categorization of blood diseases

1.3.1 Blood—Bodily Fluid

The main function of blood is to transport energy and nutrients to all the living tissues and to remove the waste products from the body. Due to this vital duty, most creatures (animals) have a complicated vascular network system. In humans, this network varies in diameter from 3 μm (for capillaries) to <3 cm (for aorta).

Blood is a complex fluid with several substances; extracellular fluid rich in protein (plasma) and formed elements including red blood cells (RBCs), white blood cells (WBCs) and platelets [24]. If a sample of blood (with anticoagulant) is spun in a cylindrical tube at 10,000 g for ~5 minutes, we have a fraction of formed elements and plasma (Fig. 1.35).

In healthy and normal adult humans, blood accounts for 6–8% of body weight [1], or about 5 L [20]. About 40–45% of blood volume is

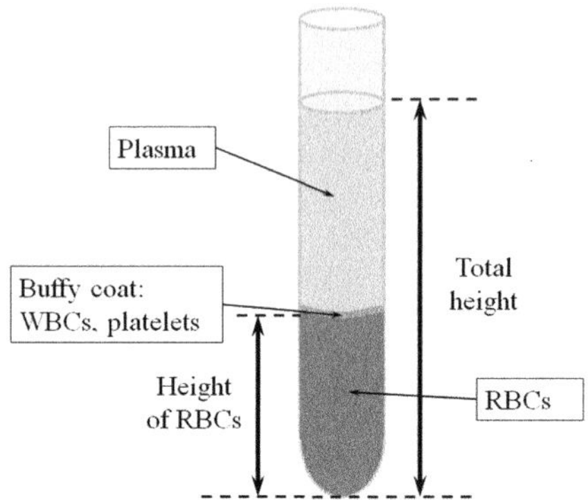

■ **FIGURE 1.35** Centrifuged blood sample.

occupied by RBCs, less than 1% by WBCs and platelets and around 55% by plasma. These percentages can vary depending on gender, weight, age and other parameters.

The density of blood is greater than the density of water (1000 kg/m^3) at approximately 1060 kg/m^3, this increased density is because of the density of the RBCs in blood [1].

1.3.2 Blood Types

Early blood transfusions were risky and many resulted in the death due to the reaction of the immune system. Two types of antigens have the main roles in this reaction: the O-A-B system of antigens and the Rh system. In a blood transfusion procedure, the blood is generally classified into four major O-A-B blood types, as illustrated in Table 1.5 [20].

Table 1.5 Blood Types With Related Genotypes, Agglutinogens and Agglutinins

Genotypes	Blood Types	Agglutinogens	Agglutinins	Prevalence
OO	O	–	Anti-A and Anti-B	~47%
OA or AA	A	A	Anti-B	~41%
OB or BB	B	B	Anti-A	~9%
AB	AB	A and B	–	~3%

The phrase for the aggregation of RBCs into clumps is agglutination, which can lead to death in severe cases. Agglutinogens (or antigens) are located on the membranes of RBCs. Agglutinins are plasma antibodies, which attempt to attack specific agglutinogens—this mechanism is recognized as an antibody–antigen reaction [25].

As the table shows, the presence or absence of the two agglutinogens (antigens), A and B, define the types of blood. The human blood is categorized into four types: O, A, B and AB. With regard to this category and Table 1.5, people with type B blood have the B agglutinogens (antigens). People with type AB blood have both the A and the B agglutinogens on their RBCs. As a final point, persons with type O blood have no A or B Agglutinogens on their RBCs [25].

Along with the O-A-B system of antigens, the Rh system is essential in the blood transfusion procedure. This system has six common types (C, D, E, c, d and e) of Rh antigens, which are called Rh factors. The type D is widely prevalent among other antigen types in the human populations. If anyone has the D type of antigen, they are said to be Rh positive, otherwise the Rh of that person is negative (it is much less common). In the negative case, other Rh antigens can still cause mild transfusion reactions [20].

1.3.3 Blood Cells

A hematocyte or blood cell is a cell is produced from an hematopoietic stem cell in the bone marrow. Blood cells are categorized by three types in mammals:

- Erythrocytes or RBCs
- Leukocytes or WBCs
- Thrombocytes or Platelets

1.3.3.1 *Erythrocytes (RBCs)*

The major function of erythrocytes (RBCs) is to transport hemoglobin, which in turn carries oxygen from the lungs to the living tissues. Besides the transport of hemoglobin, RBCs have other roles in human body, such as increasing the rate of CO_2 carrying by water in the blood and assisting in acid–base buffering of hemoglobin [20,24].

Shape and Size of RBC: a RBC is a nonnucleated biconcave cell with a diameter of 7.7 ± 0.7 μm, peripheral thickness of 2.8 ± 0.5 μm, central thickness of 1.4 ± 0.5 μm, aspect ratio ~0.4, membrane thickness ~0.02 μm, surface area ~140 μm^2 and volume of ~90 fL (90×10^{-15} L) [20,24,26]. Fig. 1.36 shows a schematic of RBC geometry.

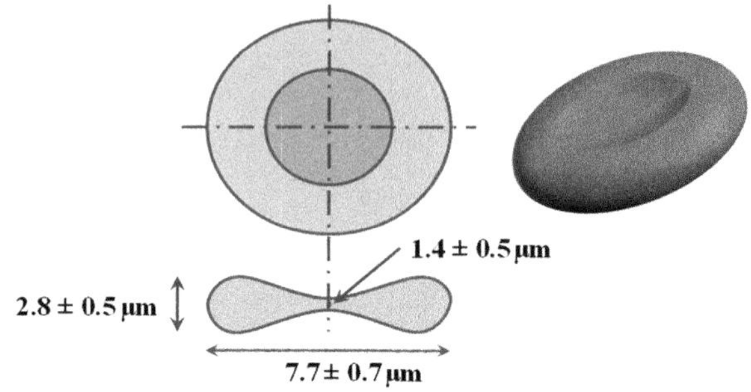

■ **FIGURE 1.36** Schematic of RBC geometry.

An average unstressed shape of normal human RBC was measured and formulated experimentally [27], the average shape equation of a RBC in Cartesian coordinates (x, y, z) is [28,29]

$$y = R\sqrt{\left(1-\frac{x^2+z^2}{R^2}\right)}\left[c_0 + c_1\frac{x^2+z^2}{R^2} + c_2\frac{(x^2+z^2)^2}{R^4}\right] \tag{1.88}$$

where $R = 3.91\ \mu m$, $c_0 = 0.0135805$, $c_1 = 1.001279$, $c_2 = -0.561381$.

The average number of RBCs per mm^3 (or μL) is 5,200,000 (± 300,000) and 4,700,000 (± 300,000) in normal men and women, respectively. RBCs normally circulate in the human body an average of 120 days before being destroyed. Also, they have the ability to concentrate hemoglobin in the cell fluid by up to about 34 g in each 100 mL of cells [20].

The RBC membrane is composed of a phospholipid bilayer and stretching of a RBC by optical tweezers approximates a shear modulus of 4–10 μN/m [26]. The RBC membrane has an elastic modulus of isotropic dilation of about 500 dyn/cm (0.5 N/m) [30].

The RBC shape (biconcave) helps to increase the cell surface area which is important for diffusion of O_2 and CO_2, also it helps to deform easily (more flexibility) to pass through smaller arterioles or capillaries.

1.3.3.2 ***Leukocytes (WBCs)***

The mobile units of the body's defensive system are called leukocytes or WBCs. WBCs are created partially in the bone marrow and partially in the lymph tissue, and then they are distributed in the blood circulation system [20].

Table 1.6 Approximate Percentages of the WBCs Types

WBC Type	Percentage
Neutrophils	62.0
Eosinophils	2.3
Basophils	0.4
Monocytes	5.3
Lymphocytes	30.0

Types of WBCs: Five types of WBCs are normally present in the blood [24,26]:

- Neutrophils (size 8–15 μm), have small plasma granules and often multilobed nucleus.
- Eosinophils (size ~15 μm), have large granules and a nucleus which frequently has two lobes.
- Basophils (size 12–15 μm), have many large basophilic granules.
- Monocytes (size 15–25 μm), are the largest leukocyte.
- Lymphocytes (size 8–15 μm), have a very large nucleus.
- Occasionally, plasma cells can be considered as sixth type [20].

The normal adult human being has about 7000 WBCs per microliter (μL) of blood; the approximate percentages of the different types of WBCs are shown in Table 1.6 [20].

1.3.3.3 *Thrombocytes (Platelets)*

Thrombocytes (also called Platelets) are minute cells 2–3 μm in diameter [24]. Platelets are produced in the bone marrow from megakaryocyte cells, their concentration in the human blood is 150,000–300,000 platelets per microliter (μL) [20]. The normal lifetime of platelets is 7–10 days [25]. Platelets have a major role in blood coagulation and the creation of a mechanical plug in response to a vascular injury in the human body.

On the surface of the platelet membrane is a coat of glycoproteins that repulses adherence to normal endothelium and yet causes adherence to injured endothelial cells, also the platelet membrane includes large amounts of phospholipids that have a major role in the blood clotting procedure [20], as we discuss in chapter "Reaction of Blood and Biomaterials."

1.3.4 Plasma

Body fluids are categorized into several components. Plasma is one of the bodily fluids and it is classified as a subsystem of extracellular fluids in the human body. About 8% of body water is found in the plasma—around 3 L in a normal adult human. Plasma represents approximately 55% of the blood volume. The plasma performs as a suspending fluid for blood cells. Plasma is composed of water (~92%), electrolytes (~0.9%), proteins (~7%), lipids (~0.6) and glucids (~0.1%) [26]. Those proteins include albumins, globulins and fibrinogens [25].

As we explained before, plasma is categorized as a Newtonian fluid. The normal range of plasma viscosity is between 1.10 and 1.30 mPas (centipoise-cP) at 37°C and it is independent of blood flow characteristics, age and gender. Capillary viscometers are the simplest method to measure plasma viscosity. Plasma viscosity is sensitive, particularly to protein changes. In parallel, the viscosity can increase with erythrocyte sedimentation rate, but it is not controlled by aggregation of RBCs, anemia, polycythemia or hemoglobinopathies. Also high plasma viscosity and fibrinogen can predict some cardiovascular disorders, such as unstable angina and stroke [31].

The term of "apheresis" refers to a filtration technology to separate and remove either plasma (plasmapheresis) or the formed elements like RBCs, WBCs, and/or platelets (cytopheresis) from whole blood. In plasmapheresis, large amounts of plasma are removed from a patient and after treatment are returned to the blood circulation system of patient.

In the plasmapheresis procedure using hollow fibers, the separation flux per unit of filter area is approximately relative to the wall shear rate (γ_w)— at least up to 7500 s^{-1}—and depends on the fiber length [32],

$$J = 10.8 \times 10^{-5} \frac{\gamma_w}{L^{2/3}} \tag{1.89}$$

where J is separation flux, γ_w is wall shear rate and L is fiber length.

1.3.5 Blood pH

Acid–base characterization is the study of the hydrogen ion (H^+). An acid is any chemical substance that can donate an H^+ and a base (alkali) is any chemical substance that can accept an H^+ [24].

The pH scale was introduced by Sørenson as a tool to measure acid-base behavior of materials,

$$\text{pH} = -\log_{10}[H^+] \tag{1.90}$$

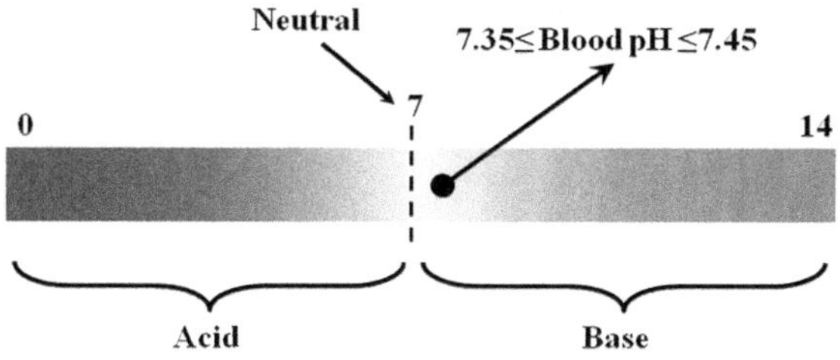

■ **FIGURE 1.37** The pH range. Acid (0–7), Base (7–14). The normal pH range of blood is from 7.35 to 7.45.

Table 1.7 Approximate pH Values for Bodily Fluids [24]

Biofluid	pH
Gastric secretions	0.7
Lysosome	5.5
Chromaffin granule	5.5
Neutral H_2O	6.81
Cytosol of a typical cell	7.2
Cerebrospinal fluid	7.3
Arterial blood plasma	7.4
Mitochondrial inner matrix	7.5
Secreted pancreatic fluid	8.1

Fig. 1.37 shows the range of the pH scale from 0 to 14. The blood pH range is narrow and for a healthy person it is between 7.35 and 7.45 (in base range). When the pH is less than 7.35, this condition is defined as acidosis.

Table 1.7 shows approximate pH values of body fluids.

Dissolving CO_2 in plasma water produces carbonic acid, which lowers blood pH. The carbonic acid–bicarbonate (mostly sodium bicarbonate) buffer helps to balance the arterial blood pH near 7.4.

$$CO_2 + H_2O \rightleftarrows H_2CO_3 \rightleftarrows H^+ + HCO_3^- \tag{1.91}$$

$$\text{Dissolving of carbon dioxide} = s.P_{CO_2} \tag{1.92}$$

where s is solubility coefficient and P_{CO_2} is partial pressure of CO_2.

The Henderson–Hasselbalch equation correctly predicts the blood pH,

$$\text{pH} = 6.1 + \log\frac{[HCO_3^-]}{s \times P_{CO_2}} \tag{1.93}$$

Table 1.8 Approximate Major Cations, Anions and Proteins of the Intracellular and Extracellular (Blood Plasma and Interstitium) Fluids

Substance	Blood Plasma	Interstitium	Intracellular
Na^+ (mM)	142	145	15
Ca^{++} (mM)	2.4	1.2	0.0001
K^+ (mM)	4.4	4.5	120
Mg^{++} (mM)	0.9	0.55	18
Cl^- (mM)	102	116	20
HCO_3^- (mM)	22	25	16
HPO_4^{2-} (mM)	1.4	0.8	0.7
Proteins (g/dL)	7	1	30

For human arterial blood, HCO_3^- is ~ 24 mM, s is ~ 0.03 mM/mmHg and P_{CO_2} is ~ 40 mmH, so the pH for arterial blood is

$$\text{pH} = 6.1 + \log\frac{24}{0.03 \times 40} = 7.40$$

1.3.6 Electrolytes and Proteins of Blood

Electrolytes refer to substances that have positively charged ions (cations) or negatively charged ions (anions). Electrolytes have fundamental roles in the performance of human organs. The most important electrolytes are

- Cations: Na^+, K^+, Mg^{++} and Ca^{++}
- Anions: Cl^-, HCO_3^-, HPO_4^{2-} and SO_4^{2-}

Electrolytes (cations and anions) are unequally distributed in biofluid compartments. Table 1.8 illustrates the key cations, anions and proteins of the intracellular and extracellular fluids. In mammals, the extracellular fluid is categorized into three major subcompartments: blood plasma, transcellular fluid and interstitium fluid. Fig. 1.38 shows the approximate water distribution in an adult human—the intracellular fluid is 60% (25 L) of total body water (TBW) and extracellular fluid (blood plasma, transcellular and interstitium) is 40% (17 L) of TBW. All biofluids have the same approximate osmolality [24].

The main families of plasma proteins (blood proteins) are:

- *Albumin*: binds (carrier protein) steroids, T_3, bilirubin, fatty acids, bile salts. Also albumin creates osmotic pressure. Its plasma concentration is 3.5–5.5 g/dL. It is synthesized by liver.

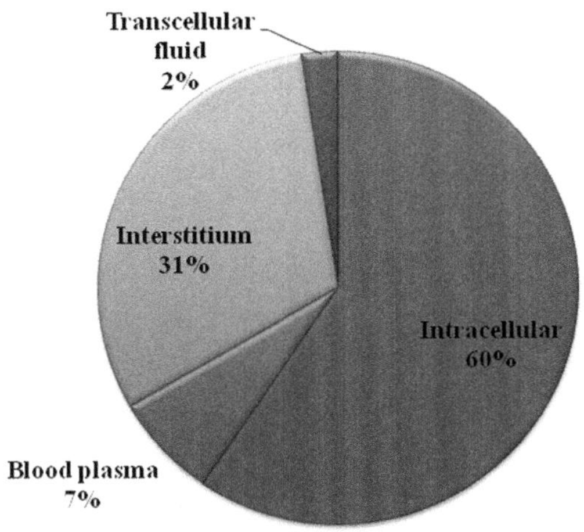

FIGURE 1.38 Approximate water distribution in an adult human.

- *Fibrinogen*: plays a key role in clotting and production of fibrin. Its plasma concentration is 150–300 mg/dL. The liver synthesizes fibrinogen.
- *Globulins*: participate in the immune system. Their plasma concentration is 2–2.5 g/dL. Most globulins are synthesized by liver.
- *Coagulation factors*: participate in the conversion of fibrinogen to fibrin. The liver synthesizes most coagulation factors.

1.3.7 Blood Viscosity and Its Aspects

One of the important assumptions in the Poiseuille equation is the Newtonian fluid condition. Generally, biofluid mechanics and medicine use this equation for cardiovascular problems and usually the viscosity of whole blood is considered constant in a Newtonian sense. This assumption is the simplest technique to solve the problems. Water and plasma have a linear relationship between shear stress and shear rate for all shear rate values. Fig. 1.39 shows this relationship for water, plasma and whole blood. As the figure explains, for shear rate less than 100 s^{-1}, the whole blood has nonlinear behavior which is called non-Newtonian fluid (dash line). Whole blood has a viscosity of $\sim$3 centipoise (cP) or $\sim$0.003 Pascal second (Pa · s) [24].

The effective whole blood viscosity depends on several factors [24]:

1. Hematocrit
2. Fibrinogen concentration

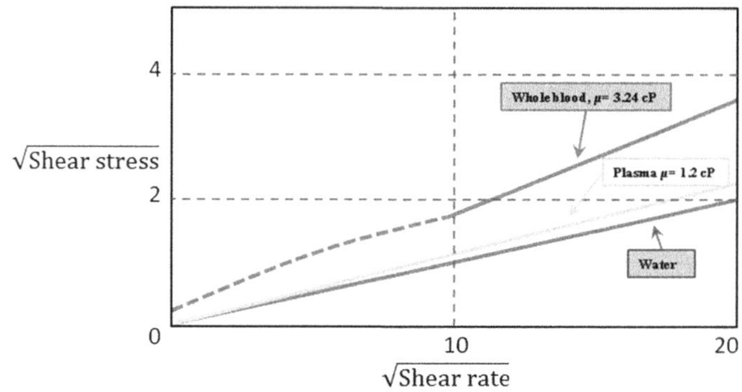

■ **FIGURE 1.39** Shear stress—shear rate diagram (or viscosity diagram) for water, plasma and whole blood.

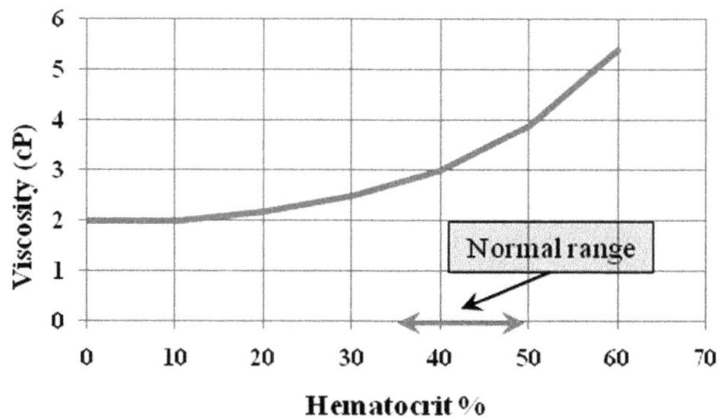

■ **FIGURE 1.40** Relation of viscosity—hematocrit.

3. Vessel radius
4. Temperature
5. Velocity

Hematocrit: Hematocrit is the volume percentage of erythrocytes (RBCs) in blood. The normal range of hematocrit is from 35% to 50% (40–50% for men and 35–45% for women). Increases in hematocrit raise blood viscosity (Fig. 1.40).

Fibrinogen concentration: One of the important reasons for the non-Newtonian behavior of blood is the fibrinogen interaction with erythrocytes (RBCs). High concentrations of fibrinogen (hyperfibrinogenemia) can raise yield shear stress in blood and affect viscosity (more shear stress means a sharper slope in Fig. 1.39).

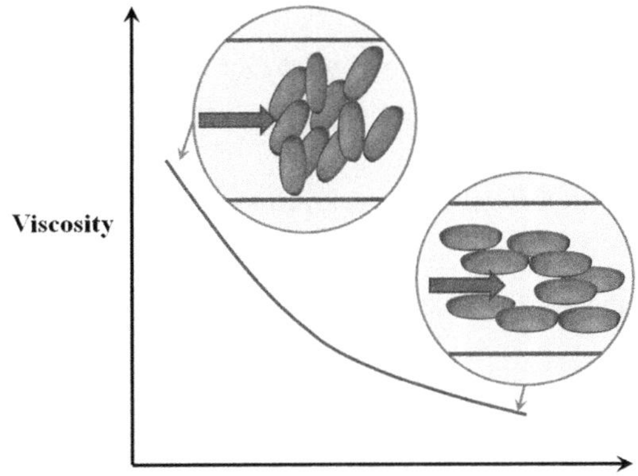

■ **FIGURE 1.41** Schematic of relation between shear rate ($\gamma - s - 1$) and viscosity (μ-Pa · s) in blood flow.

Vessel radius: It is proven that for vessels with diameters less than 2 mm ($D < \sim 2$ mm), an increase in diameter increases the viscosity.

Temperature: Cooling whole blood from normal temperature (37°C) increases its viscosity. This factor usually is negligible since human body temperature is constant.

Velocity: At low flows (it means low velocity), the shear rate is small and fluid behaves as a non-Newtonian fluid with high apparent viscosity. This aspect happens because of RBCs deformation in high velocity blood flow, RBCs will be entirely aligned in the direction of flow and it creates less flow resistance or low viscosity (see Fig. 1.41).

1.3.8 **Rheological Models of Blood**

The rheological models of biofluids are divided into two main classes: time-dependant and time-independent (as we discussed in Section 1.2). The rheological properties of biofluids and blood are time-independent in almost all flows (in vivo and in vitro). One of the realistic methods to model blood is shear stress (τ)-shear rate (γ) equation or viscosity (μ) equation.

These models must be determined and approved experimentally by the curve-fitting method. Some practical equations are given here [16]:

- *Newtonian fluid* model:

$$\tau = \mu \gamma \tag{1.94}$$

μ is constant at all shear rate.

- *Bingham fluid* model:

$$\tau = a\,\gamma + b \text{ for } \tau \geq b \tag{1.95}$$

here "a" (or μ_B) and "b" are constants, "b" is minimum shear stress (yield stress) to cause flow.

- Power law (or Ostwald-de Waele) fluid model:

$$\tau = a|\gamma|^n \tag{1.96}$$

Considering the general viscosity equation, the apparent viscosity for this model is

$$\mu_a = a\,\gamma^{n-1} \tag{1.97}$$

here "a" and "n" are constant. Also,

$n < 1$, the fluid shows pseudoplastic behavior
$n = 1$, the fluid shows Newtonian behavior
$n > 1$, the fluid shows Dilatant behavior

The following equations are based on suspensions models:

- *Einstein model*:

$$\mu_a = \mu_o(1 + k\,H) \tag{1.98}$$

where μ_o is Newtonian viscosity of suspending fluid (it is plasma in the case of blood), H is volume faction of particles in suspension (it is hematocrit in the case of blood) and k is a constant dependant on shape of particle—for solid sphere shape $k = 2.5$

- *Casson model*:

$$\tau^{1/2} = a^{1/2}|\gamma|^{1/2} + b^{1/2} \text{ for } \tau \geq b \tag{1.99}$$

$\gamma = 0$ or no flow for $\tau < b$
where "a" ($a \cong \mu_C$= fluid viscosity) and "b" are constants. For low shear rate "b" is the yield shear stress $\left(b \cong 0.05\,\frac{\text{dyne}}{\text{cm}^2} \text{ or } 0.005\text{ Pa}\right)$ [22].

- *Herschel–Bulkley fluid model*:

$$\tau = \tau_o + m\gamma^n \tag{1.100}$$

where τ_o is yield stress, "m" and "n" are constant and < 1 (see Eq. (1.96)).

- *Quemada model*:

$$\mu_a = \mu_o(1+0.5\,kH)^{-2} \tag{1.101}$$

$$k = \frac{k_o + k_\infty \gamma_r^{1/2}}{1 + \gamma_r^{1/2}}, \quad \gamma_r = \frac{\gamma}{\gamma_c}$$

where k_o, k_∞ and γ_c are constant. γ_c is critical shear rate. See Eq. (1.98) for other parameters.

1.3.9 Blood Diseases

Blood problems can affect any of the blood components such as plasma and especially the main formed elements:

- RBCs: as oxygen carriers to living tissues
- WBCs: immune system actuators
- Platelets: blood coagulation (clotting) system operators

The range of blood diseases is very extensive, around several hundred diseases are defined. The common blood diseases are:

- *Anemia*: is defined as a decrease in the number of RBCs or the amount of hemoglobin in the blood. A patient's body with anemia does not get enough oxygen and as a result the patient may feel tired or weak, dizziness, headache, or have difficulty breathing. There are more than 400 types of anemia which are categorized into three major groups: blood loss, RBC destruction and malfunction in RBC production.
- *Blood cancers*: or hematological malignancy, are types of cancers which affect the blood, lymph nodes or bone marrow. Blood cancers include three types: leukemias, myelomas and lymphomas.
  - *Leukemias*: leukemia is a form of blood cancer that usually begins in bone marrow and the WBCs become malignant and multiply inside the bone marrow. These WBCs are not fully functional/developed.
  - *Lymphomas*: this type of cancer develops in the lymph system. In lymphoma, WBCs turn malignant, increasing and spreading unusually. This disease has two main types: Hodgkin lymphoma and non-Hodgkin lymphoma.
  - *Multiple myeloma*: in this form of blood cancer, the plasma cells (WBCs family member) become malignant and increase and produce harmful materials to damage human organs, such as the kidney.
- *Thrombocytopenia*: this disease refers to a decrease in the number of platelets in the blood. A patient with thrombocytopenia has general fatigue and weakness, and also continues bleeding from injuries or wounds.
- *Hemophilia*: a genetic disorder that impairs the proteins which are responsible for blood coagulation. The disorder has three levels of active clotting factor: mild (with 5–40% active clotting factor), moderate (1–5%) and severe (less than 1%) hemophilia.
- *Von Willebrand disease (vWD)*: Von Willebrand factor (vWF) is a protein which has a key role in clotting and platelet adhesion. In this

disease, a patient's body cannot produce enough vWF protein or the existing protein has a lack of function. Patients with vWD experience excessive bleeding during surgery or after an injury.

CHAPTER SUMMARY

- Blood is a complex fluid with several substances: extracellular fluid rich in protein (plasma) and formed elements including RBCs, WBCs and platelets.
- Two types of antigens have a main role in blood reactions during blood transfusion: the O-A-B system of antigens and the Rh system. The blood is generally classified into four major O-A-B blood types: A, B, O, AB.
- Blood cells are categorized into three types in mammals: erythrocytes or RBCs, leukocytes or WBCs, thrombocytes or platelets.
- Plasma is one of the bodily fluids and it is classified as a subsystem of extracellular fluids in the human body. A normal human has around 3 L of plasma (8% of body water).
- The blood pH range is between 7.35 and 7.45 for a healthy person.
- The most important electrolytes are: Cations: Na^+, K^+, Mg^{++} and Ca^{++}; Anions: Cl^-, HCO_3^-, HPO_4^{2-} and SO_4^{2-}.
- Whole blood has a viscosity of ~3 centipoise (cP) or ~0.003 Pascal second (Pa · s).
- Among the blood rheological models, the Casson, Quemada and Herschel–Bulkley fluid models are the best and most realistic models for human blood.
- The common blood diseases are anemia, blood cancers, thrombocytopenia, hemophilia and vWD.

PROBLEMS

1. List the blood components.
2. Explain the size and shape of blood cells. Discuss the advantage of the unstressed RBCs shape.
3. Discuss the viscosity of blood, plasma and water in Fig. 1.39?
4. For each 2.5 μL of whole blood, how many RBCs do you expect to see for every platelet? How many for every neutrophil and lymphocyte?
5. Use rheological blood models to determine the viscosity of blood with hematocrit 45% and a plasma viscosity of 1.2×10^{-3} (Pa.s) (assume the blood cells are solid spheres).

6. Which of the following changes would have an effect on the viscosity of the blood?
   a. Vessel length
   b. Hematocrit
   c. Leukemia
   d. "a" and "b"
7. A peristaltic pump forces the blood through a small rigid tube with a diameter of 3 mm and length of 400 mm, the pressure drop is $3 \times 10^4\ N/m^2$. Calculate the flow rate in the tube for hematocrit 50% and plasma viscosity of 1.2×10^{-3}(Pa.s) (assume the blood cells are solid spheres).
8. Determine the pressure drop in a segmented tube, the first segment has an inside diameter of 4 mm with a length of 100 mm and the second part has a diameter of 2 mm with a length of 100 mm, the fluid inside the tube is human blood with a flow rate of 100 mm^3/min, hematocrit is 45% and plasma viscosity is 1.2×10^{-3}(Pa.s). Determine the pressure drop for each segment and the whole tube (assume the blood cells are solid spheres).
9. What is effect of anemia disease on blood viscosity? Why?

REFERENCES

[1] L. Waite, J. Fine, Applied Biofluid Mechanics, McGraw-Hill, New York, NY, 2007.

[2] Y.-C. Fung, Biomechanics: Circulation, second ed., Springer-Verlag New York Inc., New York, NY, 1997.

[3] G. Greenstone, The history of bloodletting, BC Med. J. 52 (1) (2010) 12–14.

[4] A. Zargaran, Ancient Persian medical views on the heart and blood in the Sassanid era (224–637 AD), Int. J. Cardiol. 172 (2014) 307–312.

[5] A. Holomanova, A. Ivanova, I. Brucknerova, J. Benuska, Andreas Vesalius—the reformer of anatomy, Bratisl. Lek. Listy 102 (1) (2001) 48–54.

[6] J.R. Hess, P.J. Schmidt, The first blood banker: Oswald Hope Robertson, Transfusion 40 (1) (2000) 110–113.

[7] World Health Organization (WHO). [Online]. <http://www.who.int/cardiovascular_diseases/resources/atlas/en/>, 2014.

[8] S.L. White, S.J. Chadban, S. Jan, J.R. Chapman, A. Cass, World Health Organization (WHO). [Online]. <http://www.who.int/bulletin/volumes/86/3/07-041715/en/>, 2008.

[9] F.M. White, Fluid Mechanics, seventh ed., McGraw-Hill Science, New York, NY, 2011.

[10] R.P. Chhabra, J.F. Richardson, Non-Newtonian Flow and Applied Rheology: Engineering Applications, second ed., Butterworth-Heinemann, Amsterdam and Boston, MA, 2008.

[11] G.A. Truskey, F. Yuan, D.F. Katz, Transport Phenomena in Biological Systems, Pearson Education, Upper Saddle River, NJ, 2010.

[12] F.M. White, Viscous Fluid Flow, second ed., McGraw-Hill, Inc, New York, NY, 1991.

[13] T.J. Pedley, Introduction to fluid dynamics, Sci. Mar. 61 (1) (1997) 7–24.

[14] A. Klarbring, Solid Mechanics and Its Applications, Models of Mechanics, Springer, Dordrecht, 2006.

[15] C.W. Macosto, Rheology Principles, Measurements, and Applications, Wiley-VCH, Inc, New York, NY, 1994.

[16] O.K. Baskurt, M.R. Hardeman, M.W. Rampling, H.J. Meiselman (Eds.), Handbook of Hemorheology and Hemodynamics, IOS Press, Amsterdam, 2007.

[17] A. Marcinkowska-Gapinska, J. Gapinski, W. Elikowski, F. Jaroszyk, L. Kubisz, Comparison of three rheological models of shear flow behavior studied on blood samples from post-infarction patients, Med. Biol. Eng. Comput. 45 (2007) 837–844.

[18] M. Zamir, The Physics of Coronary Blood Flow, Springer Science + Business Media, Inc., New York, NY, 2005.

[19] M. Zamir, The Physics of Pulsatile Flow, Springer-Verlag, New York, NY, 2000.

[20] J.E. Hall, A.C. Guyton, Textbook of Medical Physiology, eleventh ed., Elsevier Inc., Philadelphia, PA, 2006.

[21] N. Westerhof, N. Stergiopulos, M.I.M. Noble, Snapshots of Hemodynamics an Aid for Clinical Research and Graduate Education, Springer Science + Business Media, Inc., Boston, MA, 2005.

[22] D.R. Rubenstien, W. Yin, M.D. Frame, Biofluid Mechanics, an Introduction to Fluid Mechanics, Macrocirculation and Microcirculation, Elsevier Inc, Boston, MA, 2012.

[23] M. Ishii, T. Hibiki, Thermo-Fluid Dynamics of Two-Phase Flow, Springer Science + Business Media, Inc., New York, NY, 2006.

[24] W.F. Boron, E.L. Boulpaep, Medical Physiology: A Cellular and Molecular Approach, second ed., Saunders Elsevier, Philadelphia, PA, 2009.

[25] L. Waite, Biofluid Mechanics in Cardiovascular Systems, McGraw-Hill, New York, NY, 2006.

[26] M. Thiriet, Biology and Mechanics of Blood Flows, Part I: Biology, Springer Science + Business Media, LLC, New York, NY, 2008.

[27] E.A. Evans, R. Skalak, Mechanics and Thermodynamics of Biomembranes, CRC Press, Inc., Boca Raton, FL, 1980.

[28] I.V. Pivkin, G.E. Karniadakis, Accurate coarse-grained modeling of red blood cells, Phys. Rev. Lett. 101 (118105) (2008), pp. 118105-1-4

[29] J. Li, M. Dao, C.T. Lim, S. Suresh, Spectrin-level modeling of the cytoskeleton and optical tweezers stretching of the erythrocyte, Biophys. J. 88 (2005) 3707–3719.

[30] C. Pozrikidis (Ed.), Modeling and Simulation of Capsules and Biological Cells, Mathematical Biology and Medicine Series, Chapman & Hall/CRC, Boca Raton, FL, 2003.

[31] G. Késmárky, P. Kenyeres, M. Rábai, K. Tóth, Plasma viscosity: a forgotten variable, Clin. Hemorheol. Microcirc. 39 (2008) 243–246.

[32] B.B. Gupta, M.Y. Jaffrin, L.H. Ding, T. Dohi, Membrane plasma separation through small-area, Hollow-Fiber filters, Artif. Organs 10 (1) (1986) 45–51.

Chapter 2

Macrocirculation System

LEARNING OBJECTIVES

- Explanation of the blood circulation system.
- Categorization of blood vessels.
- Behavior of pulsatile blood flow.
- Arteries classification and properties.
- Veins classification and properties.
- Definition of vascular bifurcations and branches.
- Curved vessels aspects.
- Formulation and explanation of mechanical and elasticity properties of vessels.
- Definition of atherosclerosis.
- Description of stenosis.

2.1 INTRODUCTION

All living organs require metabolic materials (eg, oxygen, glucose, ...) to feed their living cells and also they need a mechanism to remove waste products. Organ cells can exchange metabolic materials and waste products by cellular transportation or diffusion. To transport these metabolic materials or waste products, living organs have a complicated network of blood vessels. This network has three major parts: the arteries, which carry blood and metabolic substrates from the heart; the capillaries, which have a main role of transporting metabolic substrates and waste products to living tissues; and the veins, which carry used blood (deoxygenated blood) back to the heart. Fig. 2.1 shows a schematic of these major vessels.

The blood circulation system consists of the heart, as a pump and the blood vessels. The blood vessels are classified into five types;

- Arteries,
- Arterioles,

Biofluid Mechanics. DOI: http://dx.doi.org/10.1016/B978-0-12-802408-9.00002-8

- Capillaries,
- Venules and
- Veins.

The number of arteries increases greatly from a single aorta to $\sim 10^4$ small arteries and $\sim 10^7$ arterioles [1].

The circulatory (or cardiovascular) system can be divided into three subsystems;

- *Pulmonary circulation*: Pulmonary circulation arteries transfer deoxygenated blood from the heart to the lung, and return oxygenated blood to the heart.
- *Systemic circulation*: Systemic circulation distributes oxygenated blood to all of the body's tissues, and returns deoxygenated blood back to the heart.
- *Coronary circulation*: The heart's function is unique since it can provide its own blood flow. Coronary arteries generate a blood supply to the heart muscles.

Fig. 2.2 shows schematic of cardiovascular system and its subsystems. As the figure shows the systemic and pulmonary circulation systems are placed in series in human body. Left and right circuits of diagram illustrate deoxygenated and oxygenated blood flow, respectively.

The vascular system provides two essential services: distribution and exchange. In the distribution function, the vascular system transports blood to and away from body organs. In the exchange function, gases, electrolytes and nutrients are exchanged between blood and living tissues. Additionally, some vessels, such as small arteries and arterioles, create resistance to regulate blood pressure. Capacitance is another minor function in the vascular system to regulate blood pressure and blood volume. Table 2.1 summarized the sizes and functions of human blood vessels.

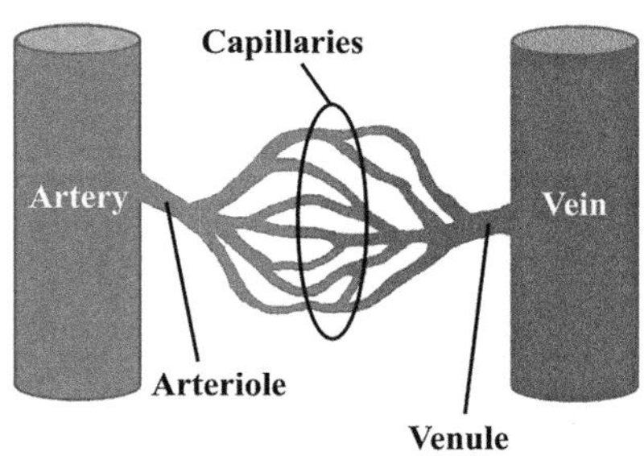

■ **FIGURE 2.1** Schematic of three major blood vessels: Artery, capillaries and vein.

The total blood volume of the body for an adult (woman 70 kg) is about 5 L and its distribution is not uniform among organs. The distribution percentage of blood volume in systemic, pulmonary and hearth circulation is ~85%, ~10% and ~5%, respectively. These percentages are not constant and the blood circulation system can be adjustable, for instance the pulmonary circulation can be much higher than 10% in some cases [1].

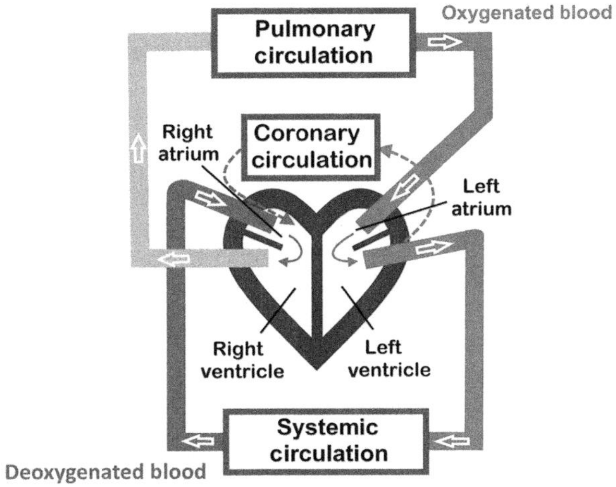

■ **FIGURE 2.2** The cardiovascular system and its subsystems: Pulmonary, systemic and coronary circulation.

Table 2.1 Approximate Geometry, Blood Velocity and Function of Human Blood Vessels

Vessel Type	Internal Diameter	Wall Thickness	Length	Average Blood Velocity	Function
Aorta	25 mm	2 mm	50 cm	45 cm/s	Distribution and pulse damping
Arteries (medium size)	4 mm	1 mm	25 cm	30 cm/s	Distribution and resistance
Arterioles (medium size)	30 μm	20 μm	1 cm	0.5 cm/s	Resistance
Capillaries (medium size)	6 μm	1 μm	0.1 cm	0.03 cm/s	Exchange
Venules (medium size)	20 μm	2 μm	0.2 cm	0.2 cm/s	Collection and capacitance
Veins (medium size)	5 mm	0.5 mm	2.5 cm	1 cm/s	Capacitance
Vena cava	30 mm	1.5 mm	50 cm	14 cm/s	Collection

The majority of the blood volume in the cardiovascular system resides in the systemic circulation veins (~3.2 L). Table 2.2 shows the distribution of the blood volume in an adult woman. For a 70 kg man, we have to add 10% to values.

The blood pressures in the systemic circulation are higher than those in the pulmonary circulation. According to Poiseuille's law, the resistance in a fluidic system depends on the pressure drop and volumetric flow rate, also it is inversely proportional to the fourth power of the tube diameter. But the steepest pressure gradient occurs for precapillary arterioles which have bigger diameters in comparison to capillaries. This feature occurs because of aggregate resistance (total resistance for parallel resistors). Table 2.3 shows the pressure drop for arteries and veins in the systemic and pulmonary circulation.

Table 2.2 Blood Volume Distribution in the Cardiovascular System [1]

Circulation Region	Volume (mL)		Percentage (%)	
Systemic circulation		4200		84
Arteries	700		14	
Capillaries	300		6	
Veins	3200		64	
Pulmonary circulation		440		8.8
Arteries	130		2.6	
Capillaries	110		2.2	
Veins	200		4	
Heart (end-diastole)	360	360	7.2	7.2
Total	5000	5000	100	100

Table 2.3 Blood Pressure Range in Systematic and Pulmonary Circulation Vessels for a Normal Adult

Region	Vessel	Pressure (mmHg)
Systemic circulation	Aorta, arteries	~80–120
	Veins	~10–15
	Vena cava	~3–5
Pulmonary circulation	Plum. arteries	~10–25
	Plum. veins	~5–7

2.2 PULSATILE FLOW PROPERTIES

The heart works as two pumps in series and each pump has two chambers; receiving and pumping chamber. Atriums (or Atria) receive blood returning from the body and ventricles pump blood to the body. Contraction of the heart muscle produces the blood flow from the right ventricle into the pulmonary artery (right side pump) and from the left ventricle into the aorta to feed systemic circulation. Each pump has valves (heart valves) to prevent the backflow of blood (eg, regurgitation problem). Contraction and relaxation of the heart muscle results in a time dependent function for blood pressure and flow rate. This mechanism creates a heartbeat which produces a pressure wave into blood vessels, which can be felt in the wrist or neck.

As has been explained in Chapter 1.2, the Womersley number (Eqs. (1.18) and (1.61)) has a key character in understanding and calculating pulsatile flow.

This number has a significant consequence on the shape of the velocity distribution [2]. The velocity profile is determined by the assumption of a Newtonian fluid with the laminar flow in the axial direction (no slip) condition in a long, circular vessel by using pulsatile flow equations. Eqs. (1.56)–(1.78) of Chapter 1.2 can be used to calculate the velocity, pressure and flow rate of a pulsatile flow in a solid or elastic tube.

Fig. 2.3 shows sequences of velocity profiles for several values of Womersley numbers [3]. For larger α values, the velocity profile is blunter near to the centerline with a maxima velocity near the wall.

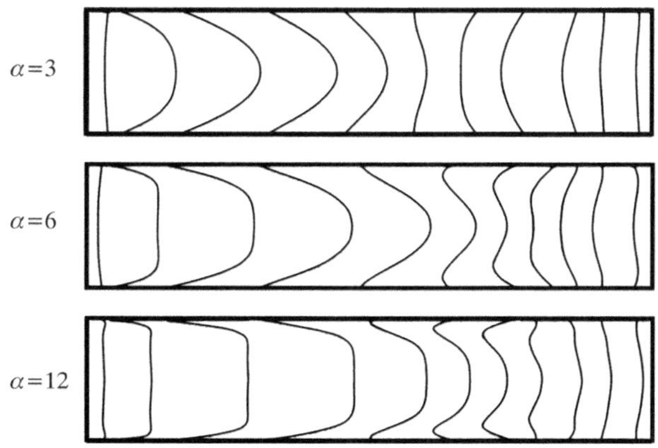

■ **FIGURE 2.3** Sequences of velocity profiles for three values of Womersley number $\alpha = 3$, 6 and 12. (The ordinate is vessel diameter and the abscissa is longitudinal cross section). *From A. Ostadfar, Design and experimental proof of selected function in implantable artificial kidney. Dissertation for the Degree of Doctor of Philosophy, 2013.*

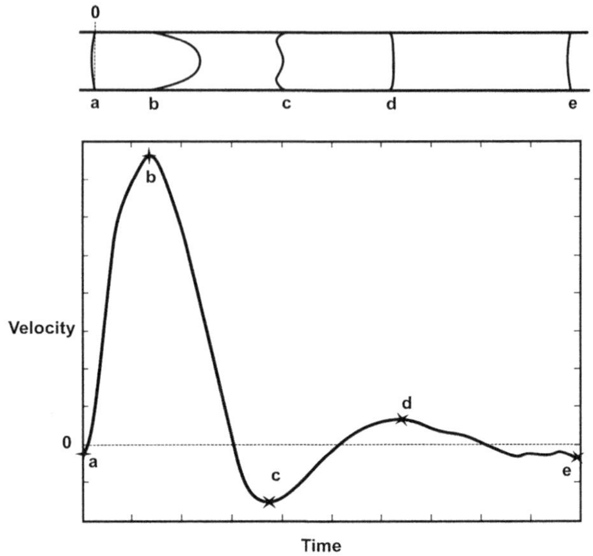

■ FIGURE 2.4 Flow velocity diagrams. Bottom: Flow velocity waveform. Top: The velocity profiles through a blood vessel.

As the Womersley number increases (for $0.2 \leq \alpha$), a phase shift (phase lag) increases between the pressure wave and the flow rate. Few methods are defined to calculate the phase shift (phase lag) between the pressure wave and flow rate in a pulsatile flow regime [2,4].

One of the useful tools in the cardiovascular system for understanding the pulsatile flow characters is the velocity waveform or flow-velocity waveform graph. This method refers to the flow velocity, which implies both speed and direction of blood flow through the vessels. Fig. 2.4 illustrates the flow-velocity waveform (bottom) and the velocity profile (top) for a pulsatile blood flow in a vessel (vessel diameter = 0.5 cm). The figure compares several points (a, b, c, d, e) between the velocity waveform graph, which numerically represents blood flow velocity, and the velocity profiles graph, which graphically represents velocity shapes of pulsatile blood flow through the blood vessel by a longitudinal cutting view.

2.3 ARTERIES

The arterial system is known as the higher pressure part of the cardiovascular system. Except for the pulmonary artery, generally arteries carry the oxygenated blood away from the heart to tissue. The human body has around 16 main arteries, Table 2.4 and Fig. 2.5 show the name and the location of these arteries in humans.

Table 2.4 Main Human Arteries

Main Arteries			
Carotid artery	Subclavian artery	Common iliac artery	Hepatic artery
Pulmonary artery	Brachial artery	Ulnar artery	Mesenteric artery
Aorta	Renal artery	Radial artery	Lateral circumflex artery
Femoral artery	Anterior tibial artery	Posterior tibial artery	Popliteal artery

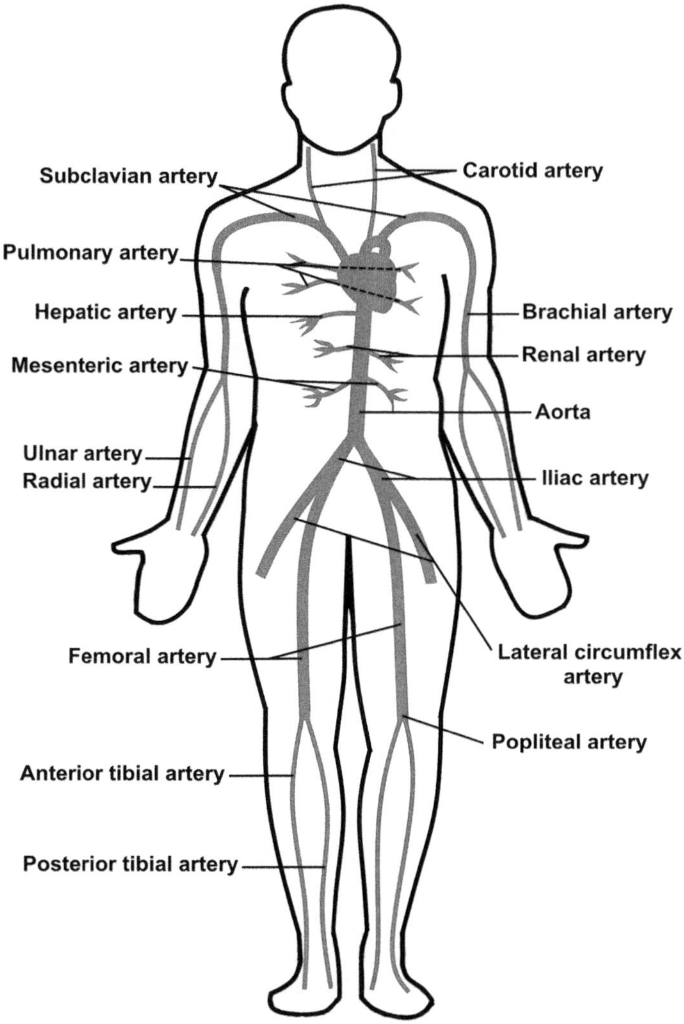

■ **FIGURE 2.5** Schematic of main arteries in human body.

Types of arteries: There are three types of arteries which can be categorized in relation to size, wall geometry and configuration.

- *Elastic arteries*: These arteries receive blood directly from the heart, such as the aorta and pulmonary arteries. The ratio of wall thickness to the diameter is high (thin wall, large diameter). The elasticity behavior (stretching and recoiling) of this type of artery helps to maintain blood pressure during systolic and diastolic function. Also their walls have a greater number of elastin (elastic fibers). Characteristically, this type of artery has a diameter of more than 10 mm.
- *Muscular arteries (or distributing arteries)*: This type of artery distributes blood to various parts of organs and living tissues. Their walls have a greater number of smooth muscles. The diameter range of the muscular arteries varies from 0.1 to 10 mm, this range is smaller than elastic arteries and larger than arterioles diameter. The radial artery and femoral artery are good examples of this type of artery.
- *Arterioles:* Arterioles are the smallest type of arteries. Arterioles work as resistors in the cardiovascular system and cause a drop in blood flow pressure. Their walls have usually only one to two layers of smooth muscle with an average vessel diameter of 30 μm. The next chapter explains more about arterioles in microcirculation system.

Typically, arteries are made of three layers [4]:

The tunica intima is the innermost layer of the artery wall. This layer is created from endothelium cells and is in direct contact with the blood.
The tunica media is the middle layer of the artery wall. This layer is made up smooth muscle cells and elastic fibers.
The tunica externa is the outermost layer of the artery wall. It is composed of fibroblast cells (collagen).

Fig. 2.6 shows the construction layers of an artery.

Blood flow properties are different in respect to the location (site) in the cardiovascular system. Fig. 2.7 illustrates a schematic of pressure-velocity waveforms at multiple sites in a cardiovascular model [5]. This cardiovascular model presents a multiscale simulation of pulsatile blood flow and flow rate within a three dimensional deformable arterial network in older individuals (>50 years of age). The average range of systolic to diastolic pressure for normal adults (young individuals) is around (120/80) mmHg but the value for people with hypertension (HTN) or older individuals, may be around (140/90) mmHg or more.

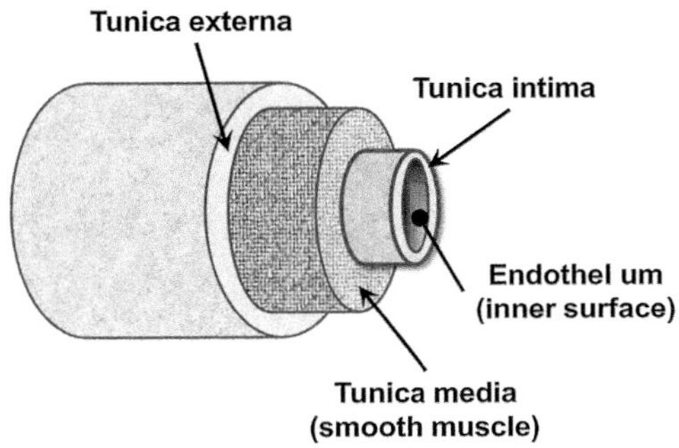

■ **FIGURE 2.6** View of an artery with three layers.

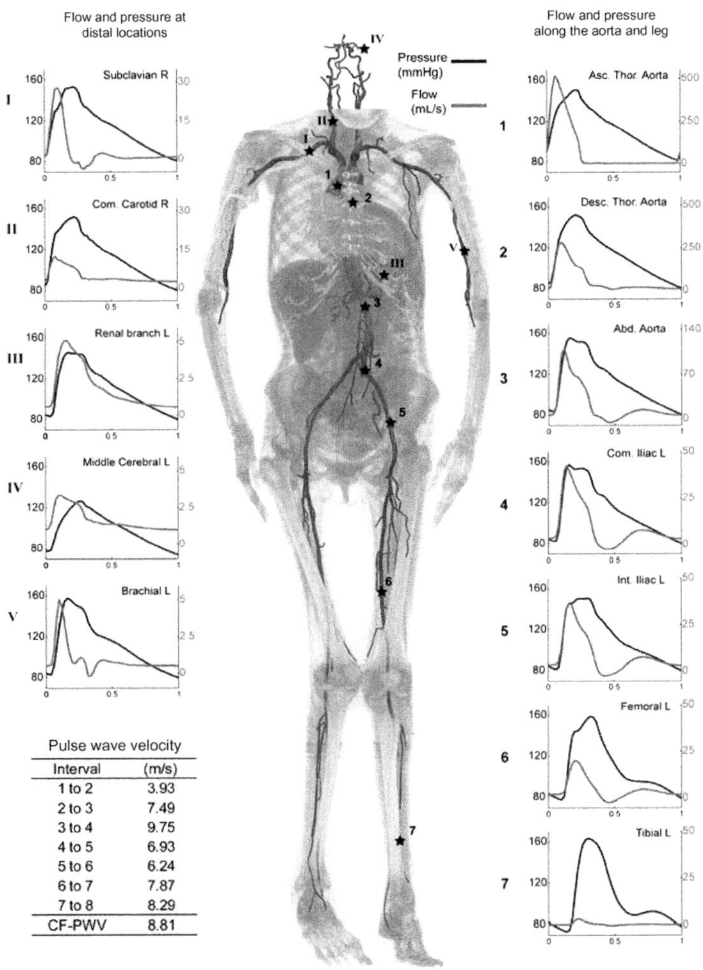

Pulse wave velocity

Interval	(m/s)
1 to 2	3.93
2 to 3	7.49
3 to 4	9.75
4 to 5	6.93
5 to 6	6.24
6 to 7	7.87
7 to 8	8.29
CF-PWV	8.81

■ **FIGURE 2.7** A three-dimensional simulation of blood pressure and flow rate waves at multiple sites of cardiovascular system for old individuals (>50 years of age). *From C.A Figueroa, J.D. Humphrey, Pressure wave propagation in full-body arterial models: a gateway to exploring aging and hypertension, vol. 10, in: Mechanics for the World: Proceedings of the 23rd International Congress of Theoretical and Applied Mechanics, ICTAM2012, 2014, pp. 382–395.*

Table 2.5 Main Human Veins

Main Veins			
Jugular veins	Subclavian vein	Iliac veins	Hepatic vein
Pulmonary vein	Brachial vein	Saphenous vein	Mesenteric vein
Vena cava	Renal veins	Radial vein	Axillary vein
Femoral vein	Anterior tibial vein	Posterior tibial vein	Popliteal vein

2.4 VEINS

The veins are known as the collection system of the cardiovascular system. In most cases this system collects deoxygenated blood from living tissues or organs and carries blood toward the heart, except the pulmonary and umbilical veins which carry oxygenated blood.

Classification of veins is based on:

- Depth (deep or superficial),
- Function (systemic, coronary or pulmonary) and
- Geometry.

The vein's walls consist of three layers, similar to the artery wall. The average diameter in the venous system is from 20 μm for venules to 30 mm for the vena cava (see Table 2.1). Table 2.5 shows the names of the major veins in humans.

Veins behave as a capacitor in the cardiovascular system. Their compliance is higher than arteries (at least in the low pressure range), a 10 mmHg increase for a relaxed vein (Vena cava) enlarges the volume by ~200% whereas this percentage can be accomplished only for 200 mmHg in an artery (Aorta). Since veins act as a volume reservoir, they are known as capacitance vessels [1]. This aspect of veins is helpful in bypass surgery especially for patients with an artery blockage problem.

Veins function: For several decades, the vein's role was assumed to be only as a return vessel to the heart. Today, several functions are defined for veins in the cardiovascular system, for instance they can store small or large amounts of blood by constricting and enlarging. Also peripheral veins can propel blood forward as a pump (*venous pump*) in the circulation system. In addition, they have a role as a controller and regulator in cardiac output. Large veins have some resistance to blood flow, and due

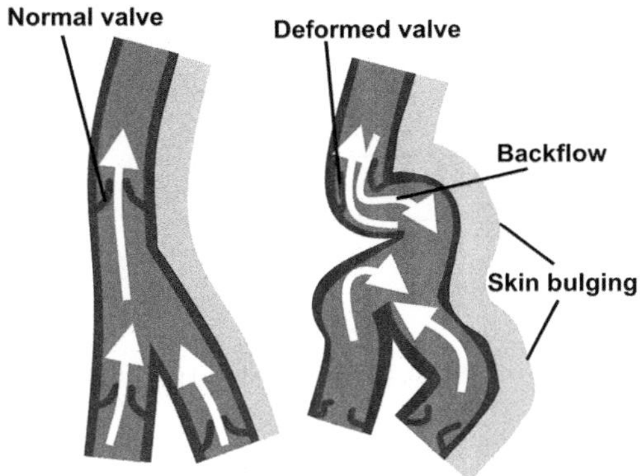

■ **FIGURE 2.8** Schematic of normal vein and valves (left) and deformed valves with a varicose problem (right).

to this aspect, sometimes (in a person lying down) the pressure in the more peripheral small veins is greater (around 4–6 mmHg) than the right arterial pressure [5].

To prevent backflow and to help in moving blood flow toward the heart, veins have *venous valves*. Dysfunction of venous valves in the leg causes "Varicose" veins. Due to overstretching of venous valves by a chronic venous high pressure, the valve will be deformed resulting in a blood backflow and a high blood pressure before the valve. The high blood pressure, blood backflow and the structure of the vein create an enlarged and twisted vein which is called a varicose vein. Fig. 2.8 shows a schematic of venous valves and a varicose problem.

2.5 VASCULAR BIFURCATIONS AND BRANCHES

Generally the arteries bifurcate several times before they become capillaries. This bifurcation involves many geometric parameters, including the bifurcation angle, the diameter ratio of daughter vessels to mother vessel, curvature radius of the bifurcation vessel and length of mother and daughter vessels.

Fig. 2.9 shows the model of a bifurcation vessel. Examples of this shape of bifurcation are carotid and iliac arteries.

The blood flow rate in the mother vessel (Q_1) has to be divided into two vessels. The lengths L_1, L_2, L_3 are affected by the branch point (O). Murray's principle presented a mathematical method to minimize the

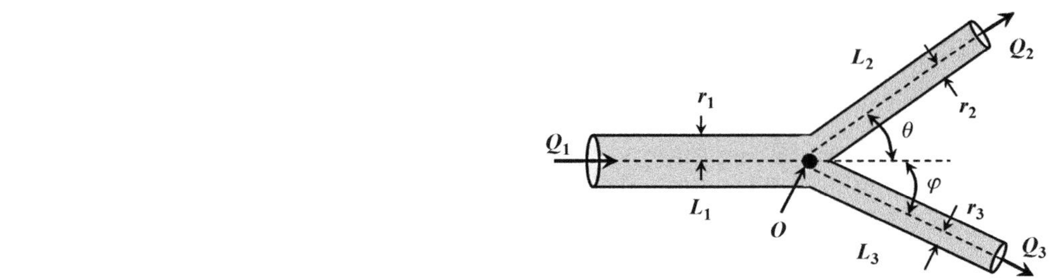

■ **FIGURE 2.9** Bifurcation of a mother blood vessel to two daughter vessels.

work in biofluid networks. There are formulae to explain the relation among mother and daughters vessels radii [6],

$$Q_1 = Q_2 + Q_3 \tag{2.1}$$

$$r_1^3 = r_2^3 + r_3^3 \tag{2.2}$$

$$\cos\theta = \frac{r_1^4 + r_2^4 - r_3^4}{2r_1^2 r_2^2} \tag{2.3}$$

$$\cos\varphi = \frac{r_1^4 - r_2^4 + r_3^4}{2r_1^2 r_3^2} \tag{2.4}$$

$$\cos(\theta + \varphi) = \frac{r_1^4 - r_2^4 - r_3^4}{2r_2^2 r_3^2} \tag{2.5}$$

Eq. (2.3) can be written as

$$\cos\theta = \frac{r_1^4 + r_2^4 - (r_1^3 - r_2^3)^{4/3}}{2r_1^2 r_2^2}, \quad \text{etc.} \tag{2.6}$$

where Q is flow rate in vessels, r is vessel diameter, θ and φ are angle of daughter vessels to mother vessel.

After bifurcation, the velocity profile will change and a flow separation occurs after entering the daughter vessel. In flow separation, the fluid flow becomes detached from the surface of the blood vessel.

In a pulsatile flow regime, the geometry of separation (separation point and size) may be different with time [6]. This geometry depends on vessel shape and pulsatile flow properties, such as the Reynolds number (*Re*) and the Womersley number (α). An increase in the Reynolds number results in an increase in the separation region, in addition it creates a recirculation flow in the separation zone. Generally, the formation of the recirculation zone and its properties are related to flow conditions. Fig. 2.10 shows the

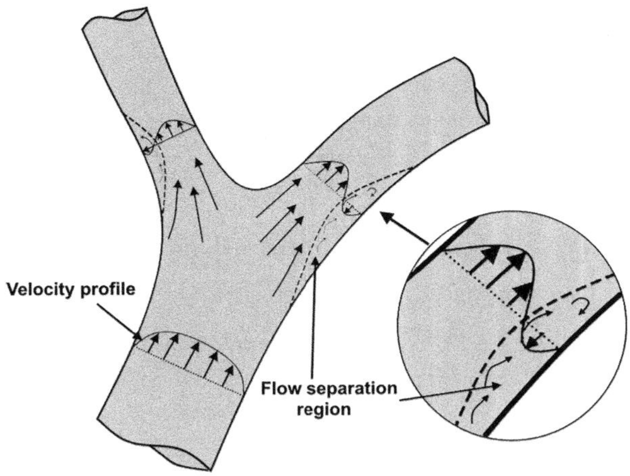

■ **FIGURE 2.10** Bifurcation of vessels and flow separation regions.

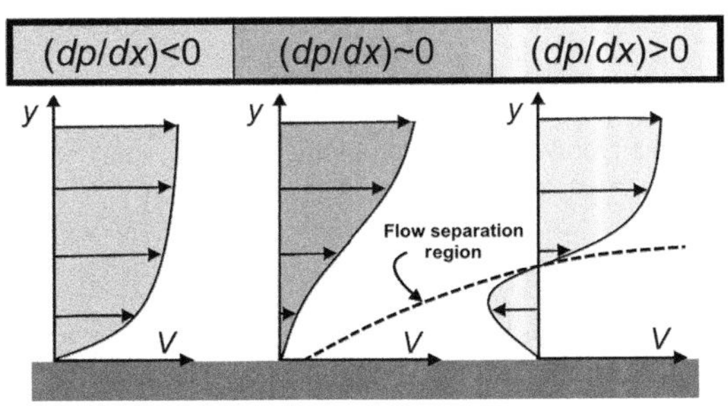

■ **FIGURE 2.11** Effect of pressure gradient on the flow velocity profile and flow separation region.

bifurcation of vessels and flow separation in each branch (daughter vessel). This aspect occurs due to momentum of fluid flow before the junction. The fluid velocity before the separation point moves the fluid toward the inner walls (V shape of daughter vessels, Fig. 2.10). After the junction, the pressure gradient will be positive $(\partial P/\partial x > 0)$ and it results in a flow separation from the wall—this character is known as the adverse pressure gradient. Fig. 2.11 shows the effects of the pressure gradient on velocity profiles. For $(\partial P/\partial x < 0)$, the velocity profile has a normal parabolic shape, with an increase in $\partial P/\partial x$ from negative to positive, a return flow occurs in the channel and the shape of velocity profile is changed by this flow. This feature causes the flow separation region in the velocity profile close to the outer wall (see Figs. 2.9 and 2.10).

The formation of the recirculation zone cannot be ignored in physiological matters since this is known as a property of viscous fluids [7]. If the angle of bifurcation between the vessel wall and the fluid flow is large (such as in a stenosed vessel), the fluid vortex within the recirculation zone will continue. Usually, the recirculation zones may act as an incentive to develop lesions such as thrombosis and atherosclerosis. Commonly, these lesions occur near the branching regions [8–10]. This problem occurs because of low endothelial wall shear stress.

2.6 BLOOD FLOW THROUGH CURVED VESSELS

Numerous blood vessels are curved and this feature affects the flow properties, especially the velocity profile and pressure. The aortic arch, circle of Willis, coronary arteries, inferior vena cava and eye arteries are examples of curvature vessels. As was explained before (Chapter 1.2), the Dean number is the ratio of the viscous force on a fluid flowing in a curved pipe to the centrifugal force. This number has a main role in the calculation of the curvature vessels. Eq. (1.21) and Fig. 1.14 explain the Dean number,

$$De = Re\sqrt{\frac{D_h}{2R}}$$

The values of the Dean number for the aortic arch, and left and right coronary arteries are 707, 47.4 and 36, respectively [2].

Generally, two flow conditions are defined within the aortic arch [11]:

- If the flow is not developed into the vessel inlet, a potential vortex in the core of the fluid can act with a velocity skewing in the direction of the inner wall in the curve. For a fully developed flow, the velocity skewing is toward the outer wall. The majority of arterial flows have velocity skewing toward the outer wall.
- Secondary flow develops as counter rotating vortices in the middle of the tube and the flow moves to the outer side. The flow can be separated by high dean numbers along the inner curve. The secondary flow can be induced by pulsatile flow through a curvature tube.

Fig. 2.12 shows the flow conditions in a curved vessel. At higher values of *De* (number 3 in the figure) the velocity develops into being more skewed. The fluid is pushed toward the outer wall by inertial forces (centrifugal forces) and secondary flow occurs to move the fluid away from the vessel wall.

In the curved vessel three velocities are developed by fluid flow: axial velocity (v_z), radial velocity (v_r) and angular velocity (v_θ). When radial velocity increases, the angular velocity decreases and the result is a secondary flow which circulates in the directions of the r and θ (see Fig. 2.12).

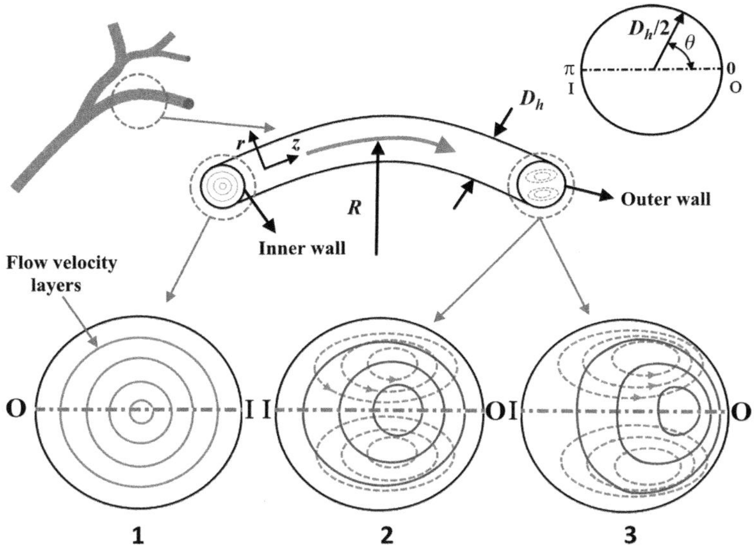

■ **FIGURE 2.12** Schematic of flow conditions through the curved vessel. Top: geometry of curved vessel. Bottom: Cross-section view of the vessel and the line of velocities at points: (1) vessel inlet; (2) vessel outlet with low Dean number (*De*); and (3) vessel outlet with high Dean number (*De*). *Dash line,* the secondary flow; *I,* inner wall; *O,* outer wall.

The conservation of mass in cylindrical coordinates is [2]

$$\frac{\partial \boldsymbol{v}_z}{\partial_z} + \frac{1}{r}\frac{\partial(r\boldsymbol{v}_r)}{\partial r} + \frac{1}{r}\frac{\partial \boldsymbol{v}_\theta}{\partial \theta} = 0 \tag{2.7}$$

The secondary flow feature has been used by heart–lung machine manufacturers to increase the performance of oxygenation of blood. The secondary flow stirs up the blood for faster oxygenation [6].

2.7 MECHANICAL AND ELASTICITY PROPERTIES OF VESSELS

As explained before, the walls of blood vessels consist of three layers: tunica intima; tunica media and tunica externa. Four structure blocks create the vessel wall [1];

- *Endothelial cells* are a thin single continuous layer that lines the interior surface of all vascular segments. This layer is very permeable and all metabolic substances can interchange quickly.

- *Elastic fibers* are very similar to rubber materials and have a key role in most of the stretch of blood vessels. They have two components: elastin, a protein that allows the tissues to recoil after deformation; and micro fibrils, which are created from glycoproteins.
- *Collagen fibers* are the key protein of the connective tissues. Under some conditions, they can be stretched by around 3–4%. They are present in all cardiovascular elements except in the capillaries.
- *Vascular smooth muscle cells (VSMCs)* are present in all blood vessels except the capillaries. Dynamically, VSMCs contract and relax the normal blood vessel wall in response to vasoactive stimuli [12].

The tunica intima consists a thin layer of endothelium cells. The tunica media, the thick layer of the vessel wall, is made up of vascular smooth muscle cells and elastic tissues. The tunica externa is mostly made up of collagen fibers.

Fig. 2.13 shows the relative abundance of these substances in the cardiovascular system. As the figure shows, the aorta and arterioles have the maximum and minimum amount of elastic fibers, respectively.

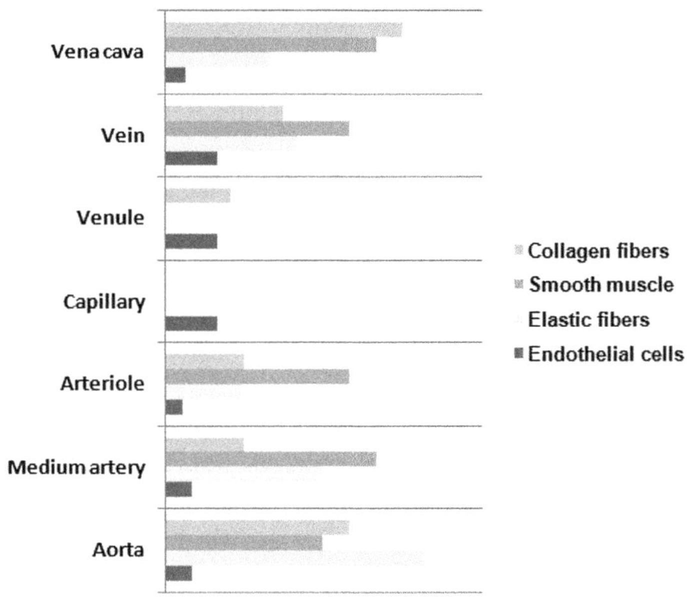

■ **FIGURE 2.13** Relative quantity of collagen fibers, smooth muscle, elastic fibers and endothelium cells in the blood vessels walls. *From W.F. Boron, E.L. Boulpaep, Medical Physiology: A Cellular and Molecular Approach, second ed., Saunders Elsevier, Philadelphia, PA, 2009.*

Fig. 2.14 shows a cylindrical blood vessel under deformation due to blood flow pressure. Two geometries are considered for the blood vessel, stress free and internal pressurization. The following equations describe the mechanical aspects of a blood vessel with internal blood pressure of P_i, tensile force (tension) T, retractive force of the wall F_R and longitudinal tether force, F_T [13]:

$$T = P_i r_i \tag{2.8}$$

$$F_R = F_T + P_i \pi r_i^2 \tag{2.9}$$

The tension is a function of stress (circumferentially directed force intensity, σ_θ) and thickness of the vessel wall (h), or

$$T = \int_{r_i}^{r_e} \sigma_\theta \, dr = \overline{\sigma_\theta} h \tag{2.10}$$

where $\overline{\sigma_\theta}$ is the mean value of circumferential stress. Also, F_T and the extending force will be balanced due to internal pressure by F_R and axial stress, σ_z, in the vessel wall,

$$F_R = 2\pi \int_{r_i}^{r_e} \sigma_z r dr = \overline{\sigma_z} \pi h (r_e + r_i) \tag{2.11}$$

where $\overline{\sigma_z}$ is the mean value of longitudinal stress. By rewriting Eqs. (2.8)–(2.11), the mean stresses are

$$\overline{\sigma_\theta} = P_i \frac{r_i}{h} \tag{2.12}$$

$$\overline{\sigma_z} = \frac{F_T}{\pi h (r_e + r_i)} + \frac{P_i r_i}{2h} \tag{2.13}$$

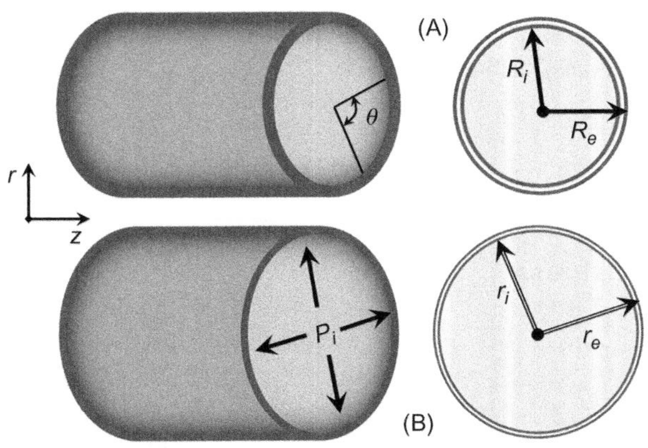

■ **FIGURE 2.14** Cylindrical geometry of a blood vessel. (A): stress free (rest). (B): internal pressurization configuration and vessel deformation.

These stresses are excellent estimates for the vessels with thin walls. Two boundary conditions and the equilibrium equation of stress distribution are defined as:

$$\sigma_r(r_i, \varepsilon) = -P_i \tag{2.14}$$

$$\sigma_r(r_e, \varepsilon) = 0 \tag{2.15}$$

$$\frac{1}{r}\frac{d}{dr}(r\sigma_r) - \frac{\sigma_\theta}{r} = 0 \tag{2.16}$$

where ε is uniform longitudinal strain. These equations demonstrate the relation between stresses in cylindrical coordinates with uniform deformation and internal pressurization.

Elasticity is a mechanical character of materials which determines the tendency of materials to return to their original shape after deformation. It can be measured by elastic modulus. The stress–strain relation is defined by Hooke's law,

$$\sigma = E\varepsilon \tag{2.17}$$

where σ is stress in N/m^2, ε is strain and E is elastic (Young's) modulus in N/m^2. Fig. 2.15 shows and compares stress–strain curves in a brittle material, a ductile material and an artery. The artery with viscoelastic behavior has a lower elastic modulus (E) due to more strain in comparison to the brittle and ductile materials.

Arteries behave as viscoelastic materials and Hooke's law (with constant elasticity) cannot be applied for these materials [4,14]. For viscoelastic materials the relation between stress and strain is as [15]:

$$\sigma = \sigma(\varepsilon, \dot{\varepsilon}) \tag{2.18}$$

where σ is stress, ε is strain and $\dot{\varepsilon}$ is strain rate (or $d\epsilon/dt$). The Maxwell, Kelvin–Voigt and standard solid model are common models for viscoelastic materials, such as arteries.

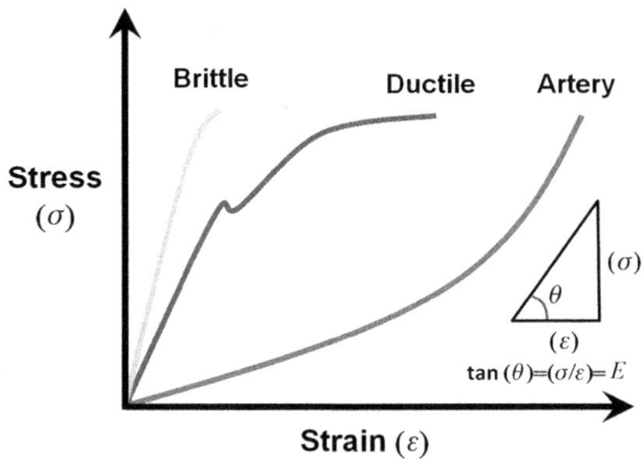

FIGURE 2.15 Stress–strain curves for a brittle material, a ductile material and a blood vessel (artery).

The Maxwell model:

$$\frac{\dot{\sigma}}{E} + \frac{\sigma}{\eta} = \dot{\varepsilon} \tag{2.19}$$

The Kelvin–Voigt model:

$$\sigma = E\varepsilon + \eta\dot{\varepsilon} \tag{2.20}$$

The standard solid model is a complex of the Kelvin–Voigt and Maxwell models [15]:

$$(E_1 + E_2)\sigma + \eta\dot{\sigma} = (E_1 E_2 \varepsilon + E_1 \eta \dot{\varepsilon}) \tag{2.21}$$

where E is elastic modulus, η is viscosity, E_1 and E_2 are elastic modules in standard solid model.

A new definition (without wall thickness) is introduced for the pressure–strain elastic modulus (E_P) [16],

$$E_P = D_d \frac{(P_s - P_d)}{(D_s - D_d)} \tag{2.22}$$

where $(P_s - P_d)$ and $(D_s - D_d)$ are pulse pressure difference and pulse diameter difference, respectively, D_d is the vessel diameter at end of diastolic, "*s*" and "*d*" denote systolic and diastolic conditions.

2.8 ATHEROSCLEROSIS CHARACTERISTICS

Atherosclerosis (hardening and narrowing of arteries) is one of the diseases of the arteries which is widespread throughout the world and especially in western and developing countries. Atherosclerosis is mostly an intimal disease of the medium and large arteries [16].

Since the main effect of arterial diseases is to limit blood flow to the organs and tissues, it is vital to consider these diseases here. The initiating process of atherosclerosis is not very clear. When the endothelium layer becomes damaged (in micro size), low density lipoproteins (LDLs) have a tendency to accumulate in the injured area. Plaque formation begins with the accumulation of LDL particles in the intima layer [17]. Oxidative stress induces cytokine proteins that help the expression of adhesion molecules, which allow white blood cells to adhere to the endothelium (intima layer) [18]. The deposition of blood substances and particles to the damaged wall will continue to initiate the formation of a thrombus and low blood flow to the organs and tissue consequently causing infarction or ischemia.

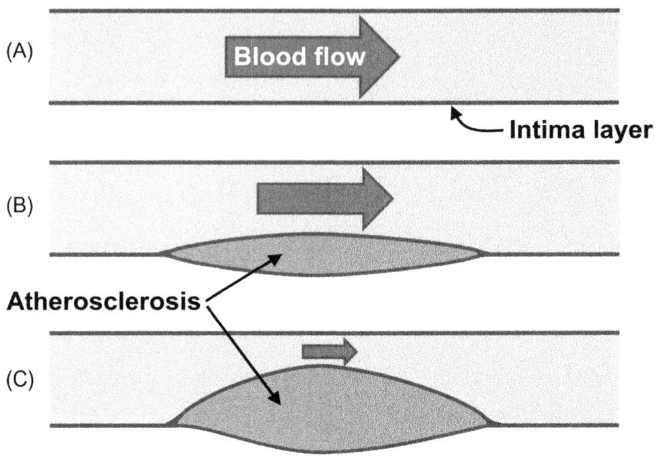

■ **FIGURE 2.16** Schematic of atherosclerosis. (A): A vessel with normal blood flow. (B): Formation of atherosclerosis with ¼ obstruction of blood vessel diameter. (C): Acute atherosclerosis with ¾ obstruction of blood vessel diameter and low blood flow rate.

Fig. 2.16 illustrates the process of atherosclerosis disease. This disease takes several decades from the point of initiation to the end stage of the disease.

Location of atherosclerosis: The main sites of symptomatic arterial diseases are in [16]:

- Carotid arteries
- Cerebral arteries
- Epicardial coronary arteries
- Femoral and iliac arteries
- Infrarenal part of abdominal aorta
- Locations with local hemodynamic flow changes such as bifurcation, dilatation, dividing and constriction.

Risk factors of atherosclerosis: Risk factors for this disease include [16]:

- Age
- Gender (male)
- Hypertension
- Hypercholesterolemia
- Cigarette smoking
- Diabetes mellitus
- Physical inactivity
- Adiposity (body fat)
- Race
- Reduced HDL cholesterol
- Elevated LDL cholesterol

- Elevated lipoprotein (a)
- Excess oxidative stress
- Elevated C-reactive protein
- Elevated homocysteine

Mechanical aspect in atherosclerosis: It is difficult to determine the particular locations and causes of atherosclerosis. Attention has been directed toward the role of a number of the mechanical factors in atherosclerosis [16]:

- Tensile stress in arterial walls
- Shear stress (low shear stress develops atherosclerosis)
- Velocity profile and flow types (fluctuation and pulsatile flow prevent the cells from aligning and it increases contacts between blood cell and damaged intima)
- Pulsatile stress and strain (they are relevant in the atherosclerosis development).

2.9 BLOOD FLOW THROUGH STENOSIS

Stenosis is an abnormal narrowing or contraction through blood vessels. It has been shown by a number of pieces of evidence that vascular biofluid dynamics play a key role in the development of arterial stenosis. An understanding of blood flow through stenosis vessels may help to solve current problems in this field of cardiovascular system. Stenosis can have an affect on the vascular system or on valves, such as stenosis of the coronary artery or the aortic valve.

A good example of stenosis is a tube with a converging-diverging throat. Fig. 2.17 shows the geometry of a stenosis case in an artery. The cylindrical coordinates (r, θ, z) are considered for the model. The angle of φ creates the possibility of different forms of stenosed artery: diverging (or diffuser) for $\varphi > 0$, straight form for $\varphi = 0$ and converging (or nozzle) shape for $\varphi < 0$ (see Fig. 2.17).

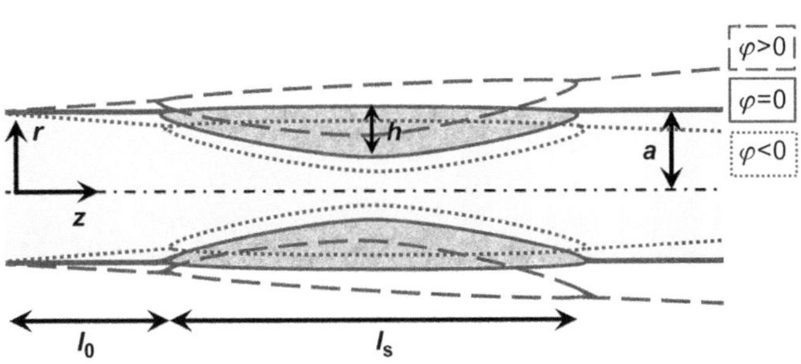

■ **FIGURE 2.17** Schematic of a blood vessel with stenosis problem.

Considering human blood vessel properties, the model has a time-variant geometry [19], so;

$$R(z,t)=\begin{cases}\left[(mz+a)-\dfrac{h\sec\varphi(z-l_0)}{h^2\sin\varphi^2-\dfrac{l_s^2}{4}}\{l_s-(z-l_0)\}\right]A(t),| & l_0\le z\le<l_0+l_s\\ (mz+a)A(t), & \text{otherwise}\end{cases} \tag{2.23}$$

where $R(z,t)$ represents the radius of stenosed region, l_s is the length of stenosis region, l_0 is the stenosis location from entrance of vessel, h is maximum height of stenosis and m is equal to tan φ which denotes the angle of vessel with respect to the z axis.

The parameter of $A(t)$ is time-variant,

$$A(t)=1-b(\cos\omega t-1)e^{-b\omega t} \tag{2.24}$$

where ω is angular velocity ($\omega=2\pi f_h$) and f_h is the frequency of pulsatile blood flow and b is a constant.

Considering two dimensional flow, axisymmetric, unsteady, fully developed and non-Newtonian blood flow for a generalized power law model, the governing equations for momentum and continuity in the cylindrical coordinate system can be represented as [19]

$$\left(\frac{\partial \mathbf{V}_r}{\partial t}+\mathbf{V}_r\frac{\partial \mathbf{V}_r}{\partial r}+\mathbf{V}_z\frac{\partial \mathbf{V}_r}{\partial z}\right)=-\frac{1}{\rho}\frac{\partial P}{\partial r}-\frac{1}{\rho}\left[\frac{1}{r}\frac{\partial(r\tau_{rr})}{\partial r}+\frac{\partial(\tau_{rz})}{\partial z}\right] \tag{2.25}$$

$$\left(\frac{\partial \mathbf{V}_z}{\partial t}+\mathbf{V}_r\frac{\partial \mathbf{V}_z}{\partial r}+\mathbf{V}_z\frac{\partial \mathbf{V}_z}{\partial z}\right)=-\frac{1}{\rho}\frac{\partial P}{\partial z}-\frac{1}{\rho}\left[\frac{1}{r}\frac{\partial(r\tau_{rz})}{\partial r}-\frac{\partial(\tau_{zz})}{\partial z}\right] \tag{2.26}$$

$$\left(\frac{\partial \mathbf{V}_r}{\partial r}+\frac{\mathbf{V}_r}{r}+\frac{\partial \mathbf{V}_z}{\partial z}\right)=0 \tag{2.27}$$

The shear stress can be written as [19]

$$\tau_{zz}=-2\left\{m\left|\left[\left(\frac{\partial \mathbf{V}_r}{\partial r}\right)^2+\left(\frac{\mathbf{V}_r}{r}\right)^2+\left(\frac{\partial \mathbf{V}_z}{\partial z}\right)^2+\left(\frac{\partial \mathbf{V}_r}{\partial z}+\frac{\partial \mathbf{V}_z}{\partial r}\right)^2\right]^{\frac{1}{2}}\right|^{n-1}\right\}\left(\frac{\partial \mathbf{V}_z}{\partial z}\right) \tag{2.28}$$

$$\tau_{rz}=-\left\{m\left|\left[\left(\frac{\partial \mathbf{V}_r}{\partial r}\right)^2+\left(\frac{\mathbf{V}_r}{r}\right)^2+\left(\frac{\partial \mathbf{V}_z}{\partial z}\right)^2+\left(\frac{\partial \mathbf{V}_r}{\partial z}+\frac{\partial \mathbf{V}_z}{\partial r}\right)^2\right]^{\frac{1}{2}}\right|^{n-1}\right\}\left(\frac{\partial \mathbf{V}_z}{\partial z}+\frac{\partial \mathbf{V}_r}{\partial z}\right) \tag{2.29}$$

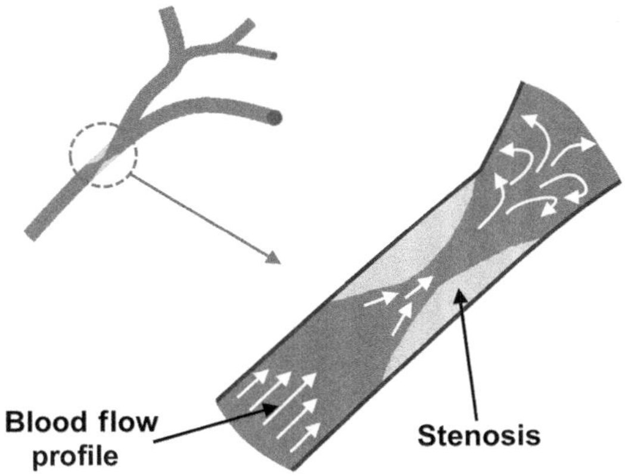

■ **FIGURE 2.18** Schematic of an axisymmetric stenosis and blood flow profiles. The blood flow profile is laminar before the stenosis and turbulent after the stenosis.

$$\tau_{rr} = -2\left\{ m \left| \left[\left(\frac{\partial \mathbf{V}_r}{\partial r}\right)^2 + \left(\frac{\mathbf{V}_r}{r}\right)^2 + \left(\frac{\partial \mathbf{V}_z}{\partial z}\right)^2 + \left(\frac{\partial \mathbf{V}_r}{\partial z} + \frac{\partial \mathbf{V}_z}{\partial r}\right)^2 \right]^{\frac{1}{2}} \right|^{n-1} \right\} \left(\frac{\partial \mathbf{V}_r}{\partial z}\right) \tag{2.30}$$

where "n" shows the behavior of fluid in the power law equation and it is $0 < n < 1$ for the case of blood (pseudoplastic fluids). There are some references [19–23] which have clarified and formulated several aspects and conditions of stenosis problems.

Fig. 2.18 shows a blood flow profile before and after a stenosis in an artery. The blood flow has laminar and turbulent profiles before and after the stenosis, respectively. The magnitude of turbulence flow depends on the blood flow properties (flow rate, pressure, etc.), artery type and stenosis conditions.

The degree of stenosis is measured by comparing the area of reduction to the normal part of the blood vessel and the percentage of the reduction demonstrates the degree of stenosis [24],

$$\text{Percentage of stenosis} = \left(1 - \frac{D_{\text{stenosis}}}{D_{\text{normal}}}\right) \times 100 \tag{2.31}$$

The best method to measure the degree of stenosis is catheterization laboratory (Cath lab). In this technique, image processing equipment is used to visualize the stenosis or other blood vessel problems. Fig. 2.19 illustrates several common shapes of stenosis which usually occur in blood vessels.

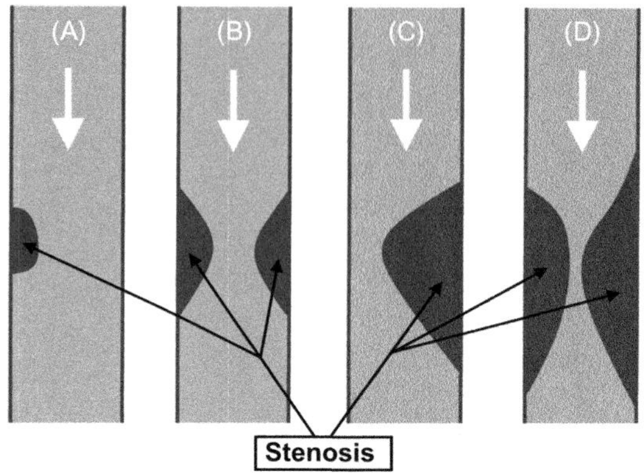

■ **FIGURE 2.19** Pattern of common stenosis types in arteries and their degree: (A): 25%, (B): 62%, (C): 72% and (D): 90%.

CHAPTER SUMMARY

- The vascular system has three major parts: *The arteries*, which carry blood and metabolic substrates from the heart; *the capillaries*, which have a main role of transporting metabolic substrates and waste products to living tissues; *the veins*, which carry used blood (deoxygenated blood) back to the heart.
- *Pulmonary circulation*: Pulmonary circulation arteries transfer deoxygenated blood from the heart to the lung, and return oxygenated blood to the heart.
- *Systemic circulation*: Systemic circulation distributes oxygenated blood to all of the body's tissues, and returns deoxygenated blood back to the heart.
- *Coronary circulation*: The heart's function is unique since it can provide its own blood flow. Coronary arteries generate a blood supply to the heart muscles.
- The Womersley number (Eqs. (1.18) and (1.61)) has a key character in understanding and calculating pulsatile flow.
- *Types of arteries*: Elastic arteries, Muscular arteries (or distributing arteries), Arterioles.
- *Arteries are made of three layers*: the tunica externa, the tunica media the tunica intima.
- *Classification of veins* is based on: Depth (deep or superficial), Function (systemic, coronary or pulmonary) and Geometry.
- The *vessel bifurcation* involves many geometric parameters, including the bifurcation angle, the diameter ratio of daughter vessels to mother vessel, curvature radius of the bifurcation vessel and length of mother and daughter vessels.

- *Dean number* has a main role in the calculation of the curvature vessels.
- *Arteries* behave as *viscoelastic* materials.
- The *Maxwell, Kelvin–Voigt and standard solid model* are common models for viscoelastic materials, such as arteries.
- *Atherosclerosis* (hardening and narrowing of arteries) is mostly an intimal disease of medium and large arteries.
- *Stenosis* is an abnormal narrowing or contraction of blood vessels and it can have an effect on the vascular system or on valves, such as stenosis of the coronary artery or the aortic valve.

PROBLEMS

1. List the blood vessels in the cardiovascular system.
2. Explain the cardiovascular components and its subsystems.
3. The majority of the blood resides in:
   (a) arteries (b) veins (c) capillaries (d) equal in all vessels
4. Which layer of the artery wall is in contact with blood?
5. Discuss blood flows in curved vessels.
6. Which vessel has a tendency to get varicose disease and why?
7. Which blood vessel (or part) is more elastic than the other vessels and why?
8. Discuss viscoelastic behavior of arteries.
9. Explain the role of mechanical factors in atherosclerosis disease.
10. What different is between atherosclerosis and stenosis?
11. Discuss blood flow regimes in an artery with stenosis?

REFERENCES

[1] W.F. Boron, E.L. Boulpaep, Medical Physiology: A Cellular and Molecular Approach, second ed., Saunders Elsevier, Philadelphia, PA, 2009.

[2] G.A. Truskey, F. Yuan, D.F. Katz, Transport Phenomena in Biological Systems, Pearson Education, Upper Saddle River, NJ, 2010.

[3] A. Ostadfar, Design and experimental proof of selected function in implantable artificial kidney. Dissertation for the Degree of Doctor of Philosophy, 2013. <http://summit.sfu.ca/item/13866>.

[4] L. Waite, J. Fine, Applied Biofluid Mechanics, McGraw-Hill, New York, NY, 2007.

[5] C.A Figueroa, JD. Humphrey, Pressure wave propagation in full-body arterial models: a gateway to exploring aging and hypertension, vol. 10, in: Mechanics for the World: Proceedings of the 23rd International Congress of Theoretical and Applied Mechanics, ICTAM2012, 2014, pp. 382–395.

[6] A.C. Guyton, J.E. Hall, Textbook of Medical Physiology, eleventh ed., Elsevier Inc., Philadelphia, PA, 2006.

[7] Y.C. Fung, Biomechanics: Circulation, second ed., Springer-Verlag New York Inc., New York, NY, 1997.
[8] D.R. Rubenstien, W. Yin, M.D. Frame, Biofluid Mechanics, an Introduction to Fluid Mechanics, Macrocirculation and Microcirculation, Elsevier Inc., Boston, MA, 2012.
[9] M. Motomiya, T. Karino, Flow patterns in the human carotid artery bifurcation, Stroke 15 (1) (1984) 50–56.
[10] Y. Louvard, M.-C. Morice, T. Hovasse, T. Lefevre, Percutaneous Coronary, Interv. Cardiol. Rev. 5 (5) (2010) 58–62.
[11] N. Arslan, Computational modeling of flow inside a diseased carotid bifurcation, Math. Comput. Appl. 11 (3) (2006) 173–180.
[12] R.P. Metz, J.L. Patterson, E. Wilson, Vascular smooth muscle cells: isolation, culture, and characterization, Methods Mol. Biol. 843 (2012) 169–176.
[13] T.R. Canfield, P.B. Dobrin, Mechanics of blood vessels, Biomedical Engineering Fundamentals, CRC-Taylor & Francis Group, LLC, Boca Raton, FL, 2006. ch. 57.
[14] L. Waite, Biofluid Mechanics in Cardiovascular Systems, McGraw-Hill, New York, NY, 2006.
[15] N. Özkaya, M. Nordin, D. Goldsheyder, D. Leger, Fundamentals of Biomechanics: Equilibrium, Motion, and Deformation, third ed., Springer Science + Business Media, New York, NY, 2012.
[16] W.W. Nichols, M.F. O'rouke, Mcdonald's Blood Flow in Arteries, Hodder Arnold, New York, NY, 2005.
[17] P. Libby, P.M. Ridker, A. Maseri, Inflammation and atherosclerosis, Circulation 105 (2002) 1135–1143.
[18] Y. Huo, K. Ley, Adhesion molecules and atherogenesis, Acta Physiol. Scand. 173 (1) (2001) 35–43.
[19] P.K. Mandal, An unsteady analysis of non-Newtonian blood flow through tapered arteries with a stenosis, Int. J. Non-Linear Mech. 40 (2005) 151–164.
[20] S. Kumar, C. Diwakar, A mathematical model of power law fluid with an application of blood flow through an artery with stenosis, Adv. Appl. Mathe. Biosci. 4 (2) (2013) 51–61.
[21] N. Padmanabhan, Mathematical model of arterial stenosis, Med. Biol. Eng. Comput. 18 (1980) 281–286.
[22] K.C. Ang, J.N. Mazumdar, Mathematical modelling of three-dimensional flow through an asymmetric arterial stenosis, Math. Comput. Model. 25 (1) (1997) 19–29.
[23] J.N. Mazumdar, Biofluid Mechanics, World Scientific Publishing Co. Pte. Ltd, Singapore, 2004.
[24] D.N. Ku, Blood flow in arteries, Annu. Rev. Fluid Mech. 29 (1997) 399–434.

Chapter 3

Microcirculation System

LEARNING OBJECTIVES

- Explanation of the blood microcirculation system.
- Arterioles and their aspects, such as blood velocity, wall shear stress and pressure drop.
- Capillaries, their categorization, blood velocity and pressure drop.
- Venules and their properties.
- Definition of Fahraeus and Fahraeus–Lindqvist effects.
- Mass transport definitions and equations such as Fick's law, Stokes–Einstein equation for diffusion, convection in porous media and biological tissues.
- Definition of porous medium properties such as porosity, tortuosity and permeability.
- Formulation and explanation of governing equations in porous media, Darcy's law and Brinkman equation.
- Explanation of fluid transport in poroelastic media.

3.1 INTRODUCTION

The first function of the blood circulation system is to provide suitable conditions for the living tissues. The microcirculation system is the last station of this mission. The arterioles, venules and capillaries are major parts of this system. The capillary flow provides nutritional needs for most tissues and nonnutritional for a few tissues [1].

Fig. 3.1 shows a schematic of microcirculation. As the figure shows, arterioles and venules are categorized into four orders, from the first order which is connected to an artery or vein to the fourth order which is connected to capillaries. The diameter of blood vessels decreases by the development of bifurcation in the cardiovascular system. Tables 2.1–2.3 and Fig. 2.13 (see chapter: Macrocirculation System) denote the

Biofluid Mechanics. DOI: http://dx.doi.org/10.1016/B978-0-12-802408-9.00003-X

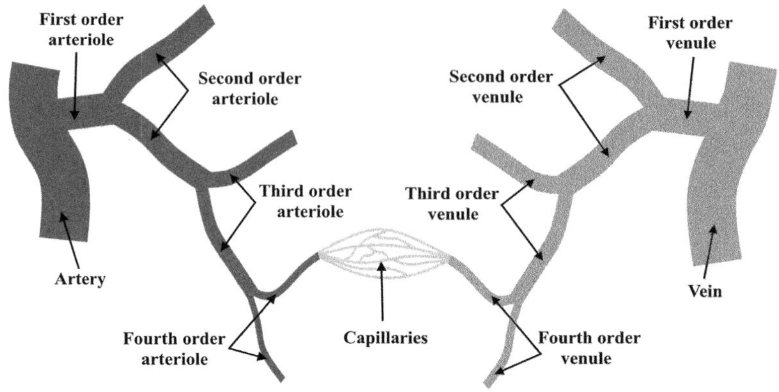

■ **FIGURE 3.1** Schematic of the microcirculation system.

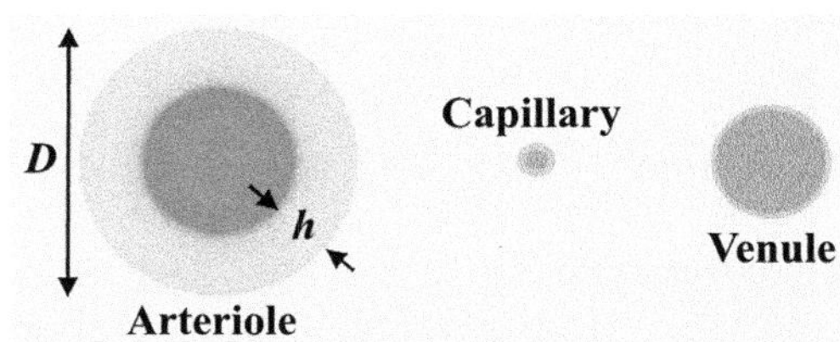

■ **FIGURE 3.2** Schematic of the cross-sections of microcirculation vessels and the approximate relation between wall thicknesses to vessel diameter (*h/D*) for average size of microvessel types.

functional information about the microcirculation system. The wall thicknesses of blood vessels are different along the cardiovascular system. Fig. 3.2 compares an approximate relation of wall thickness (h) to vessel diameter (D) in the microcirculation system. As the figure shows, the minimum and maximum relation occurs for venules and arterioles, respectively. Numerically, the approximate relation of (h/D) is 2/7, 1/8 and 1/12 for the average type of arteriole, capillary and venule, respectively.

3.2 ARTERIOLES AND BLOOD FLOW ASPECTS

One of important aspects of the microcirculation system is the maximum change in blood pressure and velocity. The blood pressure range varies from around 90 to 10 mmHg from the first order of arteriole to capillaries in the microcirculation system. The largest pressure drop in the

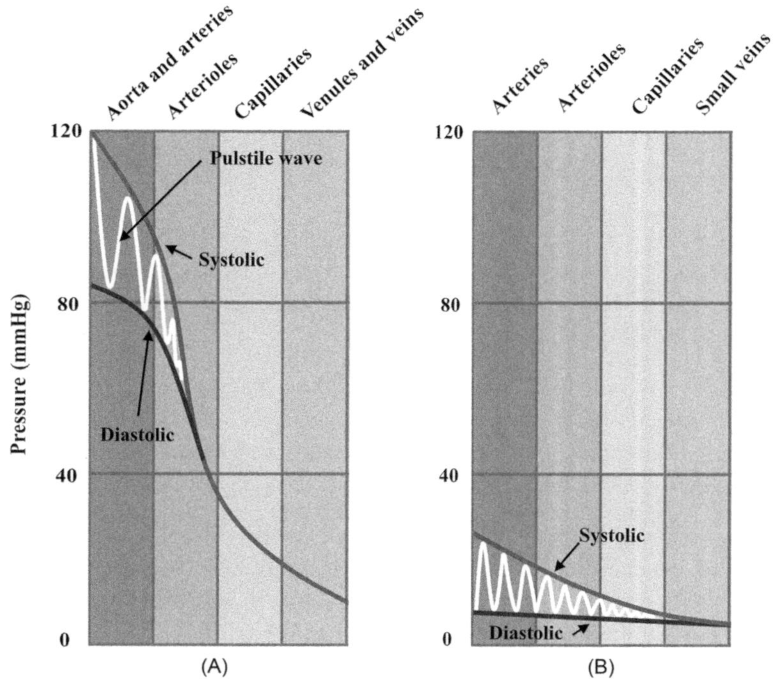

■ **FIGURE 3.3** Systolic and diastolic blood pressure ranges for (A) the systemic circulation system and (B) the pulmonary circulation system.

systematic system occurs in the arterioles, from around 90 to 15 mmHg. This means that the arterioles have the main role in the control of blood flow into living organs. In the pulmonary system, the blood pressure in arterioles ranges from 20 to less than 10 mmHg. Fig. 3.3 illustrates the blood pressure along the systemic and pulmonary systems. This figure shows the blood pressure profile for all blood vessels in the cardiovascular system. As the figure shows, the blood pressure range in the systematic system is higher than in the pulmonary system. The blood pulsatile flow can affect most of the systemic arterioles and also most of the pulmonary system in the microcirculation system.

Blood flow velocity: Flow aspects of microcirculation vary during the cardiac cycle. The velocity profile in arterioles with an internal diameter less than 60 μm is blunted [2]. The degree of blunting will be increased by the decreasing vessel diameter [3,4]. The *velocity profile* can be rewritten by the empirical relationship from Eqs. (1.23) and (1.24) in cylindrical coordinates (r, θ, z) as [2]

$$v(r,t) = v_{\max}\left(1 - \left(\frac{r}{R}\right)^{k}\right)f(t) \tag{3.1}$$

where $f(t)$ is a periodic function of pulsatile flow, R is vessel diameter, r is local radial position and k is the degree of blunting and for normal flow rate it can be described as [2,5]

$k = 2$ for parabolic profile

$2.4 \leq k \leq 4$ for arterioles with $17\ \mu m \leq$ diameter $\leq 32\ \mu m$

In a large portion of systemic and pulmonary microcirculation, pulsatile flow does not occur and the term of periodic function is close to one ($f(t) \approx 1$) in Eq. (3.1) (see Fig. 3.3).

If the size of red blood cell (RBC; $\sim 8\ \mu m$) becomes a significant fraction of the blood vessel diameter, blunting would be expected. In the first orders of arterioles, low shear rate can create the blunted velocity profile as a result of RBCs aggregation.

Considering the blood flow network and mother–daughters branching, several generation of arterioles (or venules) will exist between an artery (or vein) and a capillary. Eq. (3.2) denotes the *mean blood flow velocity* (v_n) in an arbitrary generation (g) of a blood vessel [6],

$$v_n = 2^{-(n-1)}\left(\frac{r_1}{r_n}\right)^2 v_1 \tag{3.2}$$

where r_n is the vessel radius of generation (n), r_1 and v_1 are the radius and mean flow velocity of generation number 1, respectively (see Fig. 3.4).

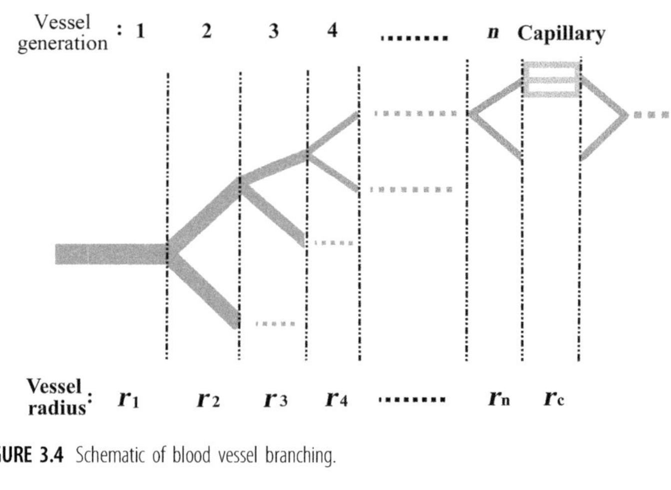

■ **FIGURE 3.4** Schematic of blood vessel branching.

Wall shear stress: Eq. (3.3) denotes the general equation of wall shear stress (WSS) measurement for a Newtonian fluid,

$$\tau_w = \frac{2\mu_p V_{max}}{R} \tag{3.3}$$

where μ_p is plasma viscosity, V_{max} is maximum velocity at center line ($V_{max} = 2V_{ave}$) and R is vessel radius.

Regarding the fluidic behavior of blood, there is empirical correction factor for Eq. (3.3) and it can be written as [7]

$$\tau_w = \frac{2\mu_p V_{max}}{R}\left(m_1 e^{m_2 H_d} \ln R + m_3 H_d + m_4\right) \tag{3.4}$$

where $m_1 = 0.0515$, $m_2 = 4.732$, $m_3 = 0.6134$, $m_4 = 1.0548$, H_d is discharge hematocrit ($0.25 \leq H_d \leq 0.55$) and the vessel radius range is $15\ \mu m \leq R \leq 70\ \mu m$.

To calculate the blood *pressure drop* in arterioles, the best method is to use the Poiseuille equation (Eq. 1.49) or

$$\Delta P = \frac{8\mu LQ}{\pi R^4}$$

3.3 CAPILLARIES

Capillaries are classified into three groups based on their degree of leakiness [1]:

- *Continuous capillaries* are the most common type of capillaries. In this type, interendothelial junctions are 10 to 15 nm wide (eg, skeletal muscle).
- *Fenestrated capillaries* have a very thin layer of endothelial cells with perforated fenestrations (eg, small intestine, exocrine glands).
- *Discontinuous capillaries* have fenestrations and also large gaps on their walls (eg, liver).

  Fig. 3.5 shows three types of capillaries.
- *Velocity of blood flow in capillary*: As Fig. 3.5 shows, a capillary diameter is around or (often) smaller than the RBC diameter and the RBCs have to deform (like parachute) when they cross the capillaries. Due to this aspect, the flow velocity in a capillary can be equal to the RBC velocity. The blood flow velocity reduces (approximately linear) in the microcirculation system from arterioles with a diameter around 110 μm to capillaries with a diameter of around 5 μm. Fig. 3.6A illustrates blood flow velocity in the microcirculation system

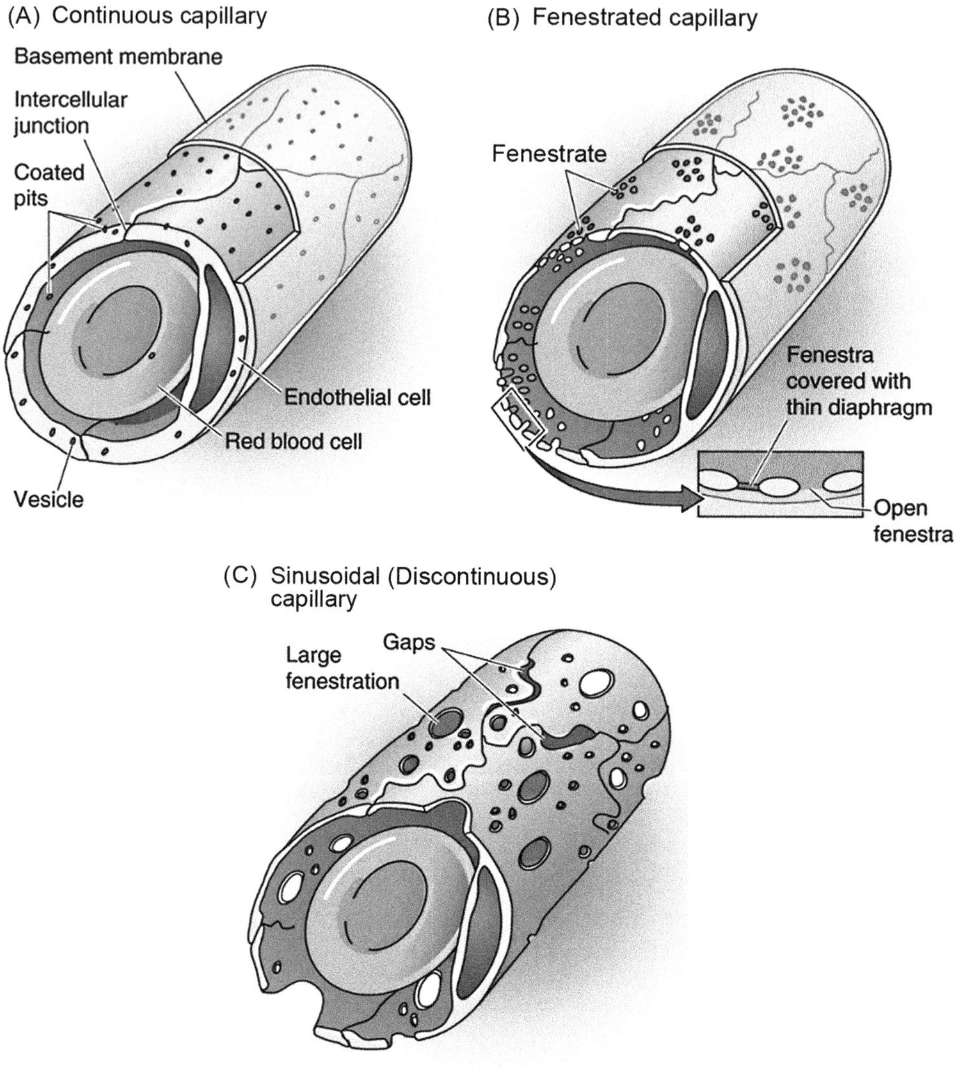

■ **FIGURE 3.5** Types of capillaries. (A) Continuous capillaries, (B) fenestrated capillaries and (C) discontinuous capillaries. *From W.F. Boron, E.L. Boulpaep, Medical Physiology: A Cellular and Molecular Approach, second ed., Saunders Elsevier, Philadelphia, PA, 2009 by permission.*

including capillaries [6]. As the figure shows, the range of blood velocity in a capillary is less than 0.2 cm/s.

- *Blood pressure in capillary*: For the capillary exchange, the blood pressure and the internal surface area of vessel have the main roles. The blood pressure drop is a function of vessel length, vessel diameter, flow rate and viscosity of blood. Intravascular pressure

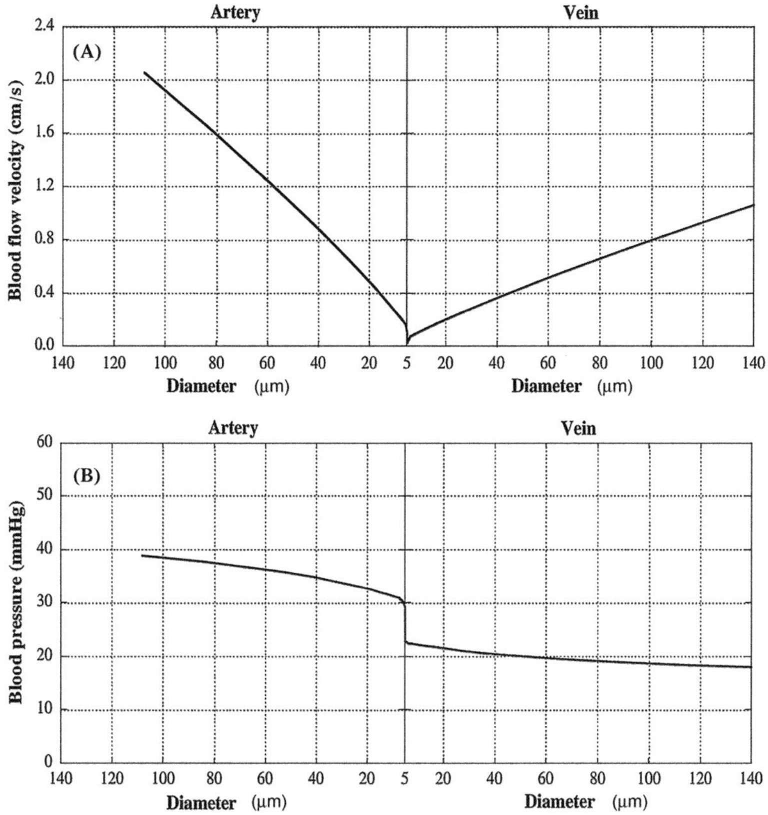

■ **FIGURE 3.6** Distribution of mean blood flow velocity (A) and blood pressure (B) in the microcirculation network. *From T. Nagaoka, H. Yanagida, T. Saitohm, A. Kamiya, T. Hein, L. Kuo, et al., A mathematical model for the distribution of hemodynamic parameters in the human retinal microvascular network, J. Biorheol., 23 (2009) 77–86 by permission.*

decreases in the microcirculation system. Fig. 3.6B shows blood pressure distribution versus blood vessel diameter in the microcirculation system. Fig. 3.7 is a mathematical model of the hemodynamic parameters in the human retinal microvascular system [6]. This model confirms the previous values of pressure drop in the microcirculation system.

- *Starling equation*: This equation shows the role of hydrostatic and oncotic pressures in the movement of fluid across the membrane of a capillary. The oncotic pressure or colloid osmotic pressure is an osmotic pressure that is applied by some blood proteins, especially albumin, to pull water into the blood circulation system.

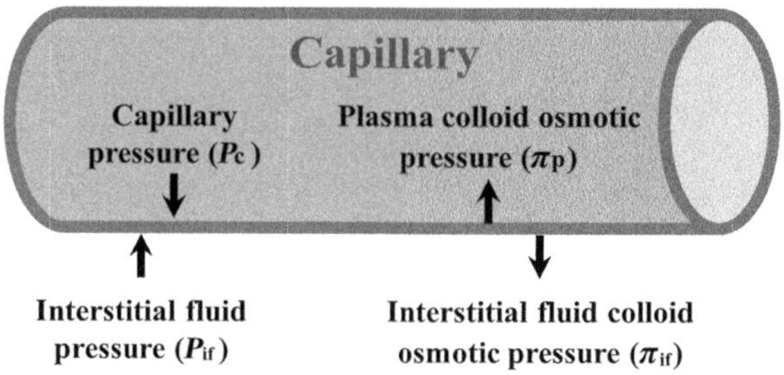

FIGURE 3.7 Procedure of fluid and colloid osmotic pressures at a capillary membrane.

The net filtration pressure (NFP) is [1,8]

$$\mathrm{NFP} = P_c - P_{if} - \pi_p + \pi_{if} \tag{3.5}$$

where

P_c: The capillary pressure, which tends to push fluid outward through the capillary wall.
P_{if}: The interstitial fluid pressure, which has a tendency to force fluid inward or outward through the capillary membrane. For $0 < P_{if}$ the flow direction is inward but outward when $P_{if} < 0$.
π_p: The capillary plasma colloid osmotic pressure, which has a tendency to cause osmosis of fluid inward through the capillary membrane.
π_{if}: The interstitial fluid colloid osmotic pressure, which has a tendency to cause osmosis of fluid outward through the capillary membrane.

The negative NFP denotes a fluid absorption from the interstitial space to capillaries. Fig. 3.7 illustrates these fluid pressures at the capillary membrane.

3.4 VENULES

Venules normally have a smaller ratio of wall thickness to diameter than arterioles [9] and their diameter range is from 10 to 200 μm [10]. Functionally, the venule wall has three layers similar to an arteriole but the middle layer is poorly developed and due to this aspect, they have a very thin wall (see Table 2.1 and Fig. 2.13).

Velocity of blood flow in venule: As Fig. 3.6A shows, with the increase in diameter of venules the blood velocity increases. Generally, the velocity profiles of RBCs are symmetric in the arteriole network (except near to bifurcation) but they are asymmetric in the venule network; also the velocity profiles are blunted and parabolic in venules and arterioles, respectively [10–12].

Blood pressure in venule: The blood pressure decreases through the microcirculation system, including in the venule network. The minimum pressure drop occurs in the venule network—less than 10 mmHg for venules with a diameter range from 10 to 140 μm (see Fig. 3.6B).

Considering Fig. 3.6A, the approximate equation ($R^2 = 0.99$) of blood flow velocity in microvessels with diameter (D) less than 140 μm is

$$V_{app} = k\,D^m \tag{3.5.1}$$

For arterioles: $k = 0.0294$ and $m = 0.9193$.

For venules: $k = 0.0148$ and $m = 0.8661$.

3.5 FAHRAEUS AND FAHRAEUS–LINDQVIST EFFECTS

These effects are named for the studies of two Swedish scientists, Fahraeus and Lindqvist, on blood behavior in microvessels (tubes).

The Fahraeus effect: Fahraeus observed the difference in RBC volume fractions (hematocrit—Hct) between micro and macrosize tubes. This effect describes that the hematocrit within a microsize vessel ($\mathrm{Hct_V}$) is lower than the hematocrit in the blood entering and the blood discharge hematocrit ($\mathrm{Hct_D}$) [13–15]:

$$\mathrm{Hct_T} < \mathrm{Hct_D} \tag{3.6}$$

$$\frac{\mathrm{Hct_T}}{\mathrm{Hct_D}} = \frac{\mathrm{V_{bulk}}}{\mathrm{V_{RBC}}} \tag{3.7}$$

$$\frac{\mathrm{Hct_T}}{\mathrm{Hct_D}} = \mathrm{Hct_D} + (1 - \mathrm{Hct_D}) \times (1 + 1.7 \times \mathrm{e}^{-0.35D} - 0.6 \times \mathrm{e}^{-0.01D}) \tag{3.8}$$

where V_{bulk} is overall mean blood flow velocity, V_{RBC} is the mean RBCs velocity within tube and D is tube diameter.

The reason for the hematocrit reduction in narrow vessels (tubes) is the formation of a cell-free layer near the vessel (tube) wall. Fig. 3.8 shows a schematic of the Fahraeus effect and the cell-free layer. The thickness of the cell-free layer is between 2 and 5 μm [16].

Fig. 3.9 shows the dependence of Eq. (3.8) on tube diameter for $\mathrm{Hct_D} = 0.45$. The minimum amount of $\mathrm{Hct_T}/\mathrm{Hct_D}$ occurs for a tube diameter around 13.5 μm and $\mathrm{Hct_D} = 0.45$.

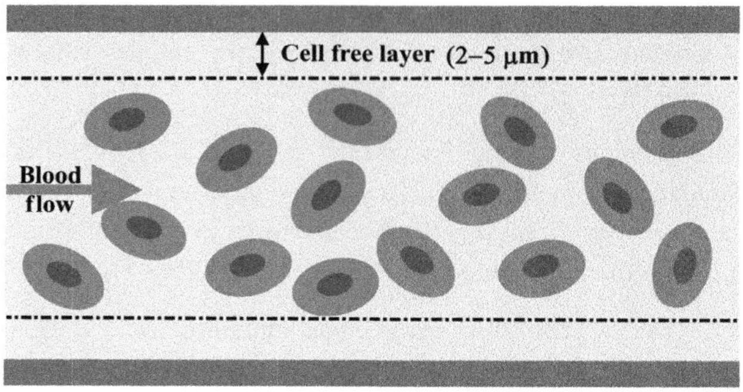

■ **FIGURE 3.8** The Fahraeus effect. Schematic of blood flow and RBCs through a microsize vessel (tube) and the formation of a cell-free layer close to the vessel wall.

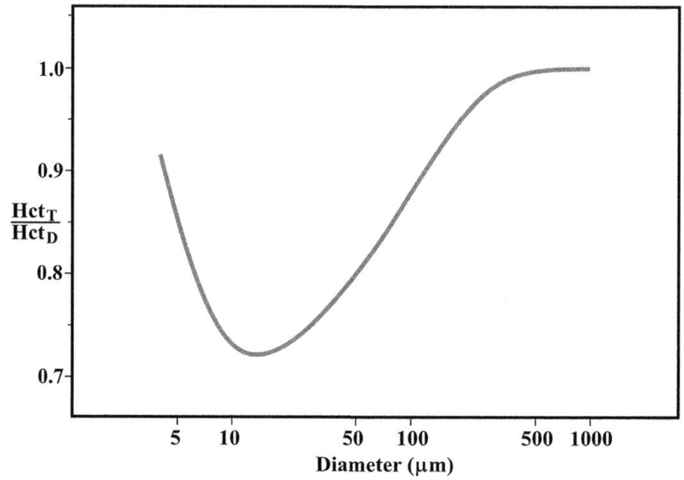

■ **FIGURE 3.9** The Fahraeus effect, based on Eq. (3.7) for $Hct_D = 0.45$.

The Fahraeus–Lindqvist effect: The apparent viscosity term is a result of resistance to blood flow. Fahraeus and Lindqvist demonstrated that the apparent viscosity (μ_{app}) is a function of the microvessel diameter for a range of $10\ \mu m \le D \le 300\ \mu m$. They also showed that apparent viscosity is reduced when the diameter is decreased. For microvessels close to the size of a capillary, this effect will reverse and a large increase in viscosity is observed.

The relative apparent viscosity is an important term in the microcirculation field and it is characterized by

$$\mu_{\text{rel}} = \frac{\mu_{\text{app}}}{\mu_{\text{p}}} \tag{3.9}$$

where μ_{p} is plasma viscosity.

Two experiments (in vitro and in vivo conditions) have been done for relative apparent viscosity [14,15,17].

In vitro: A glass tube is used for the experiment and the governing equations of the relative apparent viscosity for in vitro conditions are [14,15]

$$\mu_{\text{rel}} = 1 + (\mu_{45} - 1)\frac{(1-\text{Hct}_{\text{D}})^C - 1}{(1-0.45)^C - 1} \tag{3.10}$$

$$\mu_{45} = 220 \cdot \exp(-1.3D) + 3.2 - 2.44 \cdot \exp(-0.06D^{0.645}) \tag{3.11}$$

$$C = (0.8 + \text{e}^{-0.075D}) \cdot \left(-1 + \frac{1}{1 + 10^{-11} \cdot D^{12}}\right) + \frac{1}{1 + 10^{-11} \cdot D^{12}} \tag{3.12}$$

where D (μm) is microvascular diameter, μ_{45} is relevant apparent viscosity for a constant discharge hematocrit of 0.45 ($\text{Hct}_{\text{D}} = 0.45$) and C is a factor to show the relationship between relative apparent viscosity and hematocrit.

In vivo: The governing equations of relative apparent viscosity for in vivo conditions based on rat mesentery are [14,17]

$$\mu_{\text{rel}} = \left[1 + (\mu_{45} - 1)\frac{(1-\text{Hct}_{\text{D}})^C - 1}{(1-0.45)^C - 1} \cdot \left(\frac{D}{D-1.1}\right)^2\right] \cdot \left(\frac{D}{D-1.1}\right)^2 \tag{3.13}$$

$$\mu_{45} = 6 \cdot \exp(-0.085D) + 3.2 - 2.44 \cdot \exp(-0.06D^{0.645}) \tag{3.14}$$

The variable definitions are given by Eqs. (3.9)–(3.11) as before.

Figs. (3.10) and (3.11) illustrate this aspect of relative apparent blood viscosity and also the Fahraeus–Lindqvist effect for in vitro and in vivo conditions, respectively.

In the in vitro condition, the relative apparent viscosity declines with the reducing tube diameter (from 300 to 7 μm), it increases again for less than 7 μm, and the maximum μ_{rel} occurs for a tube diameter around 3 μm (see Fig. 3.10). In the in vivo condition, the minimum relative apparent viscosity occurs for a diameter around 45 μm (see Fig. 3.11). As the figures show, the flow resistance and viscosity in vivo (living microvessels) is higher than in vitro (glass tube). This difference is due to the presence of an endothelial surface layer (ESL or glycocalyx) bound to the inner surface of the endothelial cells [14].

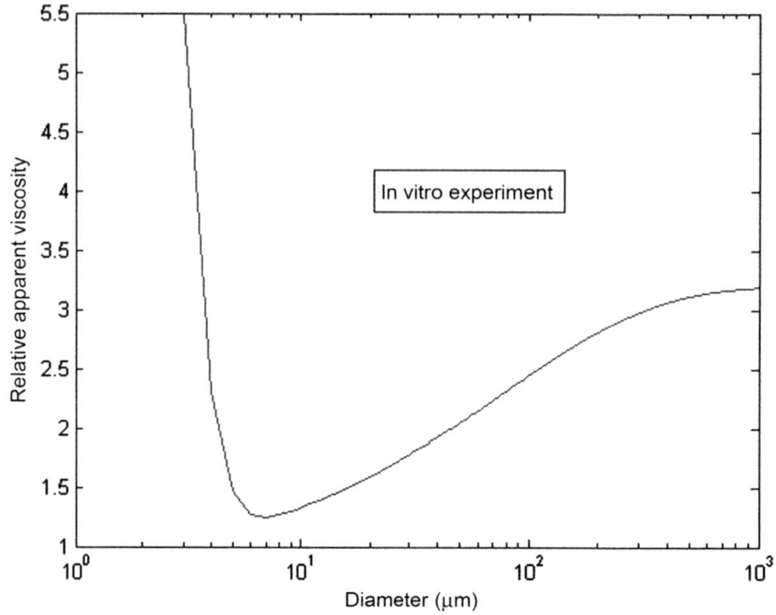

■ **FIGURE 3.10** Relation of microvessel diameter with relative apparent viscosity for $Hct_D = 0.45$ in vitro.

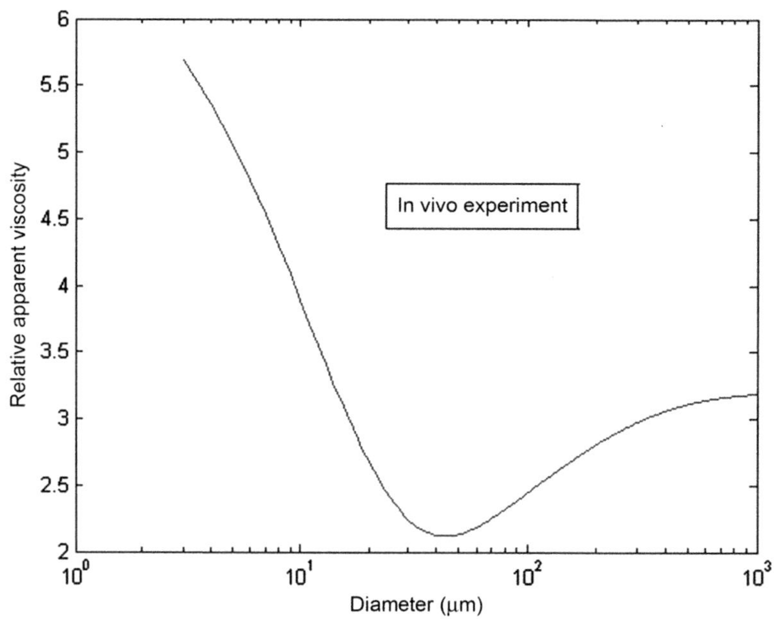

■ **FIGURE 3.11** Relation of microvessel diameter with relative apparent viscosity for $Hct_D = 0.45$ in vivo (rat mesentery).

3.6 MASS TRANSPORT IN TISSUE

All living tissues require the efficient transport of solutes and solvent. This transport provides energy and vital substances for cellular metabolism. At the cellular level, diffusion is the fastest and most efficient method to transport molecules of nutrients and metabolites to living cells over distances less than 100–200 μm. To increase transport efficiency, convection and electrical conduction are required [13].

Fick's law of diffusion: Diffusion occurs due to a concentration gradient of substances (atom, ion or molecule) between the two sides of a wall (membrane). Fig. 3.12 shows the schematic of the diffusion of a solute across a capillary wall. Fick's law demonstrates an aspect of diffusion [1]

$$J_x = P_x \cdot ([X]_c - [X]_{if}) \tag{3.15}$$

$$P_x = \frac{D_x}{a} \tag{3.16}$$

where J_x is the flux of the solute X (mol/(m^2 seconds)), P_x is permeability coefficient (m/seconds), D_x is diffusion coefficient (m^2/seconds), a is wall thickness (m), $[X]_c$ and $[X]_{if}$ are the dissolved concentration (mol/m^3) of the solute in capillary and interstitial fluid, respectively.

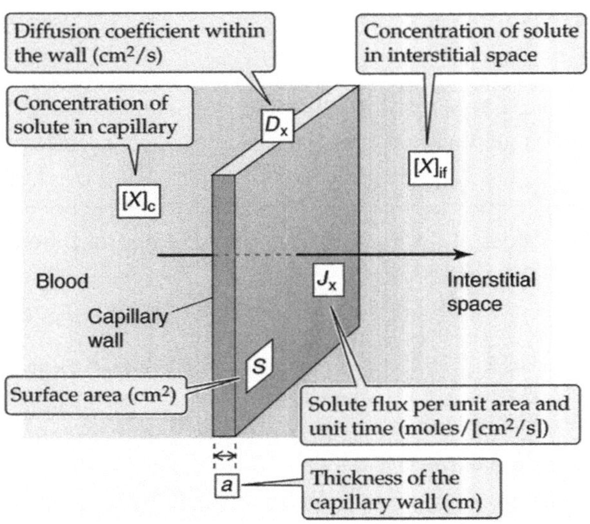

■ **FIGURE 3.12** Schematic of diffusion through a capillary wall. *From W.F. Boron, E.L. Boulpaep, Medical Physiology: A Cellular and Molecular Approach, second ed., Saunders Elsevier, Philadelphia, PA, 2009 with permission.*

Diffusion coefficient (D_x) is a function of temperature, fluid viscosity and solute size and shape. Generally, the diffusion coefficient of gases is very fast, in comparison to liquids such as water (D_x of small solutes in gases is 10,000 times larger than D_x of same solutes in water) [13].

The Stokes–Einstein equation for diffusion coefficient: This equation denotes that the diffusion coefficient is a function of temperature, frictional drag coefficient and the Boltzmann constant. For a low Reynolds number, the drag force on a particle in solution is proportional to the velocity of particle. Also, it is determined that the kinematic energy of a single particle ($m_p v^2$) is equal to the thermal energy or $K_B T$. The Stokes-Einstein equation for diffusion coefficient can be written as [13]

$$D_x = K_B T \bar{f} \tag{3.17}$$

where K_B is the Boltzmann coefficient, T is temperature and $\bar{f}$ is the frictional drag coefficient which is dependent on the geometry factors of a particle. Table 3.1 shows the values of the frictional drag coefficient for different particle shapes.

Several models can be defined for diffusion [13]:

- Diffusion in rectangular coordinates, such as liver and skin.
- Diffusion in cylindrical (radial) coordinates, such as kidney tubule.
- Diffusion in spherical coordinates, such as white blood cells, protein beads and cultured cells.

Effective diffusion factor in biological tissues:

Generally, analysis of the diffusion factor in biological tissues is complicated since it depends on both physiological and geometrical factors.

Table 3.1 Values of the Frictional Drag Coefficient for Different Particle Shapes [18,19,13]

Shape	Frictional Drag Coefficient
Sphere of radius R	$\bar{f} = 6\,\pi\,\mu\,R$
Prolate ellipsoid, $p = a/b > 1$, where a is a major axis, b is minor axis	$\bar{f} = \dfrac{6\,\pi\,\mu\,b(p^2-1)^{1/2}}{p^{1/3}\ln[p+(p^2-1)^{1/2}]}$
Oblate ellipsoid, $p = a/b < 1$	$\bar{f} = \dfrac{6\,\pi\,\mu\,b(1-p^2)^{1/2}}{p^{1/3}\tan^{-1}[(1-p^2)^{1/2}p^{-1}]}$
Thin circular disk of radius a	$\bar{f} = 16\,\mu\,a$
Cylinder of radius a and length L	$\bar{f} \approx \dfrac{4\,\pi\,\mu\,l}{\ln(L/a)+0.193}$

Approximately, the effective diffusion factor is a power function and is related to the molecular weight [13],

$$D_{\text{eff}} = b_1(M_r)^{b_2} \tag{3.18}$$

where M_r is the molecular weight of solutes, b_1 and b_2 are function of the charge, shape of solutes and tissue structure [20].

Convection: Convective transport is one of the mass transport methods which occur when a fluid characteristic, such as mass, energy or an element, is carried along with the fluid. Similar to other fluidic systems, the Reynolds number or its elements (velocity, density, viscosity and channel geometry) and the friction factor have a main role in mass transport (such as convection transport) along the blood vessels. The general equation of convection flux is

$$J_{\text{Con}} = v\,C \tag{3.19}$$

where v is velocity of particles and C is concentration of particles.

The performance of diffusion is minimized in high flow rates, however, diffusion still has a significant role in mass transportation. It is proved that when free diffusion takes place in the presence of solvent motion, the solute flux equation contains both diffusion and convection [21]. The general form of this combination is the advection-diffusion equation [22]:

$$\frac{\partial C}{\partial t} + \nabla\cdot(vC) - \nabla\cdot(D\nabla C) = g \tag{3.20}$$

where C is solute concentration, v is velocity of flow, D is diffusion, t is time and g is the source. For an incompressible fluid without any source of C ($g = 0$), Eq. (3.20) can be rewritten as [23]

$$\frac{\partial C}{\partial t} = -v\nabla C + D\nabla^2 C \tag{3.21}$$

In this equation, $\nu\nabla C$ and $D\nabla^2 C$ denote convection and diffusion transports, respectively. Also, for steady state condition, $\partial C/\partial t = 0$.

Electrical (electrolyte) transport: Many biological molecules and particles are charged and the electrical potential gradient has a main role in their transport. This potential gradient occurs due to ion concentration differences across tissues and membranes of a cell. The flux of ions in an electrolyte transport is the product of velocity and concentration of ion [13],

$$J_e = C\,v \tag{3.22}$$

The Eq. (3.19) can be rewritten as [13],

$$J_e = \frac{DCzF}{RT}\nabla\psi \tag{3.23}$$

where D is the diffusion coefficient, C is concentration, z is the net charge on the molecule, F is Faraday's constant (96,487 C/mol), R is the universal gas constant and T is temperature.

Reference [13] details comprehensive explanations of the mass transport phenomena in biological systems for a variety of coordinate systems and types.

3.7 POROSITY, TORTUOSITY AND PERMEABILITY

A porous medium is a material comprising pores. Porous materials are characterized by porosity, tortuosity, permeability and others. These are essential factors in the microcirculation, mass transport and exchange system.

Porosity: Mathematically, porosity is the fraction of pore (or void) volume to the total volume of porous material,

$$\varphi = \frac{V_\mathrm{P}}{V_\mathrm{T}} \tag{3.24}$$

where V_P and V_T are volume of pore (void) and total, respectively. Fig. 3.13 shows a porous cube with arbitrary pores.

There are three categories for pores [13]:

1. passing pores,
2. nonpassing pores and
3. isolated pores.

Fig. 3.14 shows this categorization.

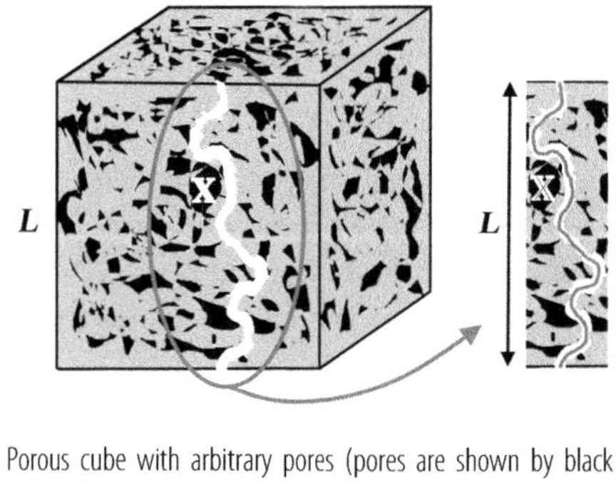

■ **FIGURE 3.13** Left: Porous cube with arbitrary pores (pores are shown by black color). Right: The tortuosity definition or (X/L), X is length of channel (pore) and L is the straight distance between the inlet and outlet of the pore which is the thickness of the membrane in this figure.

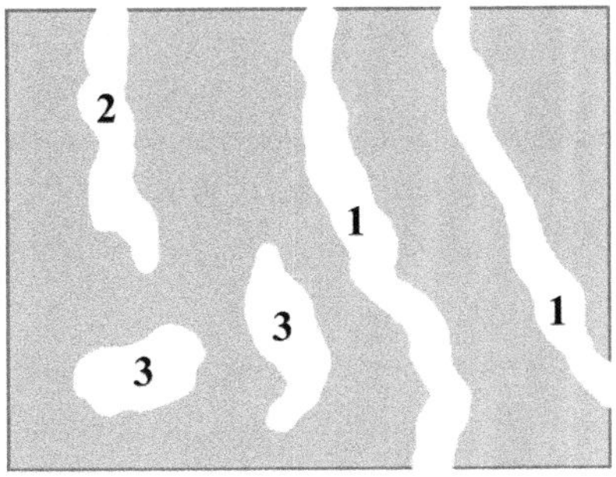

■ **FIGURE 3.14** Categorization of pores. *1*, Passing pores; *2*, nonpassing pores and *3*, Isolated pores.

Biological tissues deform under mechanical load and this deformation can change the dimensional distribution of porosity. The porosity value is larger than 0.9 for interstitial space, 0.06 for the brain, 0.3 for the skin and 0.6 for some tumor tissues [13].

Tortuosity: In a porous medium, tortuosity is the ratio of the pore channel length to the distance between inlet and outlet of channel (see Fig. 3.13),

$$\tau = \frac{X}{L} \tag{3.25}$$

where X is pore length (or channel) and L is straight distance between inlet and outlet (see Fig. 3.13). Tortuosity is always larger than one and it decreases with increasing porosity.

Tortuosity of blood vessel is also a known term in the cardiovascular system and usually it is recognized as arterial tortuosity syndrome due to the abnormal twists and turns of arteries.

Permeability: Generally, permeability is a filtration characteristic showing the ability of a porous material to pass fluids. With regards to Darcy's law (see Section 3.8), permeability or k (m^2) is

$$k = K\frac{\mu}{\rho g} \tag{3.26}$$

where K is hydraulic conductivity, μ is viscosity, ρ is density and g is gravity acceleration.

3.8 GOVERNING EQUATIONS IN POROUS MEDIA

Darcy's law is the main equation in a porous medium and describes the fluid flow through a porous material,

$$Q = \frac{k\,A\,\Delta P}{\mu\,L} \tag{3.27}$$

where Q is discharge flow rate, k is permeability, A is cross-section of porous media, ΔP is pressure drop, μ is viscosity and L is length of porous medium (see Figs. 3.15 and 3.16).

One of the closest models to the capillary is a model of a single hollow fiber. With regard to Fig. 3.16, the governing equations are [24]

$$dQ = -jD_i\pi dx \tag{3.28}$$

$$d(QC) = -jC'D_i\pi dx \tag{3.29}$$

where Q is the feed flow rate at x, D_i is the inside diameter of capillary, j is the total flux, C is the concentration of a component in the feed and C' is the concentration of the component after transport by capillary wall.

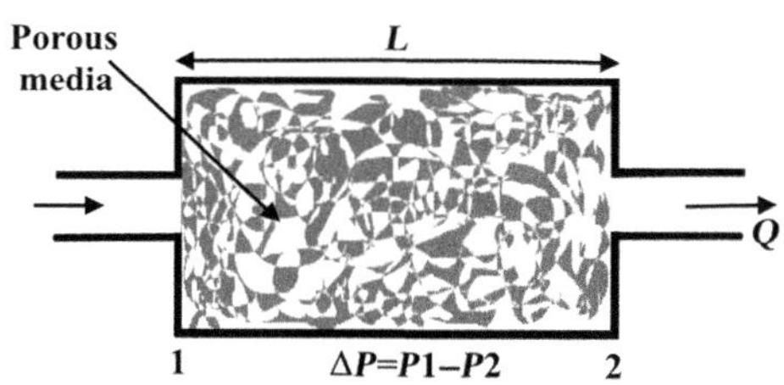

■ **FIGURE 3.15** Schematic of a porous media to show Darcy's law.

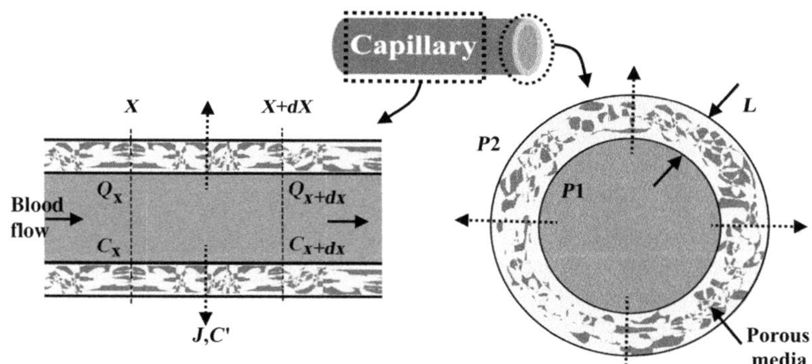

■ **FIGURE 3.16** Capillary porous wall and mass transport through the wall. $P1$ and $P2$ are inside and outside pressure, L is length or thickness of porous media, Q and C are the flow rate and the component concentration, respectively.

Furthermore, the following equation can be obtained from Eqs. (3.28) and (3.29) [24]:

$$dC = [(C - C')J]\frac{D_i \pi}{Q} dx \tag{3.30}$$

For a position x

$$Q = Q_{\rm i} - D_i \pi \int_0^x J dx \tag{3.31}$$

where $Q_{\rm i}$ is the initial flow rate at inlet. If there are several hollow fibers (or capillaries), factor n (number of fiber) will be multiplied by $D_{\rm i}$, that is, ($nD_{\rm i}$), in Eqs. (3.28)–(3.31).

Brinkman equation: This equation describes flow through a porous medium and was derived by Brinkman from Darcy's law and can be described as [25,26]

$$\nabla P = -\frac{\mu}{k} v + \mu \nabla^2 v \tag{3.32}$$

where k is the permeability, P is the pressure, μ is the viscosity and v is the velocity.

3.9 FLUID TRANSPORT IN POROELASTIC MEDIA

In mechanical engineering, a poroelastic medium is a porous material (called solid or matrix) which is filled by a viscous fluid. Biological tissues, foams, ceramics, soils and rocks are considered to be poroelastic material. Deformation behavior of poroelastic materials can be nonlinear and, usually it is neglected in analysis. Biot formulated a law for deformable porous media saturated with a fluid. Biot's law of poroelasticity has been used to predict the mechanics of biological tissues, such as bone, brain and solid tumors [27]. This theory couples Darcy's law and the strength of material (solid mechanics) and neglects some forces, such as gravity and inertial forces. Generally, Biot's law can be written as [13,27]

$$\nabla \cdot \sigma = 0 \tag{3.33}$$

$$\varphi\left(V_{\rm f} - \frac{\partial u}{\partial t}\right) = -K \nabla p \tag{3.34}$$

where σ is effective stress tensor, φ is porosity (or volume fraction of fluid), $V_{\rm f}$ is averaged fluid velocity, u is displacement, t is time, K is hydraulic conductivity and p is fluid pressure. If the tissue deformation is at steady state, then $\partial u/\partial t = 0$.

CHAPTER SUMMARY

- *Microcirculation system*: The first function of the blood circulation system is to provide suitable conditions for the living tissues and the microcirculation system is the last station of this mission.
- The largest pressure drop in the systematic system occurs in the arterioles, from around 90 to 15 mmHg; this means that the *arterioles have a main role in the control of blood flow into living organs*.
- *Velocity profile* in arterioles with an internal diameter less than 60 μm is blunted. The velocity profile can be rewritten in cylindrical coordinates (r, θ, z) as

$$v(r,t) = v_{\max}\left(1 - \left(\frac{r}{R}\right)^{k}\right)f(t)$$

- *Capillaries are classified* into three groups based on their degree of leakiness: continuous capillaries; fenestrated capillaries and discontinuous capillaries.
- For small capillaries, the flow velocity in the capillary can be equal to the RBC velocity.
- Venules normally have a smaller ratio of wall thickness to diameter than arterioles and their diameter ranges from 10 to 200 μm.
- *The Fahraeus effect* describes that the hematocrit (Hct) within a microsize vessel is less than the hematocrit in the blood entering and the blood discharge hematocrit (vessels with bigger diameter).
- *Fick's law of diffusion*: Diffusion occurs due to a concentration gradient of substances between two sides of a permeable wall.
- Several models can be defined for diffusion:
  - Diffusion in rectangular coordinates, such as liver and skin.
  - Diffusion in cylindrical (radial) coordinates, such as kidney tubule.
  - Diffusion in spherical coordinates, such as white blood cells, protein beads and cultured cells.
- *Convection*: Convective transport is one method of mass transport that occurs when a fluid characteristic, such as mass, energy or an element, is carried along with the fluid.
- *Electrical (electrolyte) transport*: Many biological molecules and particles are charged and an electrical potential gradient occurs due to the ion concentration differences across tissues and membranes of a cell.
- *Porosity*: Porosity is the fraction of the pore (or void) volume to the total volume of the porous material or $= V_P/V_T$.

- *Tortuosity*: Tortuosity is the ratio of the pore channel length to the distance between the inlet and outlet of the channel.
- *Permeability* is a filtration characteristic showing the ability of a porous material to pass fluids.
- *Darcy's law* ($Q = (k\,A\,\Delta P)/(\mu\,L)$) and the *Brinkman* equation ($\nabla P = -(\mu/k)v + \mu\nabla^2 v$) are the major equations for a *Porous Medium.*
- *Poroelastic* is a porous material (called solid or matrix) which is filled with a viscous fluid.

PROBLEMS

1. Which blood vessels have a main role in regulating blood flow through the microcirculation?
2. Which type of microvessel has a role in the exchange of gases and nutrition and waste transport? And why?
3. List the major differences among capillary types.
4. Discuss the Fahraeus effect and its reason.
5. How many models can be defined for diffusion in biological tissues?
6. Discuss the mass transport types in biological tissues.
7. Compare the types of mass transport among blood microvessels.
8. Using Eq. (3.26), find the mechanical dimensions of hydraulic conductivity.
9. Determine the porosity of a 300 g sample that contains 65 g of water when 60% of the pores are full of water.
10. Use Fig. 3.17 and discuss Darcy's law (Darcy used a similar experimental setup to valid his idea).

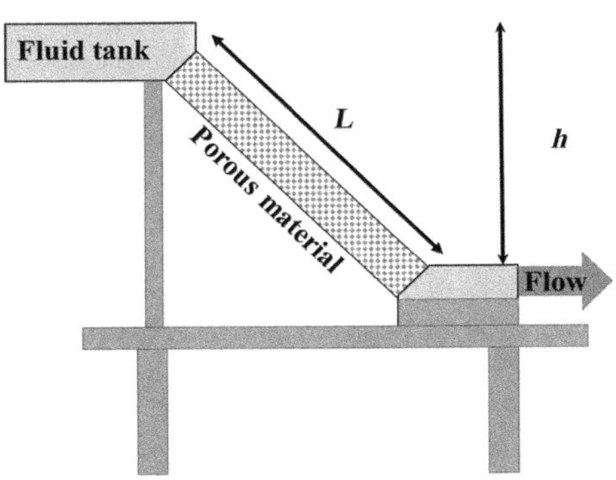

■ **FIGURE 3.17** Darcy experiment.

REFERENCES

[1] W.F. Boron, E.L. Boulpaep, Medical Physiology: A Cellular and Molecular Approach, second ed., Saunders Elsevier, Philadelphia, PA, 2009.

[2] A.S. Popel, P.C. Johnson, Microcirculation and hemorheology, Annu. Rev. Fluid. Mech. 1 (37) (2005) 43–69.

[3] A. Nakano, Y. Sugii, M. Minamiyama, H. Niimi, Measurement of red cell velocity in microvessels using article image velocimetry (PIV), Clin. Hemorheol. Microcirc. 29 (2003) 445–455.

[4] M.L. Ellsworth, R. Pittman, Evaluation of photometric methods for quantifying convective mass transport in microvessels, Am. J. Physiol. 251 (1986) H869–H879.

[5] G.J. Tangelder, D. Slaaf, T. Arts, A. Muijtjens, M. Oude Egbrink, R. Reneman, Velocity profiles of blood platelets and red blood cells flowing in arterioles of the rabbit mesentery, Circ. Res. 59 (1986) 505–514.

[6] T. Nagaoka, H. Yanagida, T. Saitohm, A. Kamiya, T. Hein, L. Kuo, et al., A mathematical model for the distribution of hemodynamic parameters in the human retinal microvascular network, J. Biorheol. 23 (2009) 77–86.

[7] K. Sriram, M. Intaglietta, D.M. Tartakovsky, Non-Newtonian flow of blood in arterioles: consequences for wall shear stress measurements, Microcirculation 21 (7) (2014) 628–639.

[8] A.C. Guyton, J.E. Hall, Textbook of Medical Physiology, eleventh ed., Elsevier, Inc., Philadelphia, PA, 2006.

[9] J.D. Bronzino, D.J. Schneck (Eds.), Biomechanics, Principles and Applications, CRC Press LLC, Boca Raton, FL, 2003.

[10] J.D. Bronzino (Ed.), The Biomedical Engineering Handbook, Biomedical Engineering Fundamentals, third ed., CRC Press, Taylor & Francis Group, Boca Raton, FL, 2006.

[11] J.J. Bishop, R. Nance, A. Popel, M. Intaglietta, P. Johnson, Effect of erythrocyte aggregation on velocity profiles in venules, Am. J. Physiol. Heart Circ. Physiol. 280 (2001) H222–H236.

[12] M. Ellsworth, R.N. Pittman, Evaluation of photometric methods for quantifying convective mass transport in microvessels, Am. J. Physiol. 251 (1986) H869–H879.

[13] G.A. Truskey, F. Yuan, D.F. Katz, Transport Phenomena in Biological Systems, Pearson Education, Upper Saddle River, NJ, 2010.

[14] O.K. Baskurt, M.R. Hardeman, M.W. Rampling, H.J. Meiselman (Eds.), Handbook of Hemorheology and Hemodynamics, IOS Press, Washington, DC, 2007.

[15] A.R. Pries, D. Neuhaus, A.P. Gaehtgens, Blood viscosity in tube flow: dependence on diameter and hematocrit, Heart Circ. Physiol. 263 (32) (1992) H1170–H1776.

[16] L. Waite, J. Fine, Applied Biofluid Mechanics, McGraw-Hill, New York, 2007.

[17] A.R. Pries, T.W. Secomb, T. Gebner, M.B. Sperandio, J.F. Gross, P. Gaehtgens, Resistance to blood flow in microvessels in vivo, Circ. Res. 75 (1994) 904–915.

[18] J. Happel, H. Brenner, Low Reynolds Number Hydrodynamics, Martinus Nijhoff publishers, Boston, MA, 1983.

[19] R. Probestien, Physiochemical Hydrodynamics, Butterworths, Boston, MA, 1989.

[20] R. Jain, Transport of molecules in the tumor interstitium: a review, Cancer Res. 47 (1987) 3039–3051.

[21] M.H. Friedman, Principles and Models of Biological Transport, Springer Science + Business Media, LLC, New York, 2008.

[22] T. Stocker, Introduction to Climate Modelling, Springer, Berlin Heidelberg, 2011.

[23] R. Abdelaziz, A.J. Pearson, B.J. Merkel, Lattice Boltzmann modeling for tracer test analysis in a fractured Gneiss aquifer, Nat. Sci. 5 (3) (2013) 368–374.

[24] N. Li, A.G. Fane, W.S.W. Ho, T. Matsuura, Advanced Membrane Technology and Applications, John Wiley & Sons, Hoboken, NJ, 2008.

[25] D.R. Rubenstien, W. Yin, M.D. Frame, Biofluid Mechanics, an Introduction to Fluid Mechanics, Macrocirculation and Microcirculation, Elsevier, Inc., Philadelphia, PA, 2012.

[26] K. Vafai, Handbook of Porous Media, Taylor & Francis Group, Boca Raton, FL, 2005.

[27] S.M. Becker, A.V. Kuznetsov, Transport in Biological Media, Elsevier, Inc., San Diego, CA, 2013.

Chapter 4

Biofluid Dynamics in Human Organs

4.1 HEART AND CARDIOVASCULAR SYSTEM

4.1.1 Introduction

The heart is a muscular organ which pumps blood through the circulation system to transport nutrients and waste products. It is located at the level of thoracic vertebra T5–T8 and its average weight is between 250 and 300 g in an adult. The heart is composed of two separate main pumps which cooperate to flow the blood in circulatory system. In this closed circulation system, the left pump transports oxygenated blood to the body and the right pump transports deoxygenated blood to the lung. Each pump has two chambers to pump the blood; the atrium (upper chamber) and the ventricle (lower chamber). By periodic muscle stimulation of these chambers, the heart pumps blood throughout the body. Passively, the atria chambers (right and left) flow the blood to the ventricles and the ventricles (right and left) actively pump the blood through the lung and body organs.

Fig. 4.1 shows a schematic of the heart structure and the blood flow through the heart [1]. Considering Fig. 4.1, the heart function phases are:

1. Oxygenated blood enters from pulmonary vein (lung) to the left atrium.
2. Oxygenated blood is transported from the left atrium to the left ventricle.
3. Oxygenated blood is pumped to the aorta for systemic and coronary systems.
4. Deoxygenated blood enters from the vena cava to the right atrium.
5. Deoxygenated blood is transported from the right atrium to the right ventricle.
6. Deoxygenated blood is pumped to the pulmonary artery (lung) to oxygenize red blood cells (RBCs).

Cardiac muscles have key role in the heart functions. The cardiac muscle is a kind of muscle which is found only in the heart. Cardiac muscle cells (or cardiac myocytes) are single nucleated. Ventricle myocytes have a

Biofluid Mechanics. DOI: http://dx.doi.org/10.1016/B978-0-12-802408-9.00004-1

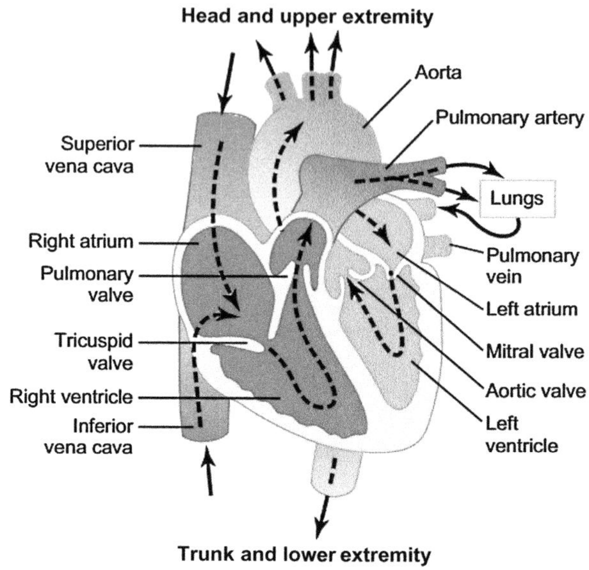

■ **FIGURE 4.1** Heart structure and blood flow through the heart. *From A.C. Guyton, J.E. Hall, Textbook of Medical Physiology, eleventh ed., Elsevier, Inc., Philadelphia, PA, 2006 with permission.*

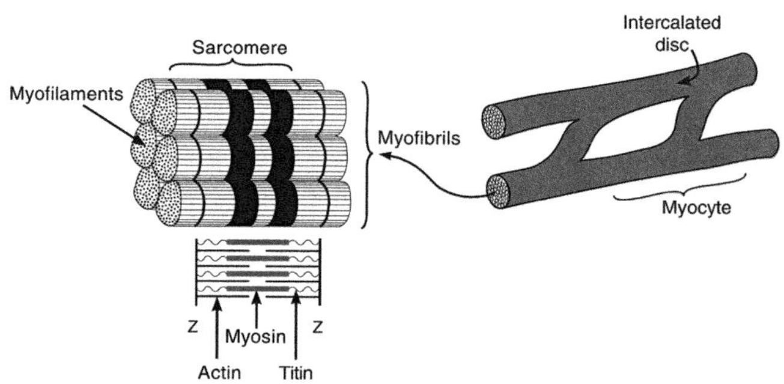

■ **FIGURE 4.2** Cardiac myocytes structure. *From R.E. Klabunde, Cardiovascular Physiology Concepts, Wolters Kluwer and Lippincot Williams & Wilkins, Baltimore, MD, 2011 with permission.*

length of about 100 μm and a diameter of around 25 μm (atria myocytes are smaller than ventricle myocytes) [2]. The cardiac myocytes are created from myofibrils that comprise myofilaments. The myofilaments are composed of thin filaments of actin (a multifunctional protein molecule) and thick filaments of myosin (a motor protein molecule). The myosin is fixed to the Z-line (a dark thin protein) by the elastic filament of titin. Fig. 4.2 illustrates the structure of cardiac myocytes. Interactions between

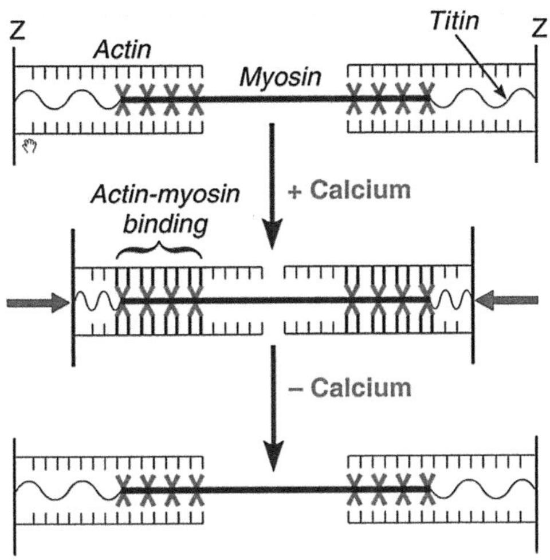

■ **FIGURE 4.3** Sliding filament theory of muscle contraction. Top: normal myocyte; middle: calcium enters the cell and results contraction; bottom: calcium removing and muscle releasing. *From R.E. Klabunde, Cardiovascular Physiology Concepts, Wolters Kluwer and Lippincot Williams & Wilkins, Baltimore, MD, 2011 with permission.*

myosin and actin result in a base of a sliding filament theory of muscle contraction. This contraction occurs due to calcium-triggered mechanism. During depolarization, calcium enters the cell and free calcium binds to troponin-C (TN-C) permitting binding of the actin-myosin. This function results in adenosine triphosphate (ATP) hydrolysis which supplies enough energy for the sliding of actin and myosin on each other and this results in muscle contraction. Removal of calcium from the cell results in a relaxing of the length of the sarcomere and this procedure is fundamental to the pumping of the heart. Fig. 4.3 shows the sliding filament theory of muscle contraction.

4.1.2 The Cardiac Cycle

The cardiac cycle describes the sequence of electrical and mechanical events that occurs with every heartbeat. The normal duration of a cardiac cycle for a heart rate of 75 beats/minutes is 0.8 seconds [3],

$$\text{Duration of cardiac cycle (seconds/beats)} = \frac{60\ \text{(seconds/minutes)}}{\text{Heart rate (beats/minutes)}} \tag{4.1}$$

The cardiac cycle may be divided into phases in any number of methods, for instance four phases [1,3] or seven phases [2]. In the four phases method, the opening and closing of the heart valves explains this method of the cardiac cycle. These phases are [1,3]:

- Phase I: Filling period—the inlet valve is opened to fill the ventricle and the outlet valve is closed. The volume of blood in the ventricle increases from about 45 mL (from previous cycle) to about 115 mL.
- Phase II: Period of isovolumetric contraction—both valves are closed, blood volume is constant but the blood pressure increases to about 80 mmHg.
- Phase III: Period of ejection—the outlet valve of the ventricle is opened and the inlet is closed and due to more contraction, the blood pressure rises.
- Phase IV: Period of isovolumetric relaxation—both valves are closed and intraventricular pressure decreases without any blood volume changes.

The period of relaxation is called diastole in which the ventricle fills with blood and the period of ventricle contraction is called systole. Fig. 4.4

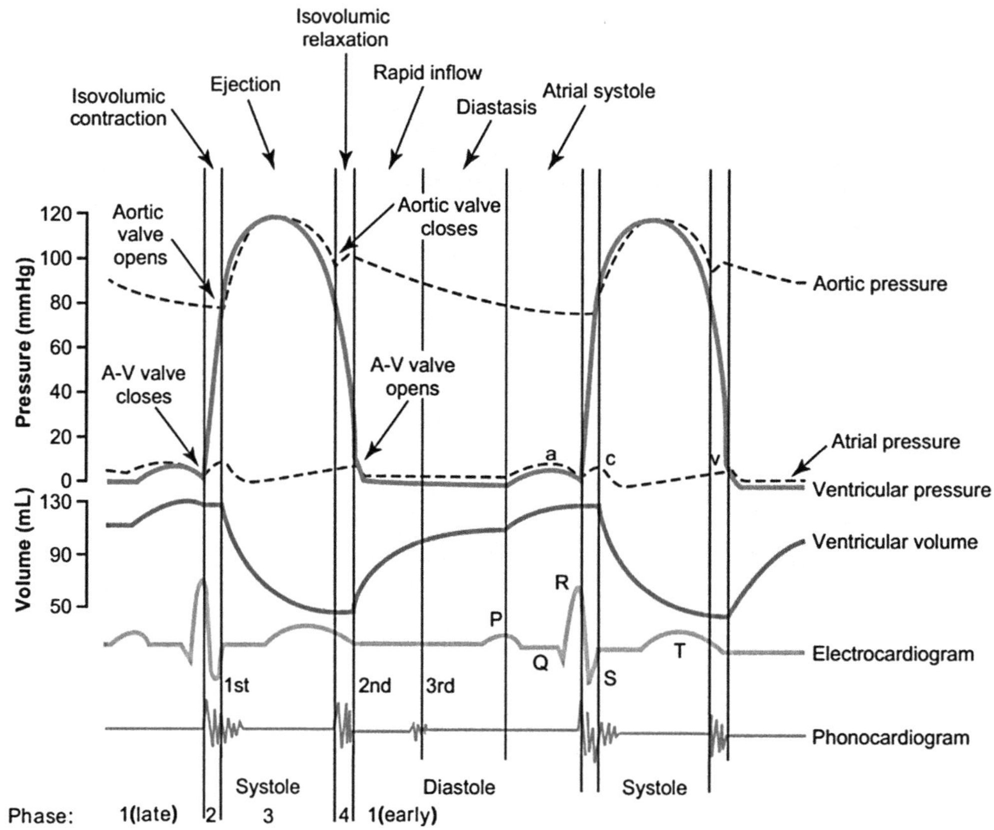

■ **FIGURE 4.4** Normal event of cardiac cycle for left ventricular function, including changes in aortic pressure, atrial pressure, ventricular pressure, ventricular volume, ECG and PCG. *From A.C. Guyton, J.E. Hall, Textbook of Medical Physiology, eleventh ed., Elsevier, Inc., Philadelphia, PA, 2006 with permission.*

shows the cardiac cycle events for the left pump of the heart for two complete cycles. This figure illustrates the pressure–volume of the events. The top three curves illustrate the pressure changes in the left pump of heart including the aorta, left ventricle and left atrium. The fourth curve from the top denotes the volume changes in the left ventricle, the fifth one is electrocardiogram curve and the sixth curve denotes phonocardiogram (PCG; heart sound).

Electrocardiogram (ECG) of cardiac cycle: The ECG is a general clinical device used to measure the electrical activity of the heart. This device records the small extracellular signals which are produced by the movement of cardiac action potential through the transmembrane ion channels in the myocytes. This cardiac potential can be measured by microelectrodes. Scientific experiments have proved that a membrane potential is about −90 mV for a resting ventricular myocyte. The ECG is recorded by an arrangement of electrodes at precise locations on the body surface. The electrodes are located on each leg and arm, and six electrodes are located on the chest to obtain a standard ECG. Fig. 4.5 shows a standard ECG and its waves (P, QRS and T). Repeating waves denote the sequence of depolarization and repolarization of the atrium and ventricles.

Depolarization of the left and right atrium muscle is reflected by the P wave and during this event the atrioventricular (AV) valve is open and ventricle will be filled with blood. The QRS denotes depolarization of

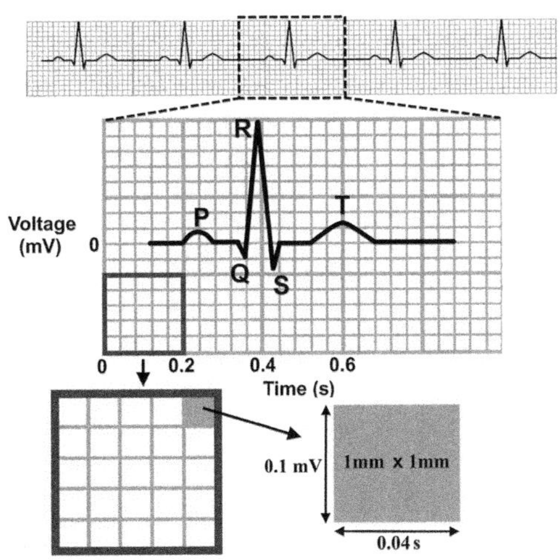

■ **FIGURE 4.5** The ECG components: the P wave; QRS complex and T wave, which denote atrium depolarization, ventricular depolarization and ventricular repolarization, respectively.

ventricular muscle and it appears 0.16 seconds after the P wave, during this event, due to ventricle contraction, the ventricular pressure rises rapidly. The T wave reflects depolarization of both ventricles and they begin to relax [3,4], see Fig. 4.4.

Normally, ECGs are recorded on paper with a vertical calibration of 1 mV/cm and a speed of 25 mm/seconds. In a normal record of ECG, a QRS follows each P wave. The shape of the ECG shows the heart function and it can reveal some heart problems, such as [2]:

- atrial flutter,
- atrial fibrillation,
- first-degree AV block,
- second-degree AV block,
- third-degree AV block,
- premature ventricular complex,
- ventricular tachycardia and
- ventricular fibrillation.

In the atrial pressure graph of Fig. 4.4, there are three minor pressure rises, *a*, *c* and *v*. The *a* wave occurs because of atrial contraction, the *c* wave occurs when the ventricles begin to contract, and the *v* wave occurs near to the end of ventricle contraction [1].

PCG of cardiac cycle: The opening and closing of the heart valves during the cardiac cycle produces sounds and these sounds are recorded by a phonocardiograph. There are two major sounds (S_1 and S_2) and two other sounds (S_3 and S_4). The sound of S_1 (sound of "lub") is for the closure of the AV valves (mitral and tricuspid valves), S_2 (sound of "dub") is for the aortic and pulmonary valves, S_3 is a normal sound for children but for adults it is considered as a problem due to ventricular dilation and S_4 is abnormal sound during atrium contraction due to changes in ventricle tissues. See Fig. 4.4 for the location of these sounds in cardiac cycle.

The frequency range of heart sound is between 10 and 500 Hz with low intensity [5]. Heart murmurs are abnormal heart sounds and they are produced because of turbulent flow of blood in the heart. Only a few heart problems produce abnormal noise, such as:

- regurgitation or back flow due to heart valve problems (mitral or aortic valves);
- stenosis or abnormal narrowing in mitral or aortic valves and
- other murmurs, such as patent ductus arteriosus, the ductus arteriosus fails to close in infants after birth.

4.1.3 Pressure, Flow and Heart Chambers Volume

Generally, the pressure and the flow rate are the two important characters for the assessment of pump function. In the heart, the pressure, the chambers volume and the flow represent the heart performance.

To measure the blood flow of the heart, cardiologists can calculate the blood flow from changes in ventricular volume by the Doppler ultrasound technique and echocardiography. Clinicians use these measurements to evaluate cardiac performance.

Cardiac volumes measurements: Generally, two methods are used by health care professionals to measure the volume of the cardiac chambers: (1) one-dimensional (M-mode) echocardiography which assesses the performance of the left ventricle in terms of ventricular wall velocities and linear dimensions, such as shortening of the ventricular circumference and (2) two-dimensional echocardiography which determines several ventricular volumes [3]:

- left ventricular end-diastolic volume (LVEDV),
- left ventricular end-systolic volume (LVESV),
- stroke volume (SV = LVEDV − LVESV) and
- left ventricular ejection fraction (LVEF = SV/LVEDV).

Generally, the SV is the difference between end-diastolic volume (or EDV around 120 mL) and end-systolic volume (or ESV around 50 mL) in the ventricle and in a normal adult is approximately,

$$\begin{aligned} SV &= EDV - ESV \\ SV &= 70\ \text{mL} \end{aligned} \tag{4.2}$$

Generally, the SV is equal for both ventricles.

The cardiac output (CO) is the volume of pumped blood by the heart. Mathematically, the CO is:

$$CO = SV \times \text{heart rate}\left(\frac{\text{beats}}{\text{minutes}}\right) \tag{4.3}$$

Ventricular pressures measurements: With regard to either side of the heart, health care professionals use two methods to measure the ventricular pressures. For measurement of blood pressure in the right side of the heart, clinicians use a Swan-Ganz catheter which consists of three parallel pipes with different lengths. The clinicians introduce the catheter through a large vein (such as femoral, subclavian, or internal jugular) and then they send it to right atrium, right ventricle and pulmonary artery. By wedging the catheter with an inflatable balloon and using probes, the blood pressure can be measured. For the left side of heart, clinicians insert a simple catheter percutaneously into an artery and then send the catheter to the left side of heart to measure the blood pressure. The pressure wave in the left ventricle is around five times larger than right ventricle [3].

Flow measurement: The blood flow in the cardiac system can be measured from changes in the volume of the ventricles by Doppler ultrasound and echocardiography. Both ventricles produce the similar amount of blood in one cardiac cycle but the blood velocity rises more rapidly in the left ventricle [3].

Table 4.1 shows some of the hemodynamic parameters for the right and left sides of the heart.

Ventricular pressure–volume loop: This loop is a powerful tool to analyze the cardiac cycle, especially ventricular performance. Fig. 4.6 shows the left ventricular pressure versus the left ventricular volume

Table 4.1 Normal Hemodynamic Parameter in Cardiac System [3,6]

Parameter	Normal Value	Unit
Right atrium pressure	Mean: 2	mmHg
Left atrium pressure	Mean: 8	mmHg
	a wave: 13	mmHg
	c wave: 12	mmHg
	v wave: 15	mmHg
Right ventricle pressure	Systolic: 15–30	mmHg
	Diastolic: 2–8	mmHg
Left ventricle pressure	Systolic: 100–140	mmHg
	Diastolic: 3–12	mmHg
Pulmonary artery pressure	Systolic: 15–30	mmHg
	Diastolic: 4–12	mmHg
Aorta pressure	Peak systolic (average): 130	mmHg
	Peak systolic (average): 80	mmHg
	Mean (average): 95	mmHg
Pulmonary capillary wedge pressure	2–15	mmHg
Cardiac output flow rate	3.5–5.5	L/min
Right ventricular outflow velocity	~45	cm/s
Left ventricular outflow velocity	~65	cm/s

during a complete cardiac cycle. In this figure the letters and definitions are [2]:

- (a): ventricular filling;
- (b): isovolumetric contraction;
- (c): ventricular ejection;
- (d): isovolumetric relaxation;
- (EDV): the maximal value at the end of ejection;
- (ESV): the minimal value at the end of ejection;
- (SV): the stroke volume of ventricle or the area within the loop, SV = EDV − ESV;
- (EDPVR): the end diastolic pressure–volume relationship. The passive filling curve for the ventricle and
- (ESPVR): the end systolic pressure–volume relationship. The highest pressure that can be developed by the ventricle at any volume of the left ventricle.

This ventricular pressure–volume loop is a reliable reference to show the normal or failure condition of the cardiac system.

4.1.4 Heart Valves Function and Mechanical Properties

Each mammalian heart has four main valves. These valves prevent the blood flowing in the reverse direction, such as a regular check valve in a fluid circuit. If the blood pressure behind the heart valves is not large enough to cause blood to flow, the valves remain closed.

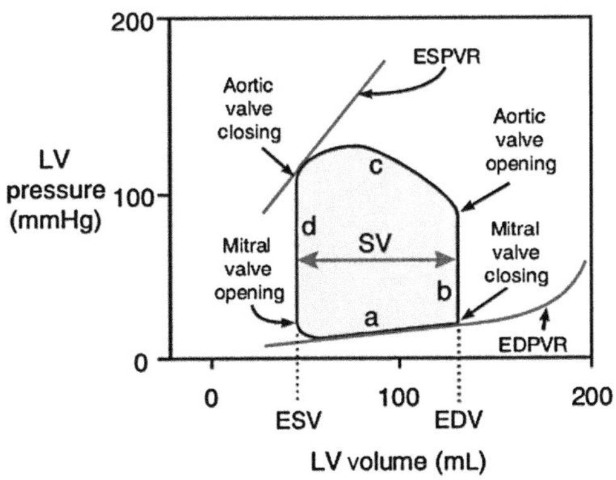

■ **FIGURE 4.6** Left ventricular pressure–volume loop. *From R.E. Klabunde, Cardiovascular Physiology Concepts, Wolters Kluwer and Lippincot Williams & Wilkins, Baltimore, MD, 2011 with permission.*

These four valve fall under two categorizations:

- *AV or inflow valves*: The mitral valve (or bicuspid valve) and tricuspid valve connect the atriums to the ventricles. The mitral valve is located in the left pump of the heart and the tricuspid valve is in the right pump of the heart, see Fig. 4.1.
- *Semilunar or outflow valves*: They are the aortic valve and pulmonary valve. The aortic valve is located between the left ventricle and the aorta and when the valve closes, it prevents back flow (reverse flow) from the aorta to the left ventricle. The pulmonary valve is located between the pulmonary artery and the right ventricle. For example, the aorta valve prevents back flow from the pulmonary artery to the right ventricle, see Fig. 4.1.

The mitral (or bicuspid valve) has two leaflets (or cusps) and the tricuspid valve has three leaflets [7]. Leaflets are thin membranes which are connected by tendons (chordate tendineae) to the ventricle muscles (papillary muscles). The papillary muscles and the chordate tendineae do not have any role in closing or opening mitral or tricuspid valves, although they prevent the leaflets from protruding into the atrium during ventricular contraction.

Fig. 4.7 shows a schematic of heart valve structure from a lateral view (open heart valve) and the top (closed heart valve).

The pressure gradient of the blood flow and the geometry of the heart chambers produces enough energy to open and close the heart valves. During diastole, blood fills the atrium and due to high pressure in the

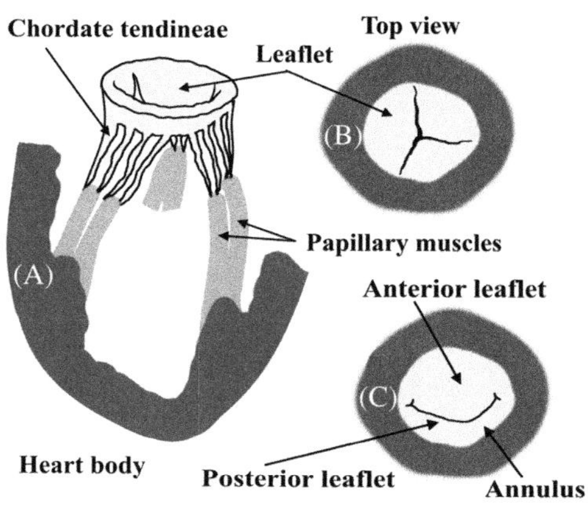

■ **FIGURE 4.7** Schematic of heart valve. (A) Lateral view of open heart valve; (B) top view of close tricuspid valve and (C) top view of mitral valve and anterior and posterior leaflets.

atrium and low pressure in ventricle, blood flow opens the heart valve, then blood begins to fill the ventricle and the result is a high pressure in the ventricle which helps to close the heart valve. Fig. 4.8 illustrates the opening–closing mechanism of the heart valve.

Mechanical properties: As previously mentioned, papillary muscles do not have any role in opening and closing the heart valves, although they do pull on the edge of the leaflets (see Fig. 4.7). The following equation shows the relation between the tension and pressure of the heart valve [4]:

$$P_i - P_o = \frac{T_i}{r_i^2} + \frac{T_o}{r_o^2} \tag{4.4}$$

where P is the pressure, T is the tension, r is curvature radius, i and o are inner and outer surface of the valve leaflet, respectively.

Studies have shown that the mitral leaflet tissue is comprised of three layers with different cellularity and collagen properties. Strength of material (tension) analysis proved that the anterior leaflet of the mitral valve would be more capable of supporting larger tensile loads than the posterior leaflet and this aspect occurs due to different material [8].

Other studies have been done on the strength of the chordae tendinae. Researchers found that the size of the chordae have a significant effect on tension, the smaller chordae with a cross-sectional area of around 0.001–0.006 cm^2 have a Young's modulus of 2×10^9 dyn/cm^2 and larger chordae with cross-sectional area 0.006–0.03 cm^2 have a Young's modulus of 1×10^9 dyn/cm^2 [9]. Also a theoretical study on stress for the mitral valve determined that the level of stress can reach as high as 2.2×10^6 dyn/cm^2 with the left ventricular pressure rising to 150 mmHg just prior to the opening of the aortic valve [10].

During past decades, several computational models of the mitral valve have developed. These studies found that the maximum stresses at pressure of 120 mmHg (a peak loading) were 5.7×10^6 dyn/cm^2 in the annular region, while the anterior leaflet stresses were between 2×10^6 and 4×10^6 dyn/cm^2 [11].

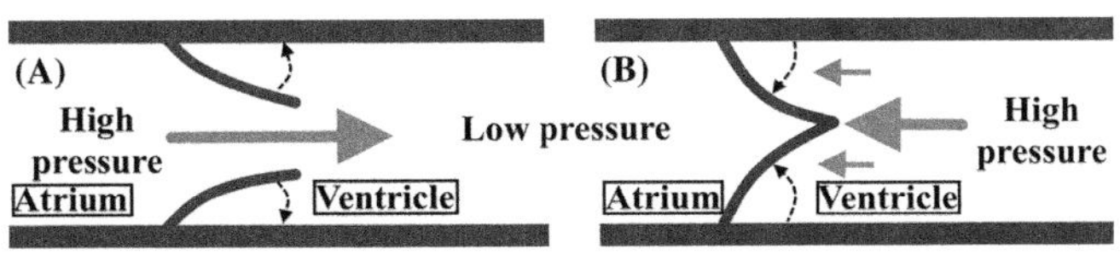

■ **FIGURE 4.8** Opening and closing heart valve due to pressure gradient. (A) High pressure in atrium helps to open the heart valve, although (B) high pressure in ventricle helps to close the valve.

4.1.5 Coronary Circulation

The coronary circulation is responsible for transporting blood to the heart muscles (myocardium). Due to the main role of the heart in the body, the coronary circulation provides an important function in the cardiovascular system and the body.

The resting coronary blood flow rate averages about 225 mL/minutes or 0.8 mL/minutes/g of heart muscle, which represents about 4–5% of the total CO [1,12]. The heart has the highest oxygen consumption per tissue mass among all human organs—its rate is about 10 mL O_2/minutes/100 g at rest [13].

Fig. 4.9 shows the heart and its coronary vessels. As the figure shows, the main coronary arteries are located on the heart surface and smaller arteries then penetrate from the surface into the heart muscles.

Generally, the left coronary artery supplies the left lateral and the anterior portions of the left ventricle, while the right coronary artery supplies most parts of the right ventricle and the posterior portion of the left ventricle. The major part (75% of total coronary blood flow) of venous blood flow from the left ventricle returns to the right atrium by the coronary

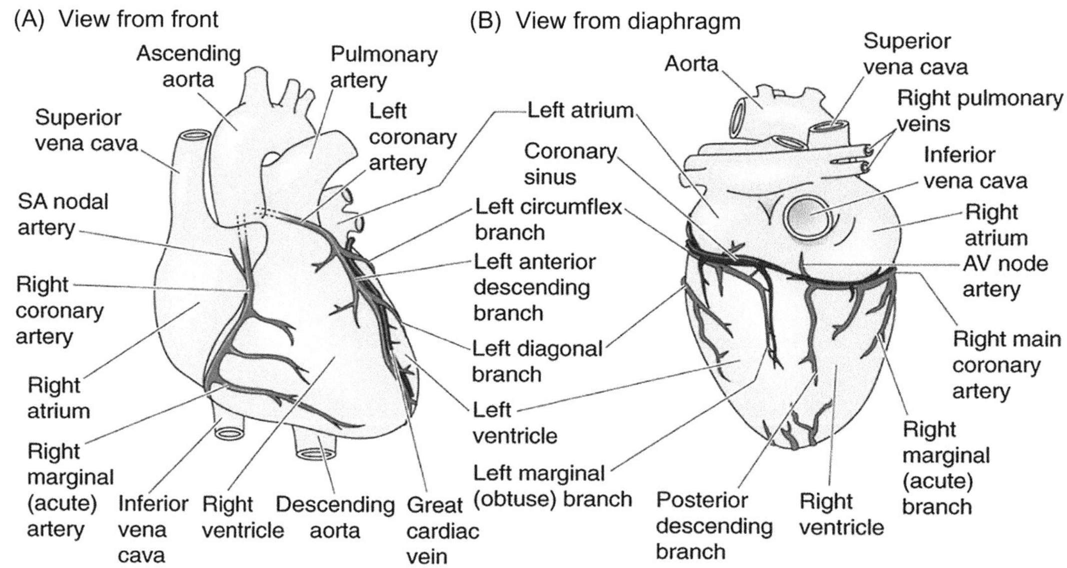

■ **FIGURE 4.9** Coronary circulation. (A) Front view and (B) diaphragm view. *From W.F. Boron, E.L. Boulpaep, Medical Physiology: A Cellular and Molecular Approach, second ed., Saunders Elsevier, Philadelphia, PA, 2009 with permission.*

sinus way, and most of the coronary venous blood flow from the right ventricular muscle returns into the right atrium, but a small amount of coronary venous blood flows back into all heart chambers [1].

The blood flow rate in the coronary vascular system is quite interesting. Considering Fig. 4.10, the blood flow within coronary arteries increases during diastole and decreases during systole. This aspect occurs due to the contraction of the cardiac muscles. This contraction compresses microvessels within the ventricular walls and it results in increasing the resistance and decreasing the blood flow rate. In the early stage of diastole, the ventricle muscles begin to relax and this relaxation removes the flow resistance in tissues, which results in an increase in the blood flow rate (see Fig. 4.10, middle panel). Consequently, most of the blood flow rate to the cardiac muscle occurs in the early stage of diastole. The compression forces affecting the coronary flow rate are greatest in the left

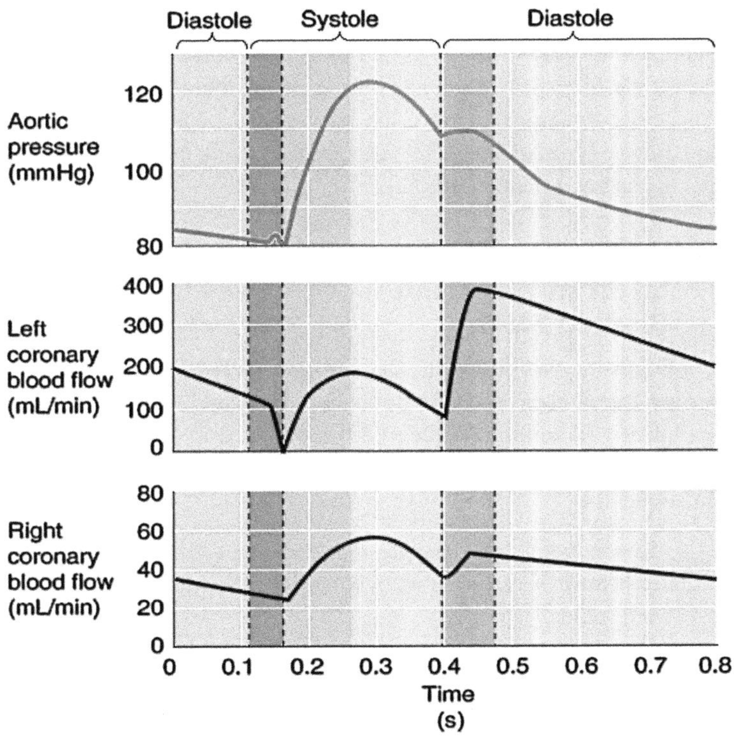

■ **FIGURE 4.10** Coronary blood flow and pressure under pulsatile nature of the heart (systolic and diastolic). *From W.F. Boron, E.L. Boulpaep, Medical Physiology: A Cellular and Molecular Approach, second ed., Saunders Elsevier, Philadelphia, PA, 2009 with permission.*

ventricle since the blood pressure in the left ventricle is several fold greater than in the right ventricle.

In the right coronary artery (Fig. 4.10, lower panel), the blood flow profile is similar to the blood pressure of the aorta, the supply vessel of the right coronary. In this area, systole contributes a greater portion of the blood flow because of the low wall tension and low resistance of the pulmonary system [3].

The collateral circulation is an alternative feature in the cardiovascular system. The collateral circulation is a standby network of blood vessels, and, normally it is not open. When a person suffers a narrowing or blockage of an artery (myocardial infarction or heart attack), collateral blood vessels become active and may enlarge. This mechanism will bypass the narrowing or blockage and supply enough blood and oxygen for the heart muscles to protect the heart from injury.

Normally, the heart consumes 70–80% of the O_2 content of arterial blood [3,7]. The blood flow in the coronary system is controlled by tissue metabolism and due to the high rate of oxygen consumption in the heart muscles, the oxidative metabolism has a major role in the heart beating. In the case of coronary artery disease, chronic narrowing reduces the blood flow oxygen delivery to heart muscles and this leads to cardiac hypoxia and heart malfunction [2].

4.1.6 Effect of Heart and Heart Valves Diseases on Biofluid Flow

According to the World Health Organization (WHO), cardiovascular diseases are the number one killer throughout the world. Cardiovascular diseases killed 17.5 million people in 2012, that is, 3 in every 10 deaths. Of these, 7.4 million people died of ischemic heart disease and 6.7 million from stroke [14].

This section examines the changes that occur in cardiovascular diseases, especially heart diseases such as hypotension, hypertension, heart failure and valve diseases.

Hypotension: Hypotension is low blood pressure in the cardiovascular system especially in the arteries and systemic circulation. Clinically, this problem occurs when the systolic arterial pressure is less than 90 mmHg ($P_{sys} < 90$), or the diastolic pressure is less than 60 mmHg $P_{dia} < 60$. Generally, a reduction in CO and systemic vascular resistance, or both,

Table 4.2 Causes and Mechanisms of Hypotension [2]

Hypotension Causes
■ *Cardiac output*: ❑ *Heart rate*: – *Arrhythmias*: Sinus bradycardia, AV nodal block, ventricular fibrillation ❑ *Stroke volume*: – *Inotropy*: Heart failure, cardiogenic shock, autonomic dysfunction – *Preload*: ■ *Hypovolemia*: Hemorrhage, dehydration ■ *Volume redistribution*: Postural changes, impaired venous return ■ *Arrhythmias*: Arterial fibrillation, tachycardia ■ *Systemic vascular resistance*: ❑ *Autonomic dysfunction* ❑ *Circulatory shock*: Sepsis, anaphylaxis, neurogenic

produces hypotension [2]. Hypotension can be caused by hypovolemia (or low blood volume), anemia, heart problems, medicine or medication side effects, hormonal changes, vessel problems, etc. Table 4.2 summarizes many causes of hypotension in the human body.

Hypertension: Hypertension is known as high blood pressure in the cardiovascular system. This term is applied to diastolic or systolic pressures above the normal amount of blood pressure. The normal ranges of blood pressures are between 90 and 120 mmHg for systolic pressure and 60–80 mmHg for diastolic pressure. Hypertension is defined as a pressure more than 140 and 90 mmHg for systolic and diastolic, respectively. This range of blood pressure can affect the cardiovascular system and even other organs, such as the renal system, visual system, brain, etc. Chronic hypertension occurs because of an increase in systemic vascular resistance and CO [2].

Chronic hypertension can be categorized into two parts: primary (essential) and secondary hypertension. Table 4.3 explains more about the causes of hypertension.

Heart failure: This disease occurs when the heart is unable to provide adequate blood flow for organs and living tissues. In general, heart failure involving the left heart (left ventricle) and right heart (right ventricle) failure occasionally occurs alone or in association with pulmonary disease or occurs secondary to left heart (left ventricle) failure [2].

Table 4.3 Hypertension Types and Causes [2]

Hypertension	
Essential Hypertension	**Secondary Hypertension**
■ Rate: 90–95% ■ Unknown causes ■ Involvement: ❑ Increasing blood volume ❑ Increasing systemic vascular resistance (vascular disease) ■ Related with: ❑ Heredity ❑ Abnormal response to stress ❑ Diabetes and obesity ❑ Age, race and socioeconomic status	■ Rate: 5–10% ■ Stenosis of renal artery ■ Hyperaldosteronism (primary) ■ Pheochromocytoma (catecholamine secreting tumor) ■ Aortic coarctation ■ Pregnancy (preeclampsia) ■ Hyperthyroidism/ hypothyroidism ■ Cushing syndrome (excessive glucocorticoid secretion) ■ Sleep apnea

Heart failure can be affected by [2]:

- end phase of coronary artery diseases;
- myocardial infarction;
- cardiomyopathies (heart muscle diseases) because of viral, bacterial, alcohol, or idiopathic origins;
- chronic arrhythmias;
- valvular disease;
- congenital disease;
- pressure load, such as hypertension;
- volume load, such as arterial venous shunt and
- high output failure, such as thyrotoxicosis or pregnancy.

One of the methods to evaluate the heart function is the Frank–Starling law (curve). This law represents that, normally, an increase (changes) in venous return results in an increase (changes) in left ventricle end diastolic pressure (LVEDP) and it increases the heart SV due to extra muscle contraction.

In heart failure, the mechanism of Frank–Starling fails because of damaged cardiac muscles which results in a failure of the heart to pump the sufficient volume of blood. Fig. 4.11 shows a Frank–Starling curve for two conditions, normal heart (*A*) and a heart with systolic failure in the left ventricle (*B*). The normal functional point (*A*) is at a LVEDP of around 8 mmHg and a SV of around 70 mL/beats.

Fig. 4.12 shows three different pressure–volume loops of the left ventricle with systolic, diastolic and both systolic-diastolic failures.

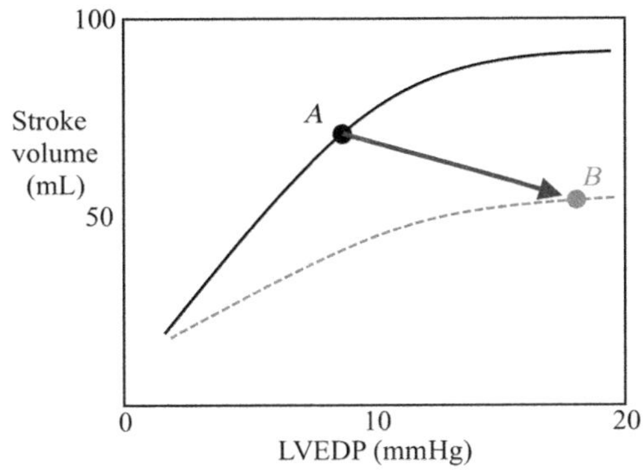

■ **FIGURE 4.11** Frank-Starling curve. *A*, Normal heart, *B*, a heart with systolic failure in the left ventricle.

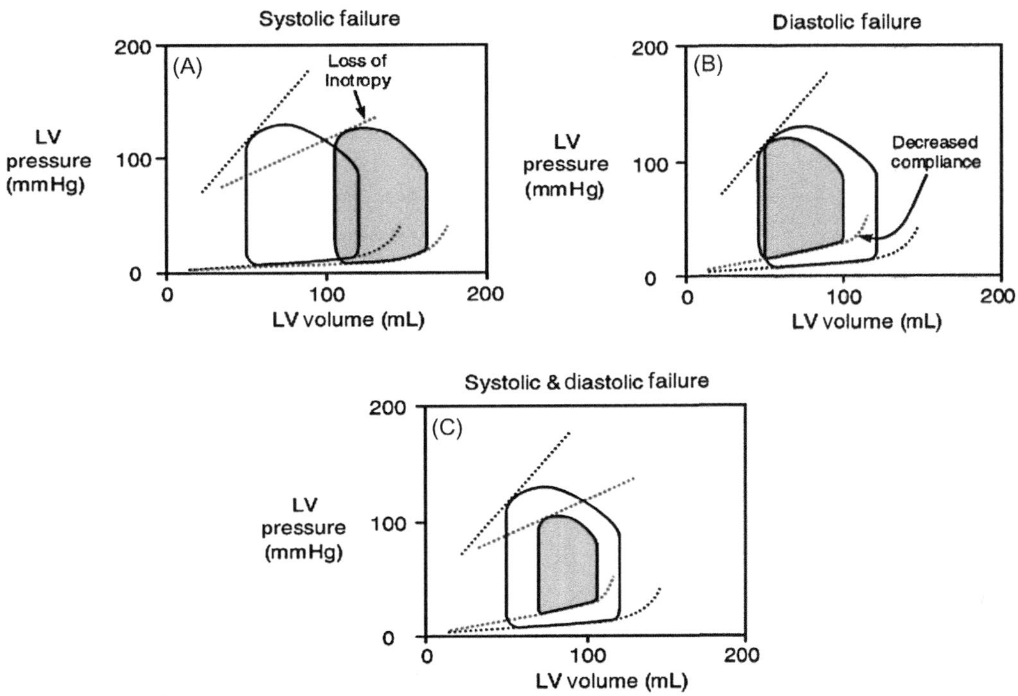

■ **FIGURE 4.12** Effects of heart failures on left ventricular pressure—volume loop. *From R.E. Klabunde, Cardiovascular Physiology Concepts, Wolters Kluwer and Lippincot Williams & Wilkins, Baltimore, MD, 2011 with permission.*

Fig. 4.12A shows that systolic failure decreases the gradient of the end-systolic pressure–volume relationship due to reduced inotropy (contractility). Because of less pressure and less ejected volume during systole, an increase occurs in ESV (see Fig. 4.12A). Diastolic failure can be characterized by either decreased ventricular compliance or impaired relaxation. Decreased ventricular compliance moves the end-systolic pressure–volume relationship to the left and up. It results in higher end diastolic pressure, decreased EDV (less ventricular filling) and finally a decrease for SV (see Fig. 4.12B). Rarely, heart failure is due to a combination of both systolic and diastolic dysfunction. This results in an increase and a decrease in the gradient of the passive filling curve and end systolic pressure–volume relationship, respectively. Because of an increase in ESV and a decrease in EDV, a dramatic reduction occurs in SV (see Fig. 4.12C) [2].

Heart valve diseases: Each heart valve has two main roles: (1) to open appropriately so the blood can empty from the heart chamber and (2) to close appropriately so the blood can flow unidirectionally. Two general diseases are defined for heat valves: insufficiency and stenosis.

Heart valve stenosis (or abnormal narrowing) can occur for all four heart valves (pulmonic, aortic, mitral and tricuspid valves) and it increases the resistance to blood flow which causes a high pressure gradient across the heart valve [2]. Considering Poiseuille's law, narrowing of the heart valve orifice (passage radius) dramatically increases the pressure gradient across the valve. Also a turbulence flow occurs after stenosis of the valve and it produces a murmur. Fig. 4.13 shows a schematic of one of the heart valve stenosis shapes.

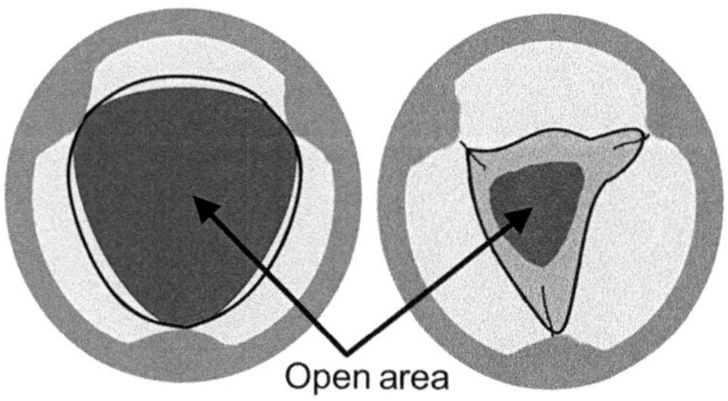

■ **FIGURE 4.13** Open heart valve. Left: Normal valve; right: stenosis valve.

The valve stenosis can be summarized for all cardiac valves as [2]:

- Aortic valve stenosis: in this disease, left ventricular pressure is increased more than normal pressure during systole. This aspect results in a high pressure gradient across the aortic valve during ejection. Turbulence flow after the stenosis valve produces a murmur. Also this problem can have an effect on the pressure–volume loop. The EDV is constant (or with small changes) while the ESV is increased, therefore, the SV is decreased.
- Mitral valve stenosis: there is an increase in the pressure gradient during filling of the ventricle and it creates an increase and a reduction in the left atrial and left ventricular filling pressures, respectively. Also, the effect of this problem on the left ventricular pressure–volume loop is a reduction on both end diastolic and ESVs and finally a decrease in SV.
- Pulmonic and tricuspid valve stenosis: this disease in the pulmonic system creates a pressure gradient across the valve during ejection of the right ventricle and a systolic murmur. A decreased right ventricular stroke reduces the left ventricular filling and SV. The right ventricular and SV are impaired by tricuspid stenosis which elevates systemic venous and atrial pressures. Also, tricuspid stenosis produces diastole murmur.

To visualize the stenosis effects of the heart valves, Fig. 4.14 is used to illustrate the left ventricular pressure–volume loop, pressure–time and echocardiogram graphs of aortic and mitral valves.

Stenotic valve orifice area (VOA): Accurate calculation of the stenotic valve orifice area A_v (cm^2) is very important for disease treatment. There are two standard formulae to calculate the area of the valve orifice: (1) the Gorlin formula and (2) the Hakki formula.

The *Gorlin formula* depends on four variables: the CO (mL/minutes), the mean pressure gradient (ΔP) (mmHg), the diastolic filling period for the mitral and tricuspid valve or the systolic filling period for other valves (seconds), the heart rate (HR) (beats/minutes) and the Gorlin constant (GC),

$$A_v = \frac{\mathrm{CO}}{(\mathrm{SEP\ or\ DEP}) \times \mathrm{HR} \times \mathrm{GC} \times \sqrt{\Delta P}} \tag{4.5}$$

The GC is 44.3 for aortic valves and 37.7 for mitral valves [14].

The second formula (*the Hakki formula*) is a simplification of the Gorlin formula. This simplification has two variables: CO (mL/minutes) and mean pressure gradient (ΔP) (mmHg),

$$A_v = \frac{\mathrm{CO}}{\sqrt{\Delta P}} \tag{4.6}$$

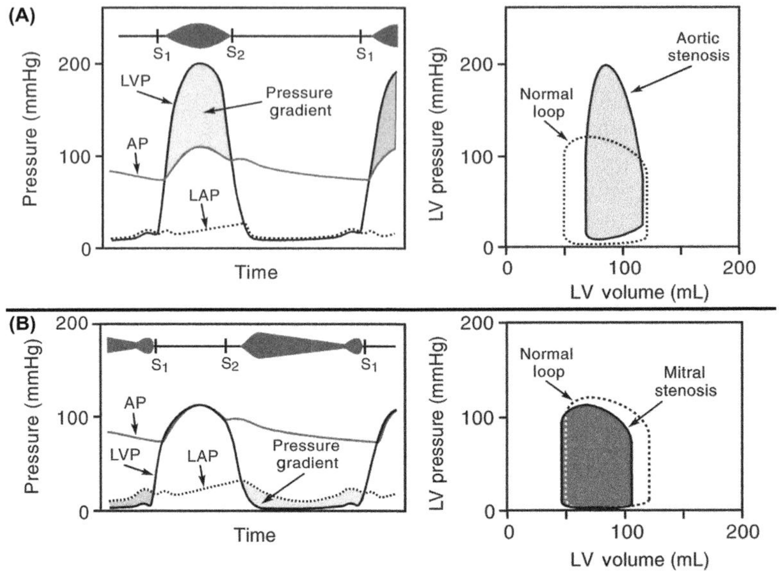

■ **FIGURE 4.14** Cardiac function with valve stenosis. (A) Chronic aortic valve stenosis in the absence of systolic failure; left: the gray area represents the pressure gradient generated by the stenosis; also echocardiogram shows a systolic murmur between S1 and S2. Right: It shows the effects of aortic valve stenosis (black loop) on the left ventricular pressure—volume loop. (B) Chronic mitral valve stenosis in the absence of systolic failure; left: the gray area represents the pressure gradient generated by the stenosis; also echocardiogram shows a diastolic murmur between S2 and S1. Right: it shows the effect (SV reduction) of mitral valve stenosis (black loop) on the left ventricular pressure—volume loop. *From R.E. Klabunde, Cardiovascular Physiology Concepts, Wolters Kluwer and Lippincot Williams & Wilkins, Baltimore, MD, 2011 with permission.*

Stenotic aortic valve resistance, R_V: This factor is one of the most important factors to classify the severity of the aorta stenosis. The equation is [15,16]

$$R_V = \frac{(\mathrm{LV} - \mathrm{Ao}) \times 80}{\mathrm{CO} \times 2.5} \tag{4.7}$$

where (LV − Ao) is the peak transaortic pressure gradient (mmHg) and CO is cardiac output (L/minutes).

There are three levels of classification of the severity of valve stenosis: mild, moderate and severe. Table 4.4 characterizes and classifies heart valve stenosis and its severity. The blood velocity and the pressure gradient increase with the increasing severity level of valve stenosis.

Valve regurgitation: In this disease, the heart valves do not close completely, which permits blood to flow backward across the valves.

Table 4.4 Characterization and Severity of Heart Valves Stenosis [17]

Classification of Aortic Stenosis Severity				
	Aortic Sclerosis	**Mild**	**Moderate**	**Severe**
Aortic jet velocity (m/s)	≤ 2.5 m/s	2.6–2.9	3.0–4.0	>4.0
Mean gradient (mmHg)	–	<20 (<30[a])	20–40[b] (30–50[a])	>40[b] (> 50[a])
Aortic valve area (cm^2)	–	>1.5	1.0–1.5	<1

Classification of Mitral Stenosis Severity			
	Mild	**Moderate**	**Severe**
Valve area (cm^2)	>1.5	1.0–1.5	<1
Mean gradient (mmHg)[c]	<5	5–10	>10
Pulmonary artery pressure (mmHg)	<30	30–50	>50

Hemodynamic of Tricuspid Stenosis	
Mean pressure gradient	≤5 mmHg
Inflow time–velocity integral	>6 cm
*T*½ (pressure half-time)	≤190 ms
Valve area by continuity equation[d]	≤1 cm^2

Grading of Pulmonary Stenosis			
	Mild	**Moderate**	**Severe**
Peak velocity (m/s)	<3	3–4	>4
Peak gradient (mmHg)	<36	36–64	>64

[a]ESC guidelines.
[b]AHA/ACC guidelines.
[c]At heart rates between 60 and 80 bpm and in sinus rhythm.
[d]Stroke volume derived from left or right ventricular outflow. In the presence of more than mild TR, the derived valve area will be underestimated. Nevertheless, a value 1 cm^2 implies a significant haemodynamic burden imposed by the combined lesion.

The characterization of heart valves regurgitation is [2]:

- Aortic valve regurgitation is categorized by an increase in aortic pulse pressure, diastolic murmur, ventricular dilation, increased SV (but reduced net aortic flow), no true isovolumetric phases, increased ventricular filling pressure and increased pulmonary vascular and left atrial pressures.
- Mitral valve regurgitation is categorized by systolic murmur, a tall v wave, ventricular dilation, increased SV (but reduced net ventricular

outflow into the aorta), no true isovolumetric phases, increased ventricular the filling pressure and increased pulmonary vascular and left atrial pressures.

- Pulmonic valve regurgitation is categorized by an increase in pulmonic artery pulse pressure, increased right ventricular EDV and pressure, diastolic murmur, no true isovolumetric phase during right ventricle systole and diastole, an increase in right atrial and systemic venous pressure and right ventricle dilation.
- Tricuspid valve regurgitation is categorized by an increase in right atrial volume and systemic venous pressure, a tall v wave in the right atrial pressure, increased SV of right ventricle and decreased pulmonary artery ejection (Table 4.5).

Table 4.5 Grading the Regurgitation Severity of Aortic, Mitral and Tricuspid Valves [18,19]

	Mild	Moderate	Severe
Grading of aortic regurgitation severity			
VC width (mm)	<3	Intermediate	>6
Pressure half-time (ms)	>500	Intermediate	<200
EROA (mm^2)	<10	10–19; 20–19	≥ 30
R Vol (mL)	<30	10–19; 20–19	≥ 60
Grading of mitral regurgitation severity			
VC width (mm)	<3	Intermediate	Large
Pulmonary vein flow	Systolic dominance	Systolic blunting	Systolic flow reversal
Mitral inflow	A wave dominant	Variable	E wave dominant (>1.5 cm/s)
TVI mit/TVI Ao	<1	Intermediate	>1.4
EROA (mm^2)	<20	20–29; 30–39	≥ 40
R Vol (mL)	<30	30–44; 45–59	≥ 60
Grading of tricuspid regurgitation severity			
VC width (mm)	Not defined	<7	≥ 7
PISA radius (mm)	≤ 5	6–9	>9
Hepatic vein flow	Systolic dominance	Systolic blunting	Systolic flow reversal
Tricuspid inflow	Normal	Normal	E wave dominant (>1.5 cm/s)
EROA (mm^2)	Not defined	Not defined	≥ 40
R Vol (mL)	Not defined	Not defined	≥ 45

4.2 LUNG AND RESPIRATORY SYSTEM

4.2.1 Introduction

The main function of the respiratory system is to supply oxygenated blood to living organs and to remove carbon dioxide. This procedure is called breathing and is divided into four major functions: (1) pulmonary ventilation, (2) diffusion of oxygen and carbon dioxide inside the lung (alveoli), (3) transportation of oxygen and carbon dioxide to and from the body's tissues and (4) regulation of ventilation.

Fig. 4.15 shows a schematic of the respiratory system and its elements. The functional elements of the respiratory system are:

- nose and nasal cavity,
- pharynx,
- trachea,
- bronchus,
- bronchiole,
- alveoli and
- blood vessels.

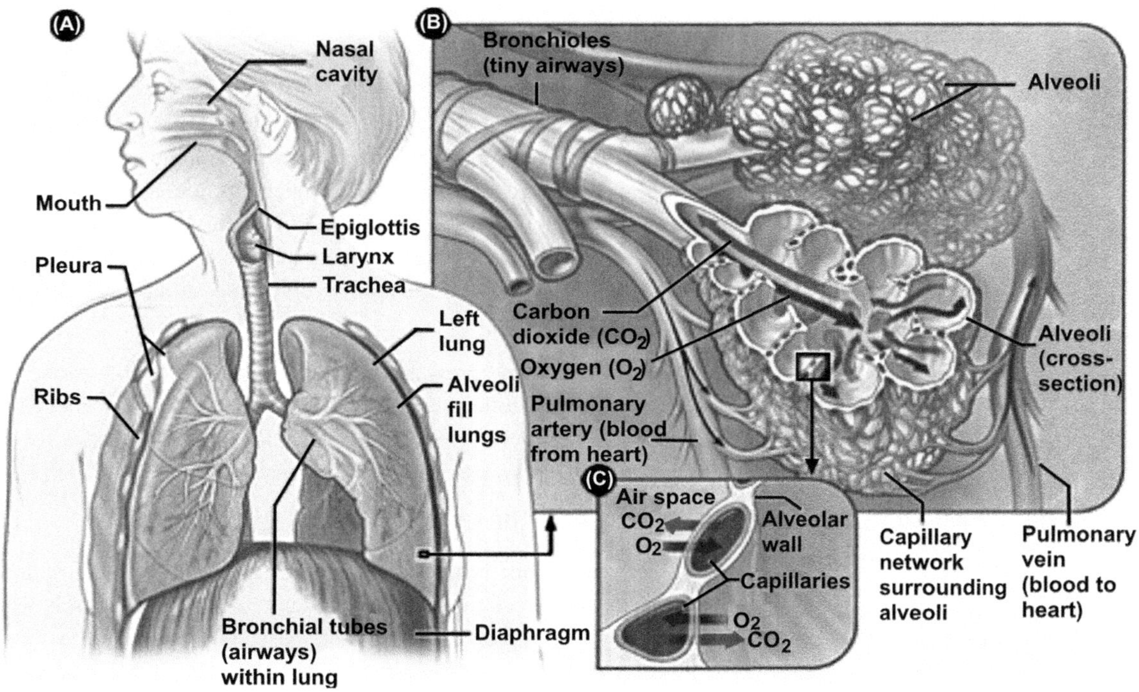

■ **FIGURE 4.15** Respiratory system and its elements. *From http://www.nhlbi.nih.gov/health/health-topics/topics/hlw/system.*

The two lungs (right and left) are elements of the respiratory system. Normally, the right lung is a little larger than the left lung due to the middle mediastinum (cardiac system and other tissues). The lungs have a (half) cone shape and concave form at the base.

The atmospheric air first enters from the nasal cavity (or the mouth), then it passes through the pharynx, the trachea, the bronchus, the bronchiole and finally the alveoli. The bifurcation mechanism in the lungs increases the surface area for gas exchange. The bifurcation mechanism has 23 generations of airways in the lung. This bifurcation begins with the zeroth generation, the trachea, then the first generation is left and right mainstem bronchi. The bronchioles will start at about 11th generation. There are around 30,000 terminal bronchioles in each lung. After the 16th generation, alveoli are present and gas exchange will be in progress via the blood capillaries. The 23rd generation is the final element of the respiratory system, that is, alveoli sacs [3].

The average internal diameter of a terminal bronchiole is around 500 μm. Each bronchiole has multiple alveoli sacs and the average number of alveoli in each lung is around 350 million (this number varies directly with lung volume). The average diameter of alveoli is around 200–300 μm and the average total surface area is around 75 m^2.

4.2.2 Pulmonary Blood Flow and Vessels

The respiratory system is fed by two separate blood circulations, the pulmonary circulation (with lower pressure) which supplies blood flow to the alveoli for gas exchange, and the bronchial circulation (with higher pressure) which supplies blood flow to the bronchial system and the trachea [2]. The pulmonary circulation receives all of the right ventricular output while the bronchial circulation receives only around 1% of the CO of the left ventricle.

The pulmonary artery pressures are 15–30 mmHg for systolic and 4–12 mmHg for diastolic pressure [6]. The mean pulmonary artery pressure (MPAP) is between 9 and 18 mmHg [20]. The new formula for predicting MPAP using systolic pulmonary artery pressure (SPAP) is [21]:

$$\text{MPAP} = 0.61 \times \text{SPAP} + 2\ \text{mmHg} \tag{4.8}$$

The vascular resistance in the pulmonary circulation is 10–15-fold lower than the systemic circulation. It is due to numerous parallel vessels which have a short length and large diameter. Also, pulmonary blood vessels are more compliant than systemic blood vessels, and because of this vital aspect, an increase in blood pressure of the right ventricle output

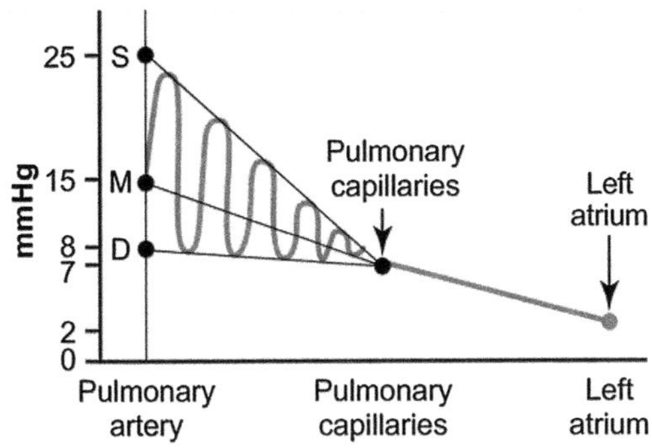

■ **FIGURE 4.16** Pressures in the lung vessels. S, systolic pressure; M, mean pressure; D, diastolic pressure and *curve line* shows arterial pulsation. *From A.C. Guyton, J.E. Hall, Textbook of Medical Physiology, eleventh ed., Elsevier, Inc., Philadelphia, PA, 2006 with permission.*

(such as exercise) does not cause a proportional increase in blood pressure of the pulmonary vessels. Increased blood pressure in pulmonary vessels can cause (1) right ventricle failure and (2) pulmonary edema because of more fluid filtration through the pulmonary capillaries [2].

Fig. 4.16 shows the blood pressure in the left atrium, the pulmonary artery, and the capillaries of the lungs. As the figure shows, in the normal adult the average systolic pulmonary arterial pressure (S) is around 25 mmHg, the diastolic pulmonary arterial pressure (D) is about 8 mmHg, and the mean pulmonary arterial pressure and the mean pulmonary capillary pressure are 15 and 7 mmHg, respectively.

The blood volume of the lungs is about 450 mL, approximately 9% of the total blood volume of the whole circulatory system. About 70 mL of this blood volume is in the pulmonary capillaries, and the rest is divided about equally between arteries and the veins in the lungs [1].

The topology of the pulmonary arteries and veins is like a tree. The most popular pattern to this tree is Strahler's method. In this method, the smallest blood vessel has order 1. When two blood vessels with order 1 meet, the next generation is named a blood vessel of order 2, etc. But, if a lower order blood vessel meets a higher order, the order number remains at the higher order and so on. The branching ratio is the ratio of the number of blood vessels of order n to that of $n + 1$, usually this ratio in the pulmonary vascular tree is close to 3. Also, the diameter ratio is the ratio of the vessel diameter of order n to that of $n + 1$. Because of some difficulties in this method, Fung et al. introduced a diameter-defined Strahler system. They defined that when two blood vessels of order n or less meet, the order number of their confluent blood vessel become $n + 1$ [22].

The alveoli is surrounded by blood capillary networks. The alveolar septum is a thin wall which separates two alveoli. This wall is similar among all adult mammalian lungs. The topology of the pulmonary capillaries is like a sheet (not treelike). A sheet flow model (or Fung's sheet flow) was introduced several decades ago to define the blood flow in the pulmonary capillaries [22]. For regular capillaries with a cylindrical tube shape, the blood flow characters can be determined the Poiseuille's equation (1.37) but because of blood elements and the porosity of the capillaries and also the shape of the sheet flow, the best formula is Darcy's equation (3.27) for porous media.

The sheet flow model of the pulmonary capillaries maximizes the performance of the lung system by maximizing the contact of erythrocytes (RBC), increasing diffusion time by increasing the contact length and notably less resistance in comparison to Poiseuille's equation in a cylindrical tube.

When the CO is normal, the length of time blood stays in the pulmonary capillaries is about 0.8 seconds and the minimum time (maximum activity) is around 0.3 seconds [1].

4.2.3 Air Flow and Pressure Aspects in the Lung

The maximum volume of the lungs and their subsystem is around 5–6 L. A spirometer is a device for measuring the air volume of the lungs and its changes during breathing. Today fully electronic spirometers have been industrialized that analyze and calculate air flowrate of the lungs.

To understand spirometer function, Fig. 4.17 shows an old model of a spirometer. This machine has a movable sliding tank (bell) that is partially merged in liquid (water). An air tube connects the patient's mouth to the tank (bell) just above the liquid (water) level. Therefore, when the patient exhales, air lifts the tank (bell) and a recording system records the movement of paper (Fig. 4.17A). The record displays the air volume that the patient exhales.

Several variables for the volumes and capacities are defined to analyze the spirometer record and the lung condition [3] (see Fig. 4.17B);

- *Tidal volume* (V_T), is the amount of air entering and leaving the lungs during normal breathing and its volume is around 500 mL for a 70 kg adult.
- *Expired lung volume* (V_E), is the volume leaving the lungs.

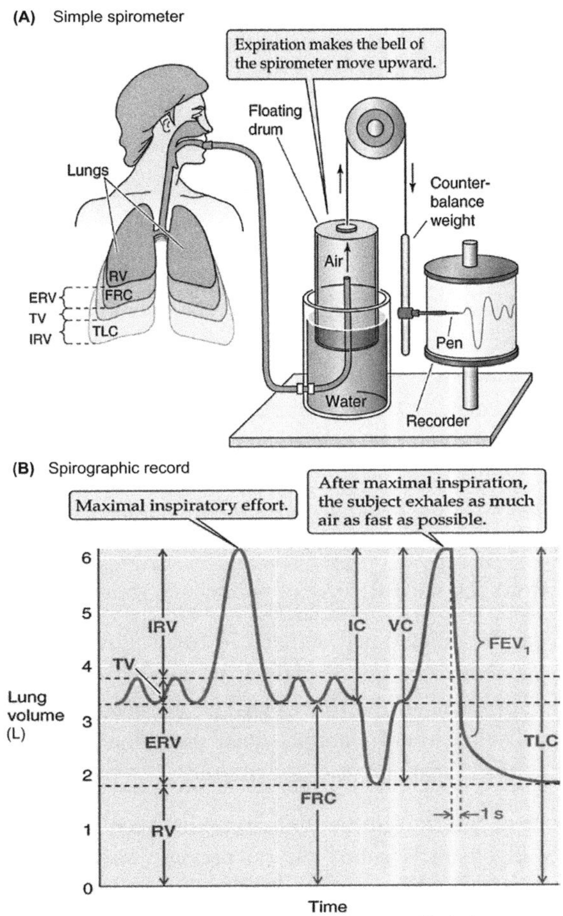

Volume and capacities	Typical ranges (L)
IRV = Inspiratory reserve volume	1.9–2.5
TV = Tidal volume	0.4–0.5
ERV = Expiratory reserve volume	1.1–1.5
RV = Residual volume	1.5–1.9
TLC = Total lung capacity	4.9–6.4
IC = Inspiratory capacity	2.3–3.0
FRC = Functional residual capacity	2.6–3.4
VC = Vital capacity	3.4–4.5

■ **FIGURE 4.17** The spirometer mechanism. (A) Spirometer and (B) the record of spirographic. *From W.F. Boron, E.L. Boulpaep, Medical Physiology: A Cellular and Molecular Approach, second ed., Saunders Elsevier, Philadelphia, PA, 2009 with permission.*

- *Inspiratory reserve volume (IRV)*, is the extra volume of air that the patient could inhale with a maximal effort at the end of quiet inspiration.
- *Expiratory reserve volume (ERV)*, is the extra volume of air that one can expire with a maximal effort after a quite expiration.
- *Residual volume (RV)*, is the remained air volume in the lungs after expiration (even maximal effort).
- *Total lung capacity (TLC)*, is the sum of all four volumes (V_T, IRV, ERV and RV).
- *Functional residual capacity (FRC)*, is the volume of air remaining inside of respiratory system after a quite expiration and mathematically, it is the sum of ERV and RV.
- *Inspiratory capacity (IC)*, is the maximal volume of air that one could still inspire after a quite expiration and is equal to the sum of IRV and TV.
- *Vital capacity (VC)*, is the maximal possible tidal volume and is equal to the sum of IRV, ERV and TV.

Table 4.6 shows the normal range of pulmonary volumes and capacities.

The lungs can be inflated and deflated in two ways: (1) by upward and downward movement of the diaphragm to change the chest cavity size and (2) by movement of the ribs to change the anteroposterior diameter of the chest cavity. Mostly, normal quiet breathing is accomplished by movement of the diaphragm (the first method).

The pressure of the fluid in the thin space between the lung pleura and the chest wall pleura is called pleural pressure and is less than atmospheric pressure, which means the pleural pressure is a slightly negative (vacuum) pressure. The second pressure factor in the respiratory system

Table 4.6 The Lung Volume and Capacity Ranges [3]

Variables	Typical Ranges (L)
Inspiratory reserve volume (IRV)	1.9–2.5
Tidal volume (V_T)	0.4–0.5
Expiratory reserve volume (ERV)	1.1–1.5
Residual volume (RV)	1.5–1.9
Total lung capacity (TLC)	4.9–6.4
Inspiratory capacity (IC)	2.3–3.0
Functional residual capacity (FRC)	2.6–3.4
Vital capacity (VC)	3.4–4.5

is alveolar pressure, and it is the pressure of the air inside the alveoli of the lungs. The third pressure factor is transpulmonary pressure which is the difference between the alveolar pressure and the pleural pressure [1]. The relation among these pressures is demonstrated in Fig. 4.18. As the figure shows, during inspiration, there is a decreasing of the pleural pressure from −5 to −7.5 cm H_2O (the lower panel) and in the upper panel an increase of 0.5 L in lung volume. During expiration, these procedures are reversed.

Air flowrate (Q_A, L/seconds) has a direct relation to the difference between atmospheric and alveolar pressure (ΔP) but an inverse relation to total airway resistance (R_{AW}) [3],

$$Q_A = \frac{\Delta P}{R_{AW}} = \frac{\text{alveolar pressure} - \text{atmospheric pressure}}{R_{AW}} \tag{4.9}$$

In normal persons, R_{AW} is around 1.5 cm H_2O/(L/seconds) but its range is between 0.6 and 2.3. In patients with lung diseases, these values are higher and even can exceed 10 cm H_2O/(L/seconds) in some extreme cases. This method to measure resistance denotes 80% of the total pulmonary resistance and the remaining 20% is the tissue resistance in the pulmonary system [3].

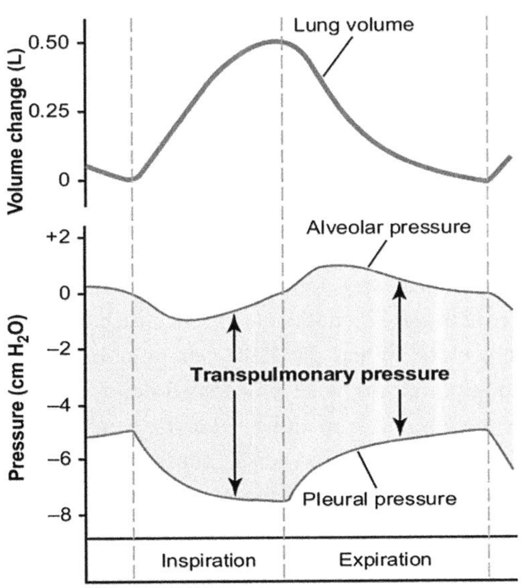

■ **FIGURE 4.18** Changes in lung pressures and volume during normal breathing. *From A.C. Guyton, J.E. Hall, Textbook of Medical Physiology, eleventh ed., Elsevier, Inc., Philadelphia, PA, 2006 with permission.*

4.2.4 Transport of Oxygen and Carbon Dioxide in the Blood

Oxygen: Blood carries oxygen (O_2) by two methods. Normally the erythrocytes or RBCs carry more than 98% of oxygen which binds to RBC hemoglobin and a small amount of O_2 dissolves in blood plasma and the cytoplasm of blood cells. According to Henry's law for the concentration of gases which dissolve in liquids, the concentration of dissolved O_2 ($[O_2]_{Dis}$) in the blood is [3],

$$[O_2]_{Dis} = k_{O_2} \cdot P_{O_2} \tag{4.10}$$

where k_{O_2} is the solubility which is $\sim$0.003 mL O_2 for 100 mL of blood and each mmHg of O_2 and P_{O_2} is partial pressure which alveolar pressure (100 mmHg). Therefore, the concentration of dissolved O_2 for arterial blood is equal to 0.3 mL O_2/100 mL blood. Furthermore, according to Fick's principle, the delivery of dissolved O_2 to tissue would be equal to 15 mL O_2/minutes, which is a result of blood flow (5000 mL/minutes) multiplied by O_2 concentration (0.3 mL O_2/100 mL blood) [3].

Hemoglobin (Hb) is the iron (Fe) containing protein molecule in the RBCs and its main role in the blood circulation system is as the transporter of oxygen to living tissues. The hemoglobin absorbs oxygen in the lung and transports it to tissues. It has two major parts, heme and globin. The heme structure is a porphyrin with an iron (Fe) ion in the center. The Fe ion can be in the states of Fe^{3+} (ferric) or Fe^{2+} (ferrous). The heme part makes the hemoglobin red in color. When the blood is saturated with oxygen the hemoglobin color is red (arterial blood) and it is a purple color when it is devoid of oxygen (venous blood).

The second part—the globin–is the protein surrounding the heme. Several globin chains are defined and named as α, β, γ, ε, etc. Individually, a hemoglobin molecule is constructed by four heme units and four globin units. When the RBC hemoglobin carries oxygen, it is called oxyhemoglobin, otherwise it is named deoxyhemoglobin. After the respiration procedure in the lung, the oxygenated blood flows in the blood circulation and its oxygen is consumed by living tissues, then the deoxygenated blood returns to the lung to take up further oxygen.

Fig. 4.19 illustrates the percentage of saturated hemoglobin for a range of P_{O_2}. The right side ordinate presents the oxygen bound to hemoglobin and the left side ordinate shows the same data in terms of oxygenated hemoglobin percentage [3].

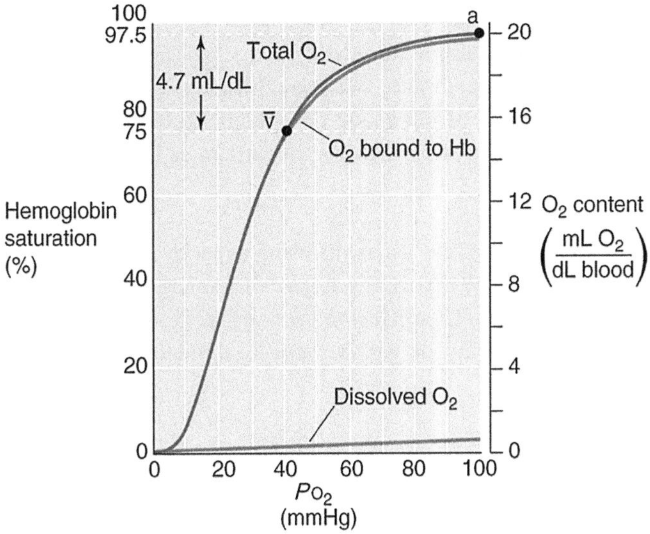

■ **FIGURE 4.19** A normal hemoglobin-oxygen dissociation curve. *From W.F. Boron, E.L. Boulpaep, Medical Physiology: A Cellular and Molecular Approach, second ed., Saunders Elsevier, Philadelphia, PA, 2009 with permission.*

Increasing the temperature shifts the hemoglobin-oxygen dissociation curve to the right—down—but decreasing the temperature moves it to the left—up.

Carbon dioxide: Blood carries carbon dioxide (CO_2) and related compounds using five methods [3]:

1. dissolved CO_2;
2. carbonic acid, H_2CO_3;
3. bicarbonate, HCO_3^-;
4. carbonate, CO_3^{2-} and
5. carbamino compounds (carbamino hemoglobin).

Transport of CO_2 by the dissolved method is a small portion of carbon dioxide transportation in the body. It is around 5–7% of CO_2 transportation or 0.2–0.3 mL CO_2 for 100 mL blood [1,3]. The transportation of CO_2 by carbonic acid is a very tiny amount, $[H_2CO_3]$ is only around 1/400 as large as of $[CO_2]$ [3]. Carbonic acid, such as carbonate, CO_3^{2-}, has a tiny role in carbon dioxide transportation. The performance of carbamino compounds is not significant for this transportation and it accounts for around 5% of total carbon dioxide [3] or less than 20% [1]. Bicarbonate, HCO_3^-, has the main role in carbon dioxide (CO_2) transportation. Bicarbonate represents around 90% of total carbon dioxide in arterial blood [3].

4.2.5 Gas Exchange in the Lung

It is clear that the transportation of blood gases (such as O_2 and CO_2) across the alveolar membrane occurs by diffusion. Diffusion is the transportation of substances from a side of high concentration to a side of low concentration. The diffusion flow can be written as [3]

$$Q_{\text{dif}} = D_L \times (P_1 - P_2) \tag{4.11}$$

where Q_{dif} is gas flow rate, D_L is diffusing capacity of the lung [mL/(min mmHg)] and $(P_1 - P_2)$ is partial pressure difference between two sides of a membrane (barrier). The diffusion capacity (D_L) is related to molecular weight (MW), solubility of gas (s), area (A) and thickness (t) of membrane (barrier),

$$D_L = k\frac{A \times s}{t\sqrt{\text{MW}}} \tag{4.12}$$

where k is a proportional constant.

Eq. (4.11) is a simplification of Fick's law. If we want to apply this law for the alveolar wall, we have to extend our model. There is a wet membrane with a film of water and air moisture on one side and blood (and blood plasma) on the other side. Figs. 4.20 and 4.21 illustrate this membrane and the diffusion of a gas across a membrane (barrier).

We can rewrite Eqs. (4.11) and (4.12) for the alveolar (A_{O_2}) and blood capillary (C_{O_2}) system,

$$Q_{O_2} = D_{L_{O_2}} \times (P_{A_{O_2}} - P_{C_{O_2}}) \tag{4.13}$$

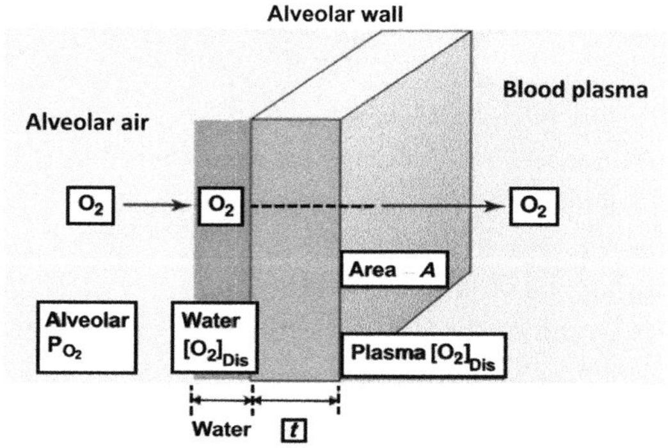

■ **FIGURE 4.20** Diffusion of O_2 across alveolar wall. *From W.F. Boron, E.L. Boulpaep, Medical Physiology: A Cellular and Molecular Approach, second ed., Saunders Elsevier, Philadelphia, PA, 2009 with permission.*

where $D_{L_{O_2}}$ is the diffusing capacity for O_2, $P_{A_{O_2}}$ and $P_{C_{O_2}}$ are partial pressure in the alveolar air and pulmonary blood capillary, these parameters are very complicated to measure. Eq. (4.13) only calculates the gas flowrate for one alveolar; for total flowrate (all lung capillaries), take the sum of all individual diffusion events added up overall alveolar walls and overall times in the pulmonary cycle [3],

$$Q_{O_2} = \sum^{\text{All alveolars}} \sum^{\text{all times in pulmonary cycle}} \left(D_{L_{O_2}} \times \left(P_{A_{O_2}} - P_{C_{O_2}} \right) \right) \quad (4.14)$$

For a CO of 5 L/minutes, total flowrate is around 250 mL O_2/minutes [3].

Two important factors in the gas transportation and exchange in alveoli capillaries are the local concentration and the total flux of gas across the capillary wall. Following equations define these factors [23].

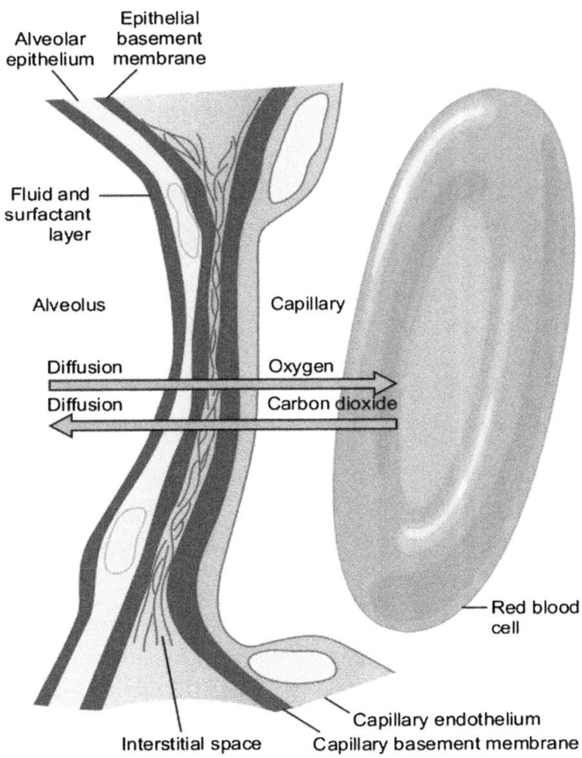

■ **FIGURE 4.21** The alveolar membrane and diffusion of CO_2 and O_2 between RBC in the capillary blood and the alveolar air. *From A.C. Guyton, J.E. Hall, Textbook of Medical Physiology, eleventh ed., Elsevier, Inc., Philadelphia, PA, 2006 with permission.*

In steady state, the concentration at position x, $C(x)$, is

$$C(x) = s \times P_g + (C_0 - s \times P_g) \times e^{-(D_M \times x/V)} \tag{4.15}$$

And total flux of gas (J_g) across the capillary with length of L is

$$J_g = V \times A \times s + (P_g - P_0) \times \left(1 - e^{-(D_M \times L/V)}\right) \tag{4.16}$$

where s is the solubility of the gas, P_g is the partial pressure of the gas, P_0 is the partial pressure of the gas at $x = 0$ or $P_0 = C_0/s$, L is capillary length and V is the velocity of fluid in the capillary. Also D_M (time^{-1}) is

$$D_M = \frac{p \times D_s}{A \times s} \tag{4.17}$$

where D_s is the surface diffusion constant and p and A are perimeter and cross-section area of the capillary, respectively.

Fig. 4.21 shows two sides of the respiratory membrane and its cross-section. It also illustrates the diffusion of O_2 and CO_2 from the alveolus into the RBC and from the RBC to the alveolus, respectively. The membrane is made from many layers and the average thickness of the membrane is around 0.6 μm (minimum 0.2 μm) [1]. These layers are:

1. A layer of fluid on the alveolus contains surfactant which reduces the surface tension of the alveolus fluid.
2. Thin alveolar epithelium layer.
3. An epithelial basement membrane.
4. An interstitial space between the epithelium cells and the capillary membrane.
5. A capillary basement membrane.
6. The capillary endothelial layer.

Four factors have an effect on gas exchange: the thickness of the membrane; the surface area of the membrane; the diffusion coefficient of the gas and the pressure difference between the two sides of the alveolar membrane.

Another definition of gas exchange in the lung is resistance to diffusion. Considering Eq. (4.13) and electrical similarity, it is possible to define the general form of the individual resistance (R_L) to diffusion as

$$R_L = \frac{1}{D_L} \tag{4.18}$$

As previously explained, the final step for O_2 transportation is binding to hemoglobin. This process is related to two factors, the binding rate of O_2

to hemoglobin or θ and the volume of blood in the lung capillaries or V_c. Eq. (4.18) can be rewritten as

$$R_L = \frac{1}{D_M} + \frac{1}{\theta \times V_c} \tag{4.19}$$

where D_M is the sum of diffusing capacities of the layers of alveolar membrane (see Fig. 4.21) or

$$\frac{1}{D_M} = \frac{1}{D_1} + \frac{1}{D_2} + \cdots \tag{4.20}$$

4.2.6 **Effect of Lung Diseases on Biofluid Flow**

Function of the lung diseases can be classified to several common types:

- obstructive,
- restrictive and
- vascular.
- *Obstructive diseases*: One of the common categories of lung diseases is obstructive diseases. All diseases in this category are characterized by the obstruction of airways. This increases the resistance to air flow in the lungs and can be classified to three subsystems: chronic obstructive pulmonary disease (COPD); asthma and localized airway obstruction. Fig. 4.22 shows the mechanism of airway obstruction. Fig. 4.22A is a partially blocked lumen by unnecessary secretions; Fig. 4.22B demonstrates a thick airway wall due to hypertrophy of wall muscles and Fig. 4.22C shows an abnormality in the tissues around the airway due to the narrowed airway because of the loss of radial traction.

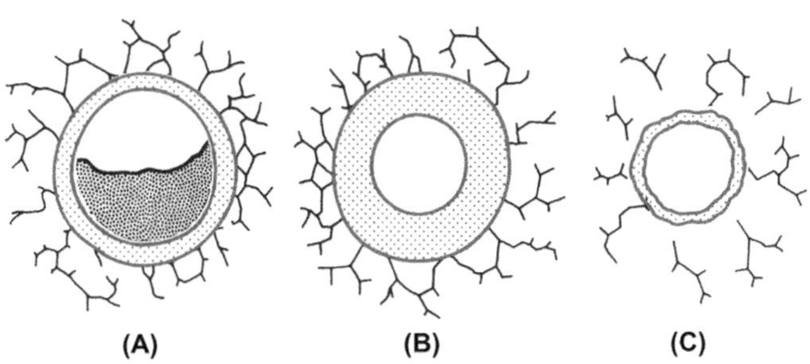

FIGURE 4.22 Three mechanisms of airway obstruction. (A) Partly blocked; (B) the thickened airway wall and (C) the abnormality of outside the airway. *From J.B. West, Pulmonary Pathophysiology, the Essentials, Lippincott Williams & Wilkins, a Wolters Kluwer Business, Baltimore, MD, 2013 with permission.*

- *COPD*: This disease is defined for patients who have emphysema, chronic bronchitis or a combination of both. Emphysema is categorized by the enlargement of the air spaces distal to the terminal bronchiole, with destruction of their walls [24]. It results in larger and fewer alveoli in the lung.
- *Chronic bronchitis*: This disease is categorized by unnecessary mucus production in the bronchial tree, enough to cause unnecessary expectoration of sputum [24].

Typically, in emphysema cases the TLC, FRC and RV are all increased. Often, the percentage of RV to TLC relation (RV/TLC %) may exceed 40% (this is less than 30% in young healthy people). Fig. 4.23 shows the relation between airway conductance (reciprocal of resistance) against static transpulmonary pressure in a series of healthy people and patients with emphysema (without bronchitis), and asthmatic patients. As the figure shows, the relationship between conductance and transpulmonary pressure for the patients with emphysema was almost normal. The small shift of the emphysema line to the right probably reflects the distortion and loss of airways in this illness [24].

Because of unknown reasons, only some patients with COPD, but especially those with severe chronic bronchitis, develop CO_2 retention since they do not increase the ventilation to their alveoli. Therefore, the O_2 cost of breathing may be enormous (Fig. 4.24) [24]. As the figure shows,

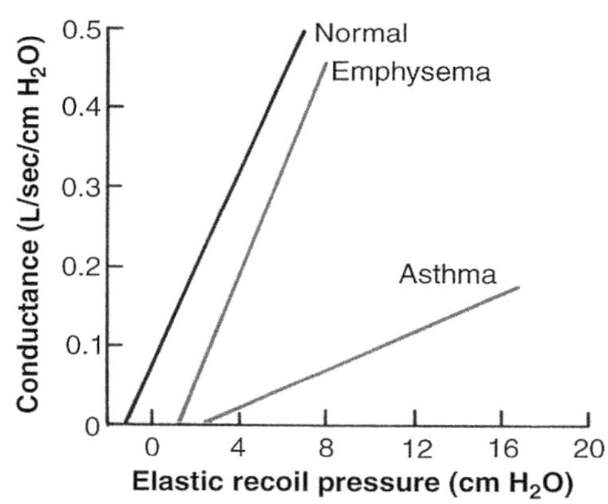

■ **FIGURE 4.23** Relationships between airway conductance (reciprocal of resistance) and elastic recoil pressure (static transpulmonary pressure) in obstructive pulmonary diseases. *From J.B. West, Pulmonary Pathophysiology, the Essentials, Lippincott Williams & Wilkins, a Wolters Kluwer Business, Baltimore, MD, 2013 with permission.*

in patients with COPD, the curve of ventilation-O_2 uptake has a sharp gradient in comparison to the curve for normal people.

- *Asthma*: Asthma is characterized by an increased sensitivity of the airways to several stimuli and is demonstrated by inflammation and extensive narrowing of the airways.

During an acute asthmatic attack, the FRC and RV of the lungs become especially increased due to the difficulty in expiring air from the lungs [1]. Fig. 4.23 shows the relation between airway conductance (reciprocal of resistance) against static transpulmonary pressure in asthmatic patients. Asthma affects bronchi (a bronchial spasm), not the alveoli. The significant abnormal behavior in Fig. 4.23 is because of effects of the bronchial spasm on all airways' sizes.

- *Localized airway obstruction*: Localized obstruction is not very common and generally causes less functional damage in the lungs. There are two categories in this field, tracheal obstruction and bronchial obstruction. Both of them can be caused by a foreign body. The tracheal obstruction occurs because of stenosis by a tracheostomy tube, or compressing mass, such as an enlarged thyroid. The right lung has a greater tendency for bronchial obstruction. The common causes are bronchial tumors and compression of a bronchus by enlarged surrounding lymph nodes [24].

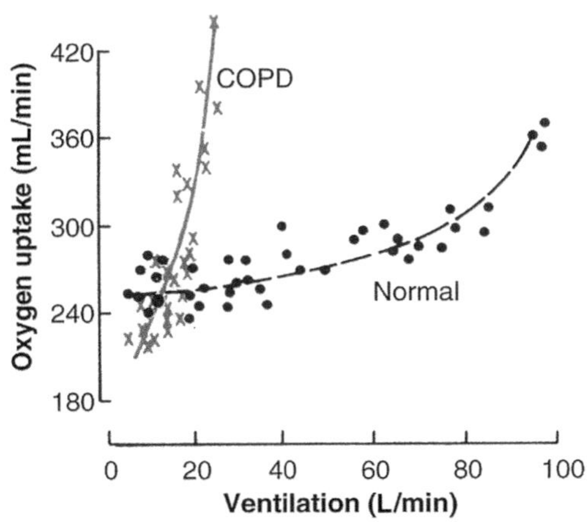

■ **FIGURE 4.24** Relation between ventilation and oxygen uptake in normal people and patients with CPOD. *From R.M. Cherniack, L. Cherniack, A. Naimark, Respiration in Health and Disease, second ed., WB Saunders, Philadelphia, PA, 1972.*

- *Restrictive diseases*: These diseases restrict the expansion of the lungs because of alterations in the lung parenchyma, or disease of the pleura, the chest wall, or the neuromuscular apparatus. Restrictive diseases are categorized by a reduced vital capacity and a small resting lung volume without any increase in the airway resistance (related to lung volume) [24].
- *Lung parenchyma disease*: This disease refers to that portion of the lung involved in gas transfer, such as the alveoli, alveolar ducts and respiratory bronchioles. This disease has several types such as [24]
  - "Diffuse Interstitial Fibrosis" with reduction in TLC, FRC, RV and low arterial P_{CO_2};
  - "Sarcoidosis" where the resting arterial P_{O_2} is low;
  - "Hypersensitivity Pneumonitis" with a reduction in lung volumes, compliance diffusing capacity and a normal or low arterial P_{CO_2};
  - collagen diseases;
  - lymphangitis carcinomatosa and
  - interstitial disease caused by drugs, poisons and radiation.
- *Pleural diseases*:
  - "Pneumothorax" is an abnormal collection of air or gas in the pleural space.
  - "Pleural Effusion" refers to fluid rather than air in the pleural space.
  - "Pleural Thickening" refers to a long-standing pleural effusion which results in a rigid, contracted fibrotic pleura that splints the lung and prevents its expansion [24].
- *The chest wall diseases*:
  - "Scoliosis" refers to lateral curvature of the spine which has an effect on the respiratory system. Clinical tests show a reduction in all lung volumes. Resistance of airways is almost normal (if related to lung volume). In progressive disease, a decreased ventilatory response to CO_2 can be demonstrated.
  - "Ankylosing Spondylitis" refers to rigidity of the vertebral joints and ribs fixation which reduces the movement of the chest wall. It reduces FVC and TLC, but resistance of the airways is normal. In this disease, the lungs remain normal and respiratory failures do not occur [24].
- *Neuromuscular disorders*: This disorders affects the muscles of the respiration system or their nerve supply. Clinical tests shows that a deep breath is reflected in a reduced FVC, TLC and IC [24].

- *Vascular diseases*: The pulmonary vascular diseases are important in respiratory function. They can be classified to three major types, pulmonary edema, pulmonary embolism and pulmonary hypertension.
  - *Pulmonary edema*: This disease is an abnormal fluid accumulation in the extravascular spaces and tissues of the lungs. A significant aspect in this disease is alveolar flooding, which reduces the affected lungs unit volume. The pulmonary edema can affect the pulmonary system by increasing capillary hydrostatic pressure, increasing capillary permeability, reducing lymph drainage, decreasing interstitial pressure, decreasing colloid osmotic pressure and uncertain etiology. Patients with this illness have rapid, shallow breathing.
  - *Pulmonary embolism*: It is a blockage of the lung main artery or one of its subsystem (branches) by a substance (usually a blood clot) that has traveled from other organs in the body through the bloodstream. Depending on the emboli size (small, medium and large), it can have an effect on cardiopulmonary factors from low blood pressure to a major hemodynamics failure with shock and chest pain. If the emboli is large and if the pulmonary artery pressure increases significantly, the right ventricle may begin to fail and increasing end diastolic pressure, evolving arrhythmias and tricuspid valve incompetence with regurgitation can follow.
  - *Pulmonary hypertension*: This disease is an abnormal increase in blood pressure in the pulmonary vascular system. High blood pressure of the pulmonary artery changes the wall structure of the pulmonary small vessels and can cause dyspnea, hemoptysis and pulmonary edema. A normal increase in blood flow rate can affect blood pressure level but pulmonary capillaries can balance the blood pressure with their adaptation capacity. One of the most common causes of pulmonary hypertension is an increase in pulmonary vascular resistance. A wide range of factors, vasoconstrictive (or the narrowing of the blood vessels), obstructive or obliterative (destroyed capillary bed) can increase pulmonary resistance.

4.3 KIDNEY AND URINARY SYSTEM

Most people are familiar with one of the important organs in their body, the kidney. Most know the first vital function of the kidneys is to get rid of waste products from the body. However, kidneys have several roles in our body and this section explains these roles and their relation to biofluid mechanics.

4.3.1 Introduction

The renal system consists of two kidneys and, in humans, the kidneys are located approximately at the vertebral level T12 to L3 (Fig. 4.25). One vital duty of the kidneys is the excretion of a variety of waste products produced by metabolism. Although this excretion is the major function of the kidney, the kidneys provide multiple functions [1] including:

- excretion of metabolic waste products and foreign chemicals;
- regulation of water and electrolyte balances;
- regulation of body fluid osmolality and electrolyte concentrations;
- regulation of arterial pressure;
- regulation of acid-base balance;
- secretion, metabolism and excretion of hormones and
- gluconeogenesis.

4.3.2 Excretion of Metabolic Waste Products and Foreign Chemicals

The waste products include urea (from the metabolism of amino acids), creatinine (from muscle creatine), uric acid (from nucleic acids), end products of hemoglobin breakdown (such as bilirubin) and metabolites of various hormones. These products have to be removed from the human body as rapidly as they are created. The kidneys also remove most toxins

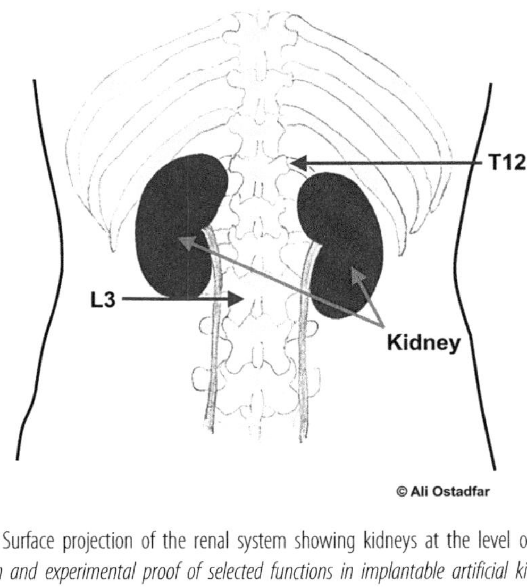

■ **FIGURE 4.25** Surface projection of the renal system showing kidneys at the level of T12 to L3. *From A. Ostadfar, Design and experimental proof of selected functions in implantable artificial kidney (thesis), Simon Fraser University Library, 2013.*

and other foreign chemicals that are either produced by the human body or ingested, such as drugs and food additives.

4.3.3 Regulation of Water and Electrolyte Balances

To maintain normal homeostasis, excretion of water and electrolytes and intake must be balanced. If the intake of a material exceeds its excretion, than the amount of that material in the body will increase. If intake is less than excretion, the amount of that material in the body will decrease.

4.3.4 Renal Blood Flow and Vessels

The blood flow in kidneys is around 22% of CO or 1100 mL/minutes. Fig. 4.26 illustrates the blood circulation in the kidneys. As the

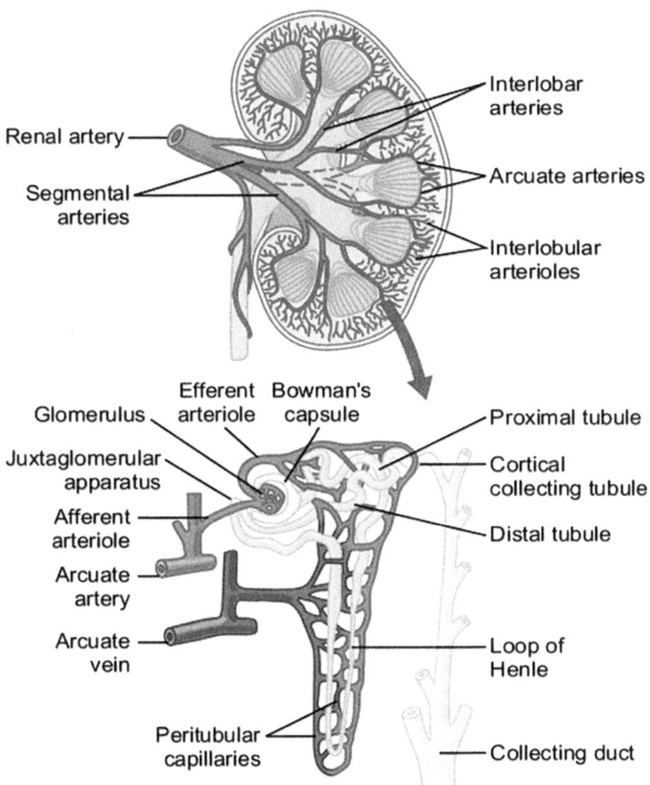

■ **FIGURE 4.26** The major vessels that provide blood flow to the kidney, and schematic of the blood circulation of each nephron. *From A.C. Guyton, J.E. Hall, Textbook of Medical Physiology, eleventh ed., Elsevier, Inc., Philadelphia, PA, 2006 with permission.*

figure shows, the blood flow enters from the renal artery, then branches to the radial arteries (interlobar, arcuate and interlobular), and the afferent arterioles provide the blood for the nephron, which is the filtration unit of kidney. Each kidney in a human includes about 1 million nephrons, which are located on the cortex of the kidney. The kidney cannot produce new nephrons. Therefore, renal disease or normal aging decrease the number of nephrons [1].

Each nephron includes glomerular capillaries (the glomerulus), through which large amounts of plasma are filtered from the whole blood, and a long tubule in which the filtered plasma is converted into waste products (urine) on its way to the pelvis of the kidney (Fig. 4.26).

The blood pressure in the renal artery is approximately equal to the systemic arterial pressure, and the average renal vein pressure is about 3–4 mmHg. Typically, the total vascular resistance through the kidneys is defined by the sum of the resistances in the arteries, arterioles, capillaries and veins [1]. Table 4.7 shows approximate blood pressures and vascular resistances in normal kidney vessels.

Mass balance in the kidney: One of key principles in the kidney is mass balance. For any solute, such as (X), which cannot be synthesized, accumulated or degraded by the kidney, three paths exist: the renal artery as an entry and two paths of exit, renal vein and ureter. Fig. 4.27 shows the

Table 4.7 Approximate Blood Pressures and Vascular Resistances in the Normal Kidney Vessels [1]

Vessel	Pressure in Vessel (mmHg)		Percent of Total Renal Vascular Resistance
	At the Beginning	At the End	
Renal artery	100	100	~0
Interlobar, arcuate and interlobular arteries	~100	85	~16
Afferent arteriole	85	60	~26
Glomerular capillaries	60	59	~1
Efferent arteriole	59	18	~43
Peritubular capillaries	18	8	~10
Interlobar, interlobular and arcuate veins	8	4	~4
Renal vein	4	~4	~0

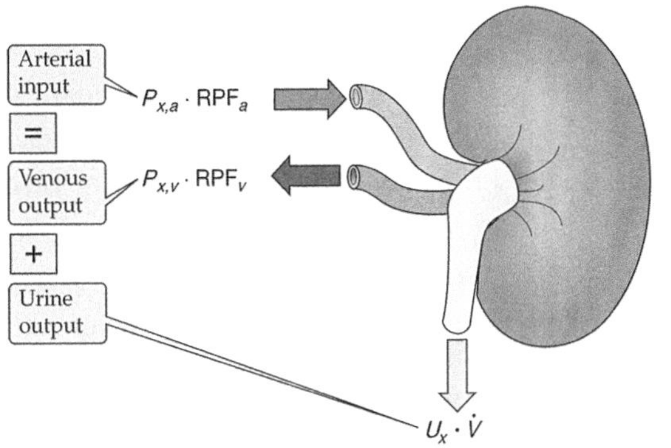

■ **FIGURE 4.27** Mass balance for solute (*X*) in the kidney. *From W.F. Boron, E.L. Boulpaep, Medical Physiology: A Cellular and Molecular Approach, second ed., Saunders Elsevier, Philadelphia, PA, 2009 with permission.*

mass balance of (*X*) solute in the kidney. From the mass balance theory, the input of (*X*) equals the output of it; therefore [3],

$$P_{X,a} \cdot \mathrm{RPF}_a = P_{X,v} \cdot \mathrm{RPF}_v + U_X \cdot \dot{V} \tag{4.21}$$

where the subscripts a and v are the renal artery and vein, respectively, P_X (mmol/mL) and RPF (mL/minutes) are the plasma concentration of the solute (*X*) and the renal plasma flow, respectively, and U_X (mmol/mL) and $\dot{V}$ (mL/minutes) are the concentration of solute (*X*) in urine and urine flow, respectively.

4.3.5 Urine Formation

The end product of the renal system is urine, and its production rate is the resultant of three renal processes, shown in Fig. 4.28: (1) glomerular filtration; (2) reabsorption of useful substances from the nephron tubules into the blood capillaries and (3) secretion of extra substances from the blood capillaries into the nephron tubules. This production rate is expressed as:

$$\begin{aligned}&\text{Urinary excretion rate } (1.5\ \mathrm{L/day}) = \text{Filtration rate } (180\ \mathrm{L/day})\\&-(\text{Reabsorption rate} + \text{Secretion rate})\ (178.5\ \mathrm{L/day})\end{aligned} \tag{4.22}$$

This equation provides a conceptual framework to examine kidney functions. In this formula, the filtration rate is determined by glomerular filtration. Reabsorption and secretion rates are the products of tubule filtration. This equation shows the base elements of the filtration procedure in the kidneys. A malfunction in any part of this procedure results in kidney disease. Each of these procedures, glomerular filtration, reabsorption and secretion, is adjusted in relation to the needs of the body. For instance, when the body has excess sodium, the filtration rate of sodium will be increased and a little fraction of the filtered sodium will be reabsorbed, resulting in an increase of sodium in the urinary excretion.

Normal balancing in the kidney filtration procedure is vital, and any unusual state can result in changes in kidney excretion. For instance, if reabsorption remains constant, a 10% increase in filtration rate (from 180 to 198 L/day) will raise urine volume around 13-fold (from 1.5 to 19.5 L/day) [1]. In the reabsorption and secretion procedures, substances are conveyed across the tubular membranes, and then they pass through the peritubular capillary membrane back into the blood (reabsorption) or back from blood capillaries to tubule (secretion), regarding body needs. The following sections explain these mechanisms.

4.3.6 Glomerular Filtration

Urine formation begins with the filtration of a large amount of plasma through the glomerular capillaries into Bowman's capsule (Fig. 4.28). The renal glomerulus system can filter 180 L of whole blood per day, and the average glomerular filtration rate (GFR) is 125 mL/minutes.

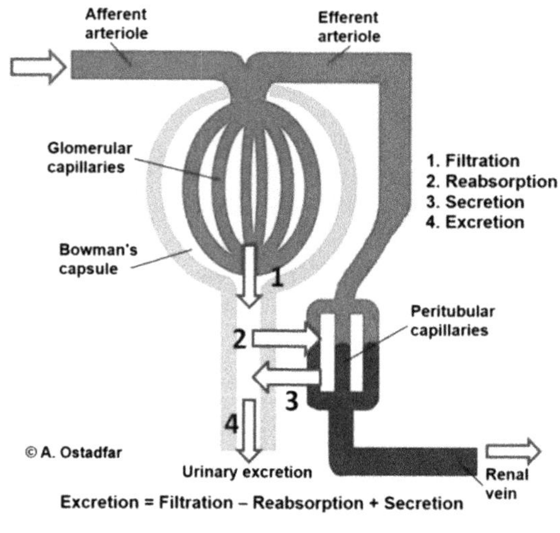

■ **FIGURE 4.28** Basic kidney processes that determine the composition of urine. *From A. Ostadfar, Design and experimental proof of selected functions in implantable artificial kidney (thesis), Simon Fraser University Library, 2013 with permission.*

The GFR is the product of glomerular capillary filtration coefficient (K_f) and the net filtration pressure (Eqs. (4.23) and (4.24)) [1],

$$\text{GFR} = K_f \times \text{Net filtration pressure} \tag{4.23}$$

The GFR can, therefore, be expressed as

$$\text{GFR} = K_f \times (P_G - P_B - \pi_G + \pi_B) \tag{4.24}$$

where P_G is glomerular hydrostatic pressure which promotes filtration, P_B is the hydrostatic pressure in Bowman's capsule, π_G is the colloid osmotic pressure of the glomerular capillary plasma proteins and π_B is the colloid osmotic pressure of the proteins in Bowman's capsule.

In a normal kidney, the net filtration pressure is

Net filtration pressure (10 mmHg)
− Glomerular hydrostatic pressure (60 mmHg)
− Bowman's capsule pressure (18 mmHg)
− Glomerular oncotic pressure (32 mmHg)

K_f can be calculated by dividing the GFR (125) by net filtration pressure (10), so a normal K_f is 12.5 mL/minutes/mmHg. This number illustrates that the glomerular capillary filtration coefficient (K_f) is 400 times greater than the K_f of most other capillary systems of the human body [1].

The diameters of the Bowman's capsule (Fig. 4.28) in different species range from approximately 100 μm (mouse) up to 300 μm (elephant); in humans, they are approximately 200 μm, in rats they are 120 μm and in rabbits they are 150 μm [26–28].

Fig. 4.29 shows a longitudinal section through a glomerulus and glomerular filtration barrier. To separate plasma from whole blood, the

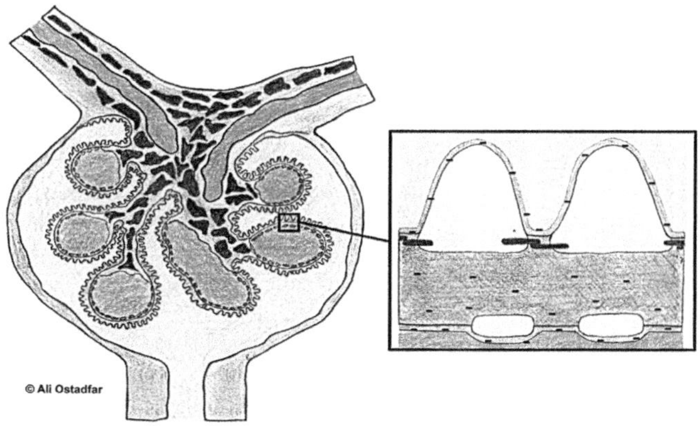

FIGURE 4.29 Left: Schematic diagram of a cross-section through a glomerulus. Right: Glomerular filtration barrier, two podocytes foot processes bridged by the slit membrane, the GBM, the porous capillary endothelium with negative charge are illustrated. *From A. Ostadfar, Design and experimental proof of selected functions in implantable artificial kidney (thesis), Simon Fraser University Library, 2013 with permission.*

glomerular capillary wall is made from three layers: endothelial cells; glomerular basement membrane (GBM) and podocytes.

Glomerular endothelial cells are large flat cells with fenestrated peripheral parts. These fenestrates are round or oval pores, and their diameters range from 50 to 100 nm. The GBM varies in width among species. In humans, the GBM thickness ranges between 240 and 370 nm [29,30], in rats and other investigational animals they vary between 110 and 190 nm [31].

Podocytes are polarized epithelial cells with a luminal and a basal cell membrane domain; the latter corresponds to the sole plates of the foot processes, which are embedded into the GBM to a depth of 40–60 nm (Fig. 4.29). The border between the basal and luminal membrane is represented by the insertion of the slit diaphragm [32].

Generally, it is accepted that the charged membrane plays a significant role in preventing polyanionic macromolecules, such as albumin, from passing from one side to the other of the glomerular barrier [33].

4.3.7 Tubular Transport

In the nephron tubule, two mechanisms perform transportation: reabsorption and secretion; these are selective mechanisms. The tubule can select ions for reabsorption or secretion from the tubule wall, depending on body needs.

Fig. 4.30 shows the reabsorption of filtered substances (water and solutes) from the tubule lumen across the tubular epithelial cells, through the

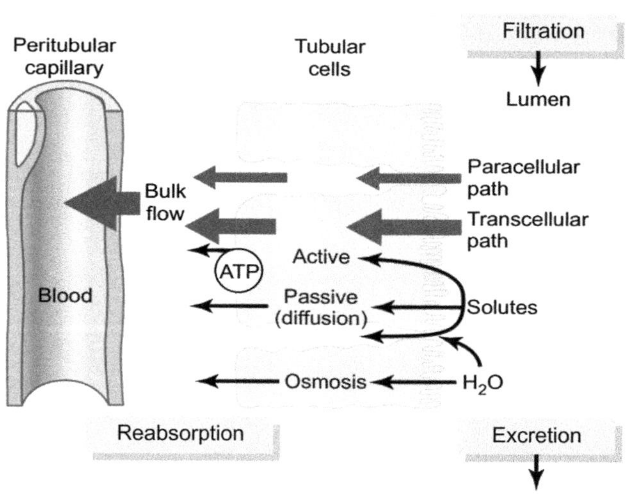

■ **FIGURE 4.30** Reabsorption of filtered water and solutes from the tubular lumen across the tubular epithelial cells, through the renal interstitium, and back into the blood. *From A.C. Guyton, J.E. Hall, Textbook of Medical Physiology, eleventh ed., Elsevier, Inc., Philadelphia, PA, 2006 with permission.*

renal interstitium (the space between cells in a tissue), and back into the blood capillaries which come from the glomerulus (efferent arteriols). Solutes are transported between the cells (paracellular route) by diffusion, or through the cells (transcellular route) by passive diffusion or active transport. The energy for this active transport comes from the hydrolysis of ATP. Osmosis transports water through the cells and between the tubular cells. Transportation of water and solutes from the interstitial liquid into the peritubular capillaries happens by ultrafiltration (bulk flow) [1].

The secretion mechanism is the reverse of reabsorption: in the tubule lumen, it is caused mainly by active transport. A secondary active transport method plays the main role in transporting the substances in the tubule. In that method, two or more substances cooperate with a specific protein in the membrane (a carrier molecule), and they are transported together across the membrane.

4.3.8 **Electrolytes Balancing**

Table 4.8 illustrates the renal operation of several substances that are all freely separated in the kidneys and reabsorbed at variable rates. As the table shows, tubular reabsorption is highly selective. Some materials, such as glucose and amino acids, are almost completely reabsorbed from the tubules. Many of the ions in the plasma fluid, such as sodium, chloride and bicarbonate, are also reabsorbed, but their rates of reabsorption are changeable, depending on the needs of the body. Waste products, such as urea and creatinine, conversely, are poorly reabsorbed from the tubules and excreted in relatively large amounts. If the concentration of these substances in the blood or urine differs from the expected value, this denotes that the kidney function is abnormal. The rate of water

Table 4.8 Rates of Filtration, Reabsorption and Excretion of Different Substances by the Kidneys [1]

	Amount Filtered	Amount Reabsorbed	Amount Excreted	Percent of Filtered Load Reabsorbed
Glucose (mol/day)	0.999	0.999	0	100
Bicarbonate (mol/day)	4.320	4.318	0.002	>99.9
Sodium (mol/day)	25.560	25.410	0.150	99.4
Chloride (mol/day)	19.440	19.260	0.180	99.1
Potassium (mol/day)	0.756	0.664	0.092	87.8
Urea (mol/day)	0.78	0.39	0.39	50
Creatinine (mol/day)	0.01591	0	0.1591	0

reabsorption varies between ~95% and 99%. Most daily variation in potassium excretion is caused by changes in potassium reabsorption in the proximal tubule and secretion in distal and collecting tubules [1].

4.3.9 Effect of Kidney Diseases on Biofluid Flow

Kidney diseases are among the most significant causes of death and disability in many countries throughout the world.

Severe kidney diseases can be categorized into two main groups: (1) acute renal failure (ARF) and (2) chronic renal failure.

- *Acute renal failure*
  ARF is a condition characterized by a rapid decrease of the GFR and, consequently, an increase in blood nitrogen products, such as urea and creatinine. It is associated with oliguria (the low output of urine) in about two thirds of cases. Depending on the localization or the nature of the renal insult, ARF is categorized as prerenal, parenchymatous or obstructive (postrenal) [34].
- *Chronic kidney disease*
  In general, chronic renal failure is similar to ARF, but in chronic kidney disease (CKD) patients lose large numbers of healthy nephrons permanently, and this situation is referred to as end-stage renal disease (ESRD).

The GFR is the best measurement method to recognize the level of kidney disease. Table 4.9 illustrates a classification and staging system based on the level of GFR and it acts as a guiding measurement tool for renal patients. It shows the staging system and an action plan for people with CRF [35]. Defining CKD using this staging system presents a regular standard for communication between the various health care systems and researchers. Regarding this staging (Table 4.9), five stages explain the level of Chronic renal failure. In stage 1, the GFR is equal or a little more than 90 mL/minutes, and the patient should be under observation also control of blood pressure. In stage 5 of the CKD ranking, GFR is less than 15 mL/minutes, and this stage is the ESRD. In this condition, the patient must be under hemodialysis machine treatment and be on a transplant waiting list.

4.4 GASTROINTESTINAL SYSTEM

The alimentary tract is responsible for providing the body with a continual supply of the required nutrients, water and electrolytes. To achieve this duty, the blood circulation carries away the absorbed substances through the gastrointestinal organs. Fig. 4.31A illustrates the entire alimentary tract.

Table 4.9 Staging System and Action Plan for Chronic Kidney Disease

Stage	Description	GFR mL/min/1.73 m^2	Action
—	At increased risk for CKD	≥90 with risk factors[a]	Screening CKD risk reduction
1	Kidney damage[b] with normal or increased GFR	≥90	Diagnosis and treatment, slow progression of CKD, treat comorbidities, cardiovascular risk reduction
2	Mild decrease in GFR	60–89	Estimate progression
3	Moderate decrease in GFR	30–59	Evaluate and treat complications
4	Severe decrease in GFR	15–29	Prepare for renal replacement
5	Kidney failure	<15 or dialysis	Replacement if uremic

[a]Risk factor: hypertension, dyslipidemia, diabetes mellitus, anemia, chronic analgestic ingestion.
[b]Kidney damage as manifested by abnormalities noted on renal pathology, blood, urine, or imaging tests [35].

The gastrointestinal (GI) tract consists of the glands and hollow organs from the mouth to the anus. These organs and their function are [3]:

- The *mouth*, responsible for chopping food into small pieces, lubrication, initiation of carbohydrate and fat digestion.
- The *esophagus*, acts a channel to the stomach.
- The *stomach*, temporarily stores food and also secretes acid and protease to initiate digestion. The average capacity of the human stomach is around 1 L.
- The *small intestine*, the first place for reabsorption of nutrients and minerals. Its length is around 7 m in an adult human.
- The *large intestine*, reabsorbs electrolytes (Cl, Na, Mg, etc.) and huge amount of liquids (90% of the absorbed water), it also stores fecal matter before expulsion from the body. Its length is around 1.5 m in an adult human.
- The glands and organs include the *salivary glands*, *liver* and *pancreas*, which secrete digestive enzymes.

Fig. 4.31B shows a characteristic cross-section of the intestinal wall. This wall has five layers, including the serosa, a longitudinal muscle, a circular muscle, the submucosa and the mucosa layer.

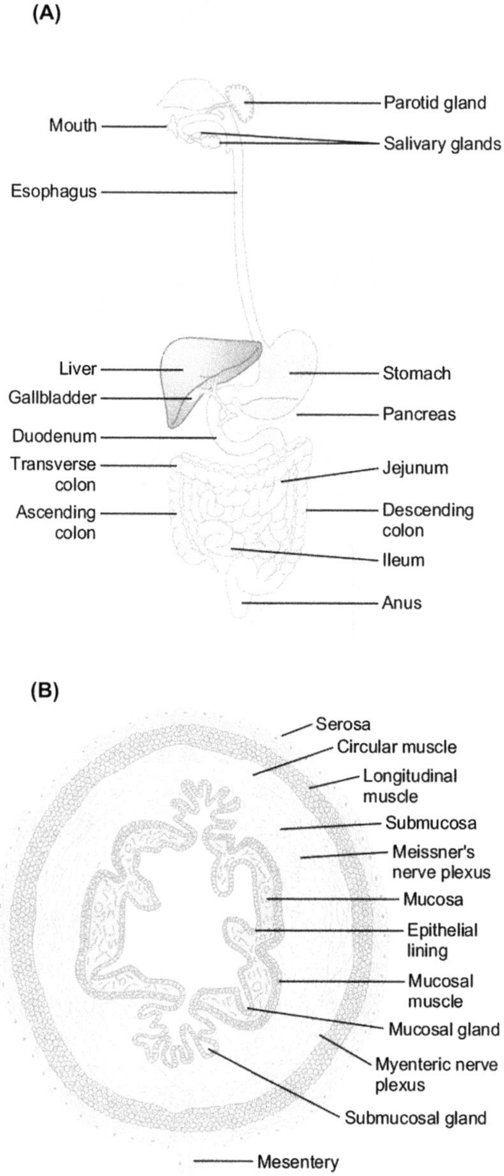

■ **FIGURE 4.31** Gastrointestinal system. (A) Alimentary tract organs and (B) typical cross-section of the gut and its wall layers. *From A.C. Guyton, J.E. Hall, Textbook of Medical Physiology, eleventh ed., Elsevier, Inc., Philadelphia, PA, 2006 with permission.*

Naturally gastrointestinal contractions occur rhythmically, and this rhythm is determined mainly by the frequency which is called "slow waves." These waves occur because of the electrical potential of smooth muscle membrane (between 5 and 15 mV). Usually, the contraction rhythm (frequency) of the stomach is around 3 per minute, of the duodenum around 12 per minute and of the ileum 8–9 per minute [1].

4.4.1 Gastrointestinal Blood Flow and Vessels

The main function of the gastrointestinal blood circulation is the delivery of the oxygen required for the absorptive and secretory functions of the gastrointestinal tissues, especially the mucosa and other gut layers. At rest, the GI tract uses 25% of total body oxygen consumption and CO, and this demand is increased postprandially. Summarizing, the physiologic character of the blood and lymph circulation supports the overall absorptive, secretory and motor function of the gastrointestinal system, such as stomach, small intestine and large intestine (colon), with an emphasis on the microcirculation of the mucosal [36].

The blood vessels of the gastrointestinal system are called the splanchnic circulation, Fig. 4.32 shows a schematic of this circulation. This system includes the blood flow through the intestines and stomach plus blood flows through the liver, pancreas and spleen. All the blood that courses through the stomach, intestines, spleen and pancreas then flows directly

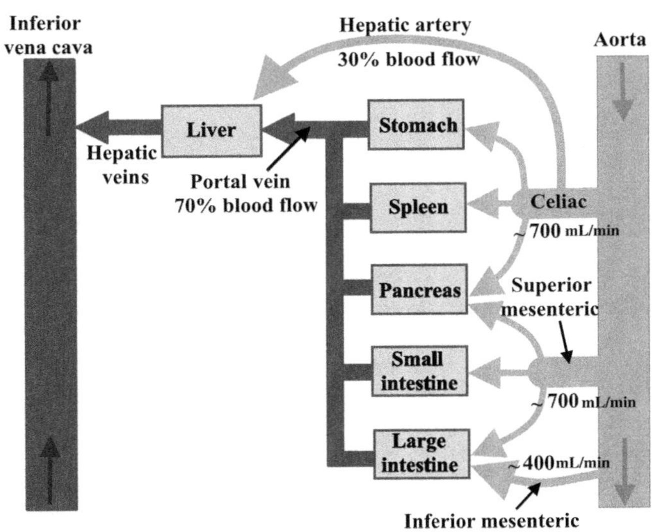

■ **FIGURE 4.32** Schematic of the splanchnic circulation.

into the liver via the portal vein. In the liver, the blood passes through millions of tiny sinusoids of liver and as a final point, it leaves the liver by the hepatic veins to the vena cava and then the heart [1]. As the figure shows, the gastrointestinal system relates to an unusual blood circulation system. The oxygenated blood flow is arranged as parallel fluidic circuits to the stomach, spleen, small intestines, large intestine, pancreas and liver, then the blood from these organs drains via the portal vein to the liver. The blood from the liver drains by the hepatic veins to the inferior vena cava. The liver receives around 30% of the CO by the celiac, superior mesenteric and inferior mesenteric arteries. The liver receives about 1300 mL/minutes blood from the portal vein and 500 mL/minutes oxygenated blood from the hepatic artery (during fasting) [37].

The gastrointestinal blood vessels can be classified as three main arteries: the celiac artery; the superior mesenteric artery and the inferior mesenteric artery. These three major branches are separate and cannot substitute for one another. Table 4.10 explains these vessels, their branch arteries, zone and function.

Fig. 4.33A shows the microvascular network in the small intestine. Small arteries branch into arterioles through the various muscle layers. Some of these arterioles develop into the mucousa including the villi. Venules developing from mucosa layers link to veins. Fig. 4.33B shows microvasculature of the villus. The structure of microvessels within a villus is similar to a fountain. The incoming arteriole develops to the center of villus

Table 4.10 The Gastrointestinal Circulation Vessels, Regarding Name, Branch Location and Function

Vessel	Branching Zone	Function
Celiac artery	A major branch of the abdominal aorta at level of T12 vertebra	It supplies the oxygenated blood to liver, stomach, spleen, esophagus, pancreas and duodenum
Left gastric artery	Branch of the celiac artery	It supplies oxygenated blood to the lower curvature of the stomach
Splenic artery	Branch of the celiac artery	It supplies oxygenated blood to the spleen
Common hepatic artery	Branch of the celiac artery	It supplies oxygenated blood to the liver, pancreas, pylorus and duodenum
Hepatic artery proper	Branch of the common hepatic artery	It supplies systemic blood to the liver
Right gastric artery	Branch of the common hepatic artery	It supplies oxygenated blood to the stomach
Superior mesenteric artery	Branch of the abdominal aorta at level of L1 vertebra	It supplies oxygenated blood to the small intestine (duodenum, jejunum and ileum) and the pancreas, also to majority of the large intestine
Inferior mesenteric artery	Branch of the abdominal aorta at level of L3 vertebra	It supplies oxygenated blood to the distal segment of the large intestine and rectum

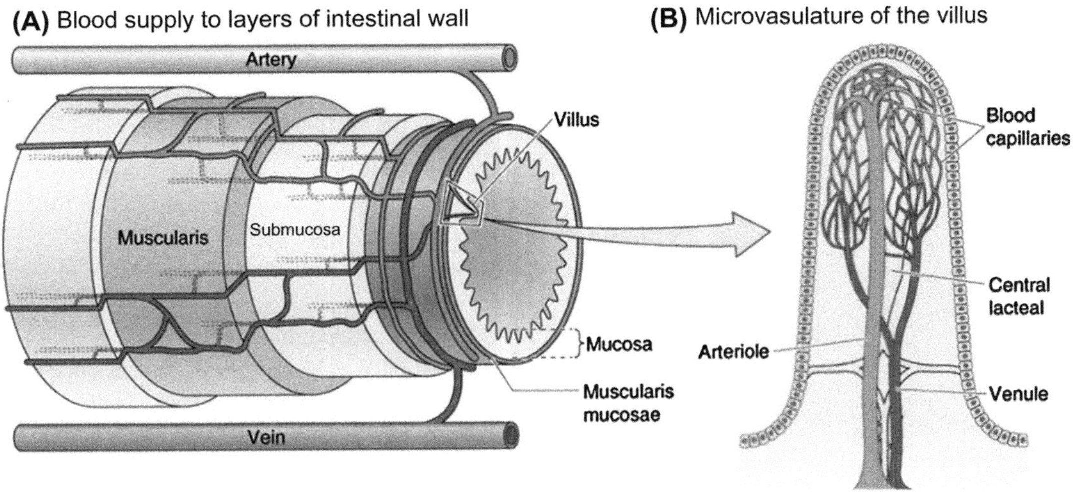

■ **FIGURE 4.33** The microvascular network in the small intestine. *From W.F. Boron, E.L. Boulpaep, Medical Physiology: A Cellular and Molecular Approach, second ed., Saunders Elsevier, Philadelphia, PA, 2009 with permission.*

and it branches into several capillaries close to the tip of villus. These microvessels are very permeable to high lipid solubility or solutes of low molecular weight. These capillaries have a large surface area due to fenestrated walls and help to absorb nutrients from intestine lumen [3]. The flow rate in each layer of the gut correlates with the local metabolism and it can be from 30 mL/minutes for each 100 g of tissue during fasting state to 250 mL/minutes during peak hyperemia after a meal [3].

The splanchnic circulation contains around 15% of the total blood volume and has a vital role in the cardiovascular system. It can regulate blood volume as a major reservoir and also as an adjuster in the human blood circulation. Significantly, body activities, such as exercise, reduce the blood flow in the splanchnic circulation.

4.4.2 Gastrointestinal Tract Functions

The GI tract can be simulated as a continuous tube that stretches from the mouth to the anus. Several procedures occur in this tube to digest the foods and absorb the body needs [1]:

- Ingestion of food, has two mechanical aspects:
  - Mastication (Chewing), by a jaw muscle force on the teeth, it is around 55 pounds on the incisor and 200 pounds on the molar. Also it aids digestion by increasing the surface area of the chewed food.
  - Swallowing (Deglutition).

- Motor functions of the stomach, are threefold:
  - Storage function, is storage of large quantities of food until the food can be processed.
  - Mixing function, is mixing of the food with stomach secretions until it forms a semifluid blend called chyme.
  - Stomach emptying, is slow emptying of the chyme from the stomach into the small intestine at an appropriate rate for better digestion and absorption by the small intestine.
- Movements of the small intestine are classified by the following processes:
  - Mixing Contractions (Segmentation Contractions), is the contractions of the circular and longitudinal muscles due to electrical impulses. This contraction includes a small segment of the intestinal wall, around 1–2 cm, and contracts the lumen, tending to divide its contents. The maximum frequency of this segmentation is 12 per minute.
  - Propulsive movements are the movements of chyme through the small intestine by peristaltic waves. These movements can occur in all parts of the small intestine, and they push the contents toward the anus at a velocity range of 0.5–2.0 cm/seconds. The movements are faster and slower in the proximal and the terminal intestines, respectively.
- Movements of the large intestine, are normally very slow because of its major role in the absorption of water and electrolytes. Similar to the small intestine, the large intestine movements can be classified once again into mixing and propulsive movements.
- Secretory functions are provided by several glands to produce the digestive enzymes and mucus for lubrication and protection of the GI tract. The glands, such as single-cell mucous glands, pits, tubular glands, the salivary glands, pancreas and liver, provide the different kinds of the GI tract secretions for proper digestion and absorption.
- Chemical digestion functions are the processes which use enzymes to break carbohydrates, fats and proteins into small enough compounds (molecules) and also water, electrolytes and other materials for absorption.
- Absorption is a kind of nutrients transportation between the intestine wall and the circulatory and lymphatic capillaries.

Table 4.11 Daily Water Turnover (mL) in the Gastrointestinal Tract [37]

Function		
Ingested		2000
Endogenous secretions		7000
Salivary glands	1500	
Stomach	2500	
Bile	500	
Pancreas	1500	
Intestine	1000	
	= 7000	
Total input		9000
Reabsorbed		8000
Jejunum	5500	
Ileum	2000	
Long intestine	1300	
	= 8800	
Balance in stool		200

Table 4.11 shows the overall water balance in the gastrointestinal system in an adult. Daily water turnover by intestines is around 2000 mL of ingested fluid plus 7000 mL of secretions from the mucosa of the gastrointestinal system and related glands. Around 98% of this fluid is reabsorbed, with a daily fluid loss of around 0.2 L in stools [37].

4.4.3 Effect of Gastrointestinal System Diseases on Biofluids

There are more than fifteen gastrointestinal system diseases and this section describes some of the common diseases around the world, such as acid reflux (heartburn), vomiting, diarrhea (acute and chronic) and peptic ulcer disease. Effects of these diseases on biofluids are miscellaneous.

- *Acid reflux* is a physical condition in which the stomach acid flows backward up into the esophagus. The best method for diagnosing the disease is esophageal pH monitoring. The acid reflux is defined as esophageal pH dropping below 4.
- *Nausea and vomiting*: Nausea is the unpleasant impulse to vomit. Vomiting is the powerful ejection of stomach contents through the

mouth. Both are not diseases but they are symptoms of around fifty eight diseases, such as food poisoning, infection, some form of cancers, overeating, brain trauma, etc. Severe vomiting can affect the biofluid system by electrolyte imbalances along with dehydration in the human body.

- *Peptic ulcer disease*: This disease occurs by an injury to the mucosal lining of the stomach or small intestine via acidic juice of the stomach. This disease can have an effect on the biofluid aspect of the body by gastrointestinal bleeding (most common sign), perforation of the stomach, which causes catastrophic bleeding and infection of abdominal organs, and obstruction of the gastric outlet, which causes vomiting without bile in patients.
- *Diarrheal diseases (acute and chronic)*: According to the WHO definition, diarrhea is the passage of abnormally loose or watery stools at least three times in a 24 hours period [38]. Diarrhea is one of the most commonly reported illnesses around the world especially among kids. If it lasts less than 2 weeks it is termed acute diarrhea, whereas chronic diarrhea continues longer than 4 weeks. Most acute cases are caused by viruses. There are many different causes of diarrhea, such as a virus, a bacterium or a parasite.

Diarrhea results from rapid movement of fecal matter through the large intestine [1] and the intestine does not have enough time to absorb enough water, ions and electrolytes. One of the most important treatments in diarrhea is management of water and electrolytes in patients. Similar to vomiting, severe diarrhea can lead electrolyte imbalances along with dehydration in our body. One useful method to prevent dehydration is to drink oral rehydration solution (ORS). Table 4.12 shows the composition by weight and molar concentrations of reduced (low) osmolarity ORS solution.

Table 4.12 Composition of Low Oral Rehydration Solution (ORS) [38]

Reduced Osmolarity ORS	G/L	Reduced Osmolarity ORS	mmol/L
Sodium chloride	2.6	Sodium	75
Glucose, anhydrous	13.5	Chloride	65
Potassium chloride	1.5	Glucose, anhydrous	75
Trisodium citrate, dihydrate	2.9	Potassium	20
		Citrate	10
		Total osmolarity	245

4.5 LIVER

4.5.1 Introduction

The liver is one of largest organs (after skin) in the human body. Its weight range is between 1200 and 1500 g, or 2–5% of body weight in an adult [3]. As Fig. 4.32 shows, the circulation system in the liver is special, since it receives the portal blood or drained blood from the intestines, the stomach, the pancreas and spleen. The liver has important roles in the human body including:

- metabolizing, detoxifying and inactivating endogenous compound and exogenous substances;
- filtering mechanism to remove foreign particulate matters such as endotoxins, bacteria, aging RBCs and parasites;
- conversion of vitamins and hormones into a new active form;
- storage of carbohydrate, lipids, vitamins and minerals;
- synthesize of carbohydrates, protein and intermediary metabolites and
- secretion of bile.

4.5.2 Liver Blood Flow

The liver is supplied by two blood sources, the portal vein and the hepatic arteries which provide around 75% and 25% of liver blood, respectively. These flows combine in the network of hepatic sinusoids. Fig. 4.34 illustrates the blood supply and circulation in the liver. The network of hepatic sinusoids is supplied by blood from the combination of hepatic arterioles and portal venules (Fig. 4.34B). Blood flow from these sinusoids converges on central veins, which, sequentially, connect to form the hepatic veins (Fig. 4.34C). All these branches (the portal vein, hepatic artery, bile duct, lymphatics and nerves) travel together as a portal tract. Generally, the arterial blood supply arises from the right hepatic artery designed for the bile ducts (Fig. 4.34C). These arterioles give rise to an extremely rich plexus of blood capillaries that surround the bile ducts as they pass through the portal tracts. Blood flowing through this peribiliary plexus drains into the sinusoids via branches of the portal vein so that this blood may transport solutes from the bile ducts and return them to the hepatocytes [3].

The blood flow in the livers is 27% of the blood circulation or 1350 mL/minutes (95 mL/minutes/100 g) [1]. In a normal adult, the portal venous pressure is around 7.5 ± 2.5 mmHg and the hepatic artery pressure is 90 ± 10 mmHg.

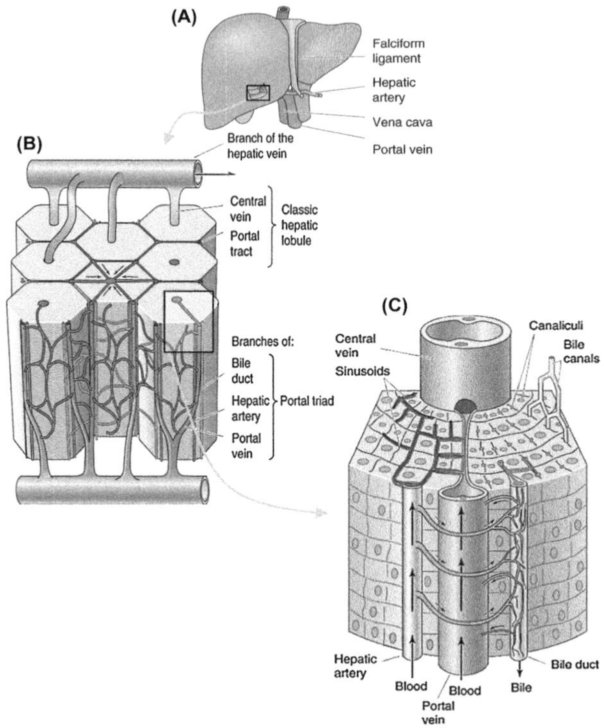

■ **FIGURE 4.34** Blood supply to the liver. (A) Liver, (B) liver lobules and (C) portion of liver lobule. *From W.F. Boron, E.L. Boulpaep, Medical Physiology: A Cellular and Molecular Approach, second ed., Saunders Elsevier, Philadelphia, PA, 2009 with permission.*

4.5.3 Biliary Flow

The gallbladder is a small organ which is located under right lobe of the liver. This organ stores bile produced by the liver. The maximum volume that the gallbladder can store is only 30–60 mL. After a meal, food begins to be digested in the upper part of the GI tract, then the gallbladder drains the bile, particularly when fatty foods reach the duodenum [1].

Table 4.13 shows the composition of the bile after secretion by the liver. As the table shows, most substances in the bile are bile salts. The gallbladder can concentrate the bile up to 20-fold (the normal amount is around five-fold). In the concentrating procedure, the gallbladder mucosa membrane reabsorbs water and large portions of the electrolytes (except calcium ions). Also bilirubin, cholesterol, lecithin and the plasma electrolytes are secreted or excreted in large concentrations. Some elements, particularly the bile

Table 4.13 Composition of Bile [1]

Substance	Liver Bile	Gallbladder Bile
Water	97.5 g/dL	92 g/dL
Bile salts	1.1 g/dL	6 g/dL
Bilirubin	0.04 g/dL	0.3 g/dL
Cholesterol	0.1 g/dL	0.3 to 0.9 g/dL
Fatty acids	0.12 g/dL	0.3 to 1.2 g/dL
Lecithin	0.04 g/dL	0.3 g/dL
Na^+	145.04 mEq/L	130 mEq/L
K^+	5 mEq/L	12 mEq/L
Ca^{++}	5 mEq/L	23 mEq/L
Cl^-	100 mEq/L	25 mEq/L
HCO_3^-	28 mEq/L	10 mEq/L

salts and the lipid substances cholesterol and lecithin, are not reabsorbed and, consequently, become highly concentrated in the bile.

4.5.4 **Effect of Liver Diseases on Biofluid Flow**

There are more than a hundred types of liver (hepatic) disease. The indications related to liver malfunction include a variety of indications related to digestive problems, coagulopathies (clotting disorder and bleeding disorder), immune system malfunction, blood sugar problems, renal problem, abnormal fat absorption and metabolism issues. Among these diseases, the coagulopathies impacts the blood more than others.

The coagulopathies: The liver has a key role in hemostasis, as it is the location of the synthesis of clotting factors, inhibitors of coagulation and fibrinolytic proteins. The most common coagulation disorders occurring in liver disease include thrombocytopenia and impaired humoral coagulation [39]. The thrombocytopenia is a disorder which refers to decreasing platelets due to lack of thrombopoietin (THPO) production. Platelets have a vital role in the clotting and coagulation mechanism and a reduction in the number of platelets increases the probability of bleeding in the body.

4.6 **BRAIN**

4.6.1 **Introduction**

The brain is one of the most complex and largest organs in the human body. The average weight of the human brain is about 1.5 kg with an average volume of about 1200 cm^3. The brain receives around 14–15%

of the resting CO [3]. The brain consumes about 20% of oxygen intake and the cerebral circulation supplies high blood flowrate—around 50–60 mL/minutes/100 g tissue weight—for the brain. This flowrate is high in relation to the weight fraction of the brain to body which is around 2% [2].

4.6.2 Brain Blood Flow

The blood flow in the brain is 14–15% of the blood circulation or 700 mL/minutes (50 mL/minutes/100 g) [1]. Four main arteries supply the brain circulation:

- the left and right carotid arteries (main blood supply) and
- the left and right vertebral arteries.

The vertebral arteries join to the basilar artery, then the basilar artery travels up and splits to form the right and left posterior cerebral arteries as a part of the circle of Willis. Fig. 4.35 shows major arteries in the brain. The circle of Willis and its arteries (anterior, middle and posterior) distribute blood flow to different sections of the brain from the brainstem to pial arteries over the surface of the brain and then to the brain tissues by arterioles and capillaries. The blood circulation of the brain is more reliable than other organs because of parallel supply circulation. The brain veins drain the blood radially in a centrifugal direction and all of the venous blood leaves the cranium by the internal jugular vein. The brain capillary only transports water, O_2, CO_2 (or other gases) and glucose by diffusion from the capillary membrane (barrier) [3]. Table 4.14 shows the blood flow rate in carotid arteries and cerebral circulation. The mean arterial blood pressure range in internal and external carotid arteries is around 110–120 mmHg.

Poiseuille's equation is the best equation to represent pressure drop and flow rate in cerebral arteries, it can be written from Eq. (1.49):

$$Q = \frac{\Delta P \, \pi \, R^4}{8 \, \eta \, L}$$

The pressure gradient between the arteries and the veins is called the cerebral perfusion pressure (CPP) which is important in the brain perfusion. Mathematically CPP is

$$\mathrm{CPP} = \mathrm{MAP} - \mathrm{ICP} \tag{4.25}$$

where MAP is mean arterial pressure and the ICP is intracranial pressure. Because of the solid aspect of the cranium, changes in the ICP can affect cerebral perfusion. It is as such an hermetic chamber with some elastic tubes inside the chamber; increasing pressure inside the chamber affects

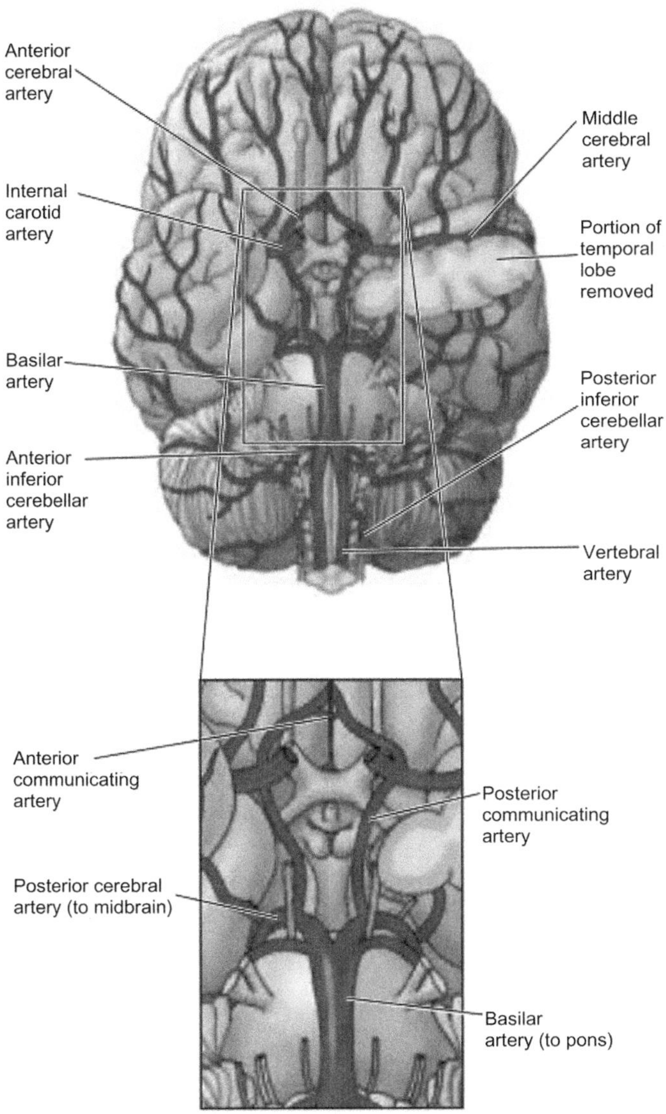

■ **FIGURE 4.35** The main arteries in the brain. *From S.M. Williams, G.J. Augustine, D. Fitzpatrick, W.C. Hall, A.-S. Lamantia, J.O. McNamara, et al., Neuroscience, third ed., Sinauer Associates, Inc., Sunderland, MA, 2004 with permission.*

the elastic tubes performance. Fig. 4.36 clarifies this aspect of the brain circulation. As the figure shows, the veins have more tendency to deform than the arteries because of the properties of the vein wall (thin wall) and also the low blood pressure in the venous circulation.

Table 4.14 The Blood Flow Rate in Carotid Arteries and Cerebral Circulation [40]

Vessel	Blood Flow Rate
Common carotid arteries	360–557 mL/min (mean ± standard deviation, 465 mL/min ± 52)
Internal carotid arteries	132–367 mL/min (mean, 265 mL/min ± 60)
External carotid arteries	113–309 mL/min (mean, 186 mL/min ± 51)
Cerebral circulation	546–931 mL/min (mean, 774 mL/min ± 134)
Vertebral arteries	133–308 mL/min (mean, 244 mL/min ± 43)

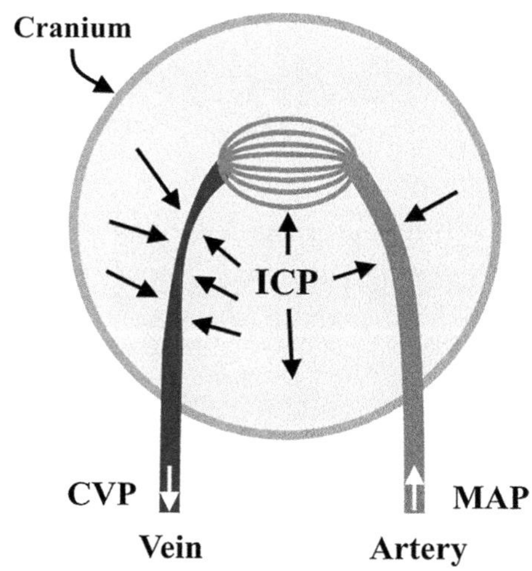

■ **FIGURE 4.36** The ICP effects on the brain blood circulation. MAP, mean arterial pressure; CVP, central venous pressure.

The normal range of the ICP is from 0 to 10 mmHg [2]; an increase (systemic hypertension) significantly reduces blood flow rate and the CPP. Also, the normal range of the MAP is between 60 and 130 mmHg.

The ICP is decreased by: intracranial bleeding; cerebral edema and tumor. Increasing ICP can lead to veins collapsing and a blood flow reduction due to decreasing CPP [2].

4.6.3 Effect of Brain Diseases on Biofluid Flow

Brain diseases have several different forms. The major brain diseases can be categorized as trauma, stroke and tumors.

- *Trauma* is one of the most common brain diseases around the world. It occurs when an external force creates a stress concentration on the cranium and injures the brain. Indications of this disease are related to grade of the brain injury. Some of common symptoms are vomiting, headache, nausea, dizziness, bleeding (as a result of damaged vessel and tissues) from the ear and balancing problem. After an accident, the first damage occurs at the time of impact, such as bleeding and blood clots from ear and point of impact. The second damage develops over time after the accident, such as increased blood pressure within the cranium and swelling of the brain which results in and increase in the ICP.
- *Stroke* is a malfunction in the blood transportation to the brain tissues. This disease has two major categories: ischemic and hemorrhagic. In the ischemic type, the problem occurs as a result of an obstacle within a blood vessel transporting blood to the brain tissues. There two types of obstruction, thrombosis because of developing an obstacle (eg, blood clot, etc.) within a vessel, and embolism which is a moving obstacle (eg, blood clot, fat particle, etc.). In the hemorrhagic disease, the blood accumulates in the skull because of the rupture of a weakened vessel in the brain. Acute stroke, naturally involves significant variations in blood pressure, and management (optimization) of blood pressure before and after stroke plays an important role in the treatment of patients.
- *Tumor* is an abnormal grow of cells within the brain. One of the indications in this disease is an increase in the ICP which normally occurs in large and edema tumors. Also some small tumors can increase ICP due to the blocking of the passage of cerebrospinal fluid. The abnormal ICP can produce vomiting, headache, anisocoria (the pupil dilation) and coma.

4.7 JOINTS

4.7.1 Joints Lubrication

The locations where two skeletal elements come together are called joints. There are two general classifications of joints [42]:

- synovial joints which are separated by a cavity and
- solid joints which are connected by a connective tissue.

The synovial joint is the most common joint in mammals. Fig. 4.37 shows a synovial joint. The synovial joint has several parts including, hyaline cartilage, joint capsule (consisting of a synovial membrane and a fibrous membrane) and articular cavity. The synovial produces synovial fluid, which permeates into the articular cavity and helps to lubricate the articulating surfaces [42].

Biotribology is a branch of tribology (the study of rubbing) and embraces basic concepts in biology, physics and chemistry. It includes three areas: friction; wear and lubrication. Friction may be defined as the tangential resistance force that is presented to the sliding of one body over another. Wear is the loss of material from the working surface of a body as a result of relative motion at the contact surface. Lubrication is a procedure of decreasing friction and/or wear between moving surfaces by the application of a liquid, gaseous, or solid material [43].

The lubricant performance depends on several factors, including the lubricant viscosity, roughness and surface aspects.

4.7.2 **Synovial Fluid**

Synovial fluid is a non-Newtonian fluid (pseudo plastic) with a biomechanical function. The main role of synovial fluid is to lubricate the cartilage of a synovial joint to reduce friction and wear and it also absorbs movement shocks. Synovial fluid is a viscous fluid (liquid), clear and straw colored. In a healthy human knee joint, the synovial fluid

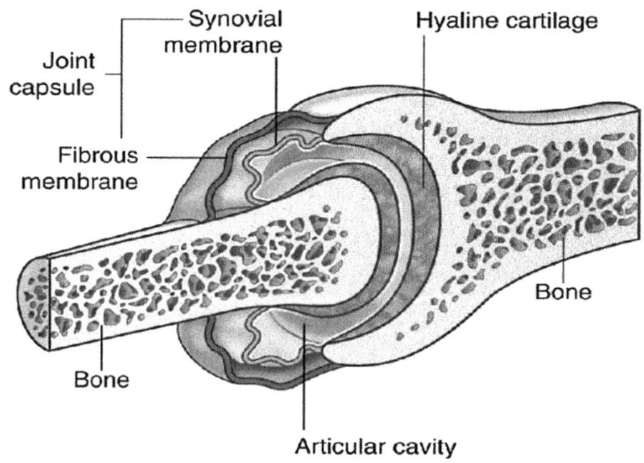

■ **FIGURE 4.37** Synovial joint. *From R. Drake, A.W. Vogal, A.W.M. Mitchell, Gray's Anatomy for Students, third ed., Churchill Livingstone, London, 2014 with permission.*

volume is around 1–4 mL [44,45]. Biomechanical experiments have shown that the viscosity of synovial fluid decreases with increasing shear rate [46]. Viscosity of normal synovial fluid of a knee joint decreases approximately from more than 10,000 to 100 mPa seconds as the shear rate increases from around 0.1 to 450 per seconds.

The range of shear module (G) is from 20 to 45 dyn/cm^2 when the shear rate range is from 0 to 100 per seconds, and increases moderately for high shear rates. The relaxation time ratio (high shear viscosity to low shear module) of synovial fluid is very long—around 3–10 seconds [47].

The shear thinning is due to the alignment of the hyaluronic acid molecules as shear rate rise. Degradation of synovial fluid in rheumatoid arthritis leads to loss of non-Newtonian behavior and synovial fluid losses its lubricant properties [48].

Synovial fluid is an ultrafiltrate or plasma dialysate and contains glucose, acids and proteins. Chemically, synovial fluid elements are [45]:

- *Protein*: Synovial fluid contains all proteins found in blood plasma, excluding proteins with high-molecular weight.
- *Glucose*: Normally, glucose level of synovial fluid is less than 10 mg/dL lower than serum levels.
- *Uric acid*: Uric acid of synovial fluid normally ranges from 6 to 8 mg/dL.
- *Lactic acid*: Generally, lactate level in synovial fluid is less than 25 mg/dL.
- *Lactate dehydrogenase*: Lactate dehydrogenase (LD) can be raised in synovial fluid, while level of serum remain normal.
- *Rheumatoid factor*: Rheumatoid factor (RF) is an antibody to immunoglobulins.

To demonstrate the synovial membrane, the best model is a semipermeable membrane. The governing equation of mass balance for the semipermeable membrane can be written as [49]:

$$\frac{\partial[V^{SF}(t,\alpha)\cdot c_i^{SF}(t,\alpha)]}{\partial t} = r_i^{SYN}(t,\alpha)\cdot A^{SYN}(t,\alpha) + r_i^{AC}(t,\alpha)\cdot A^{AC}(t,\alpha) - d_i(t,\alpha)\cdot J_i^{SYN}(t,\alpha) \tag{4.26}$$

where V^{SF} is the volume of synovial fluid (mL), c_i^{SF} is the concentration of i in synovial fluid (mg/mL), t is time (seconds), r_i^{SYN} is secretion rate of i by synoviocytes (SYN) (mg/cm^2 seconds), A^{SYN} is the area of synovial membrane (cm^2), r_i^{AC} is secretion rate of i by articular cartilage (AC) (mg/cm^2 seconds), d_i is the degradation rate of i in synovial fluid (mg/seconds), J_i^{SYN} is the total flux of i through membrane area (mg/seconds) and α is environmental factor.

The reference [49] explains more about the synovial fluid and its aspects, such as permeability, lubricant diffusion coefficient, partition coefficient, volume fraction and tortuosity of the collagen matrix and the fluxes from the synovial membrane.

4.7.3 Effect of Joint Diseases on Biofluid Flow

Joint diseases are categorized into five groups including septic, hemorrhagic, noninflammatory, inflammatory, and crystal-induced. Chemistry changes to normal joint and cell counts can lead bacterial, mechanical or chemical damage to the joint. Alterations of membrane and capillary permeability can create a variety of inflammatory responses [45]. Table 4.15

Table 4.15 Categorization Synovial Fluid and Joint Diseases [45]

Group	Category	Visual	Viscosity	Mucin Clot	Cell Count	Glucose Blood: SF	Other
	Normal	Colorless—straw clear	High	Good	<150 WBCs, <25% neutrophils	0–10	
I	Noninflammatory	Yellow; slightly cloudy	Decreased	Fair	<1000 WBCs, <30% neutrophils	0–10	
II	Inflammatory	White, gray, yellow; cloudy, turbid	Absent	Poor	<100 k WBCs, >50% neutrophils	0–4	
III	Septic	White, gray, yellow, or green neutrophils cultures; cloudy, purulent	Absent	Poor	50–200 k WBCs, >90% neutrophils	20–100	Positive cultures
IV	Crystal induced	White; cloudy, turbid, opaque, milky	Absent	Poor	500–200 k WBCs, <90% neutrophils	0–80	Crystal present
V	Hemorrhagic	Sanguinous, xanthochromic, red or brown cloudy	Absent	Poor	50–10 k WBCs, <50% neutrophils	0–20	RBCs present

categorizes the laboratory experiments for major joint diseases, and shows the blood components reactions to the joint diseases.

4.8 INTRAOCULAR SYSTEM

4.8.1 Eye Structure

The human eye is a slightly asymmetrical sphere, and its diameter is about 25 mm. The major parts of the human eye includes:

- the iris, the pigmented element;
- the cornea, a clear cover over the iris;
- the pupil, the black circular opening in the center of the iris that allows light into strike the retina;
- the sclera, the white part of eye;
- the lens, to focus inlet light on the retina;
- the retina, a light sensitive layer;
- optic nerves to transmit visual information to the brain and
- blood vessels to supply oxygen and nutrients for eye tissues.

Fig. 4.38 illustrates the human eye structure with all functional parts.

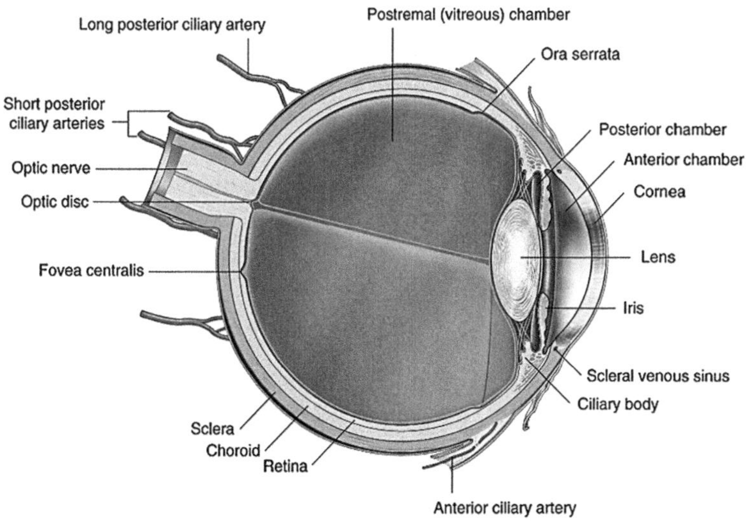

■ **FIGURE 4.38** The human eye structure and its elements. *From R. Drake, A.W. Vogal, A.W.M. Mitchell, Gray's Anatomy for Students, third ed., Churchill Livingstone, London, 2014 with permission.*

4.8.2 Eye Blood Flow Properties

The ophthalmic artery supplies the blood flow to the structures in the orbit. This artery is a branch of the internal carotid artery, and passes into the orbit through the optic canal with the optic nerve. The ophthalmic artery gives off many branches in the orbit [42]:

- The *lacrimal artery*, supplies the lacrimal gland, muscles, the anterior ciliary branch to the eyeball and the lateral sides of the eyelid.
- The *central retinal artery*, which enters the optic nerve, and obstruction of this artery or of the parent artery leads to blindness.
- The *long* and *short posterior ciliary arteries*, supplies structures inside the eyeball.
- The *muscular arteries*, deliveries the blood for the intrinsic muscles of the eyeball.
- The *supraorbital artery*, supplies the scalp and forehead as it passes across these parts to the vertex of the skull.
- The *posterior ethmoidal artery*, supplies the ethmoidal cells and nasal cavity.
- The *anterior ethmoidal artery*, supplying the lateral wall and septum, and ending as the dorsal nasal artery.
- The *medial palpebral arteries*, supplying the medial area of the lower and upper eyelids.
- The *dorsal nasal artery*, supplying the upper surface of the nose.
- The *supratrochlear artery*, supplying the forehead.

The orbit has two venous networks [42]:

- The *superior ophthalmic vein*, receiving tributaries from the companion veins to the branches of the ophthalmic artery and veins draining the eyeball posterior part.
- The *inferior ophthalmic vein*, receives several tributaries from muscles and the posterior part of the eyeball.

Figs. 4.39 and 4.40 show the arterial supply and venous drainage of the orbit and eyeball, respectively.

To create and simulate the biofluid model of the intraocular system, the normal values of the structure parameters are important. Table 4.16 summarizes the normal values for an adult human eye.

A model was done by using thermally driven flow in the anterior chamber of the eye to analyze biofluid mechanics parameters including the pressures and velocities. The model considered flow driven by thermal effects in the front portion of the eye between the cornea and the pupil.

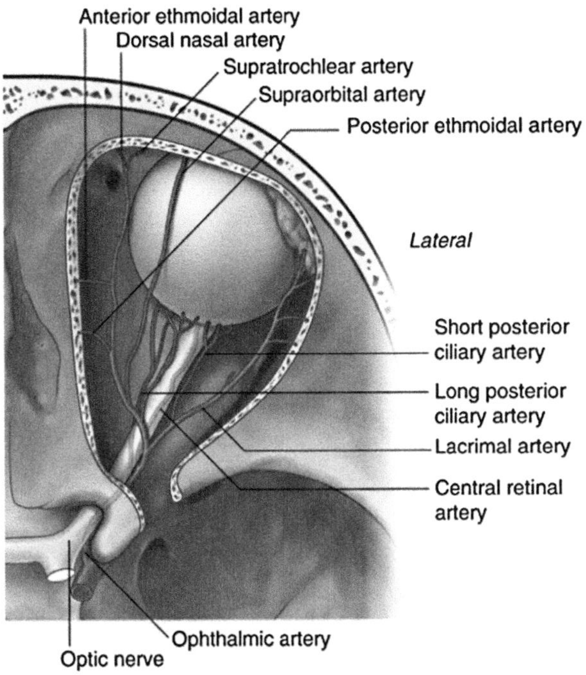

■ **FIGURE 4.39** Arterial supply to the orbit and eyeball. *From R. Drake, A.W. Vogal, A.W.M. Mitchell, Gray's Anatomy for Students, third ed., Churchill Livingstone, London, 2014 with permission.*

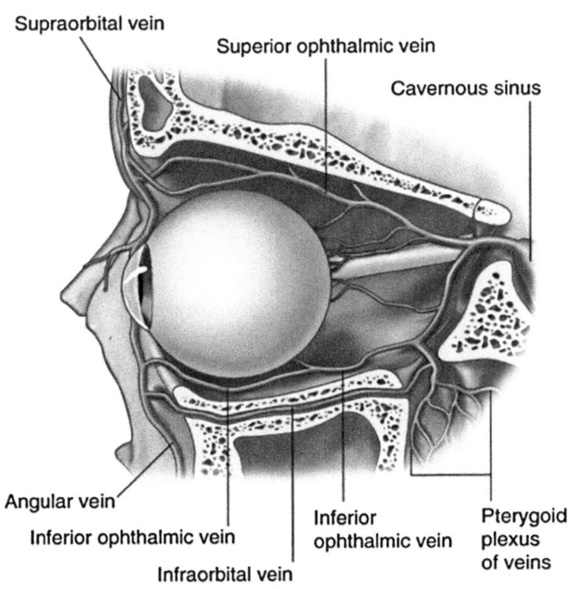

■ **FIGURE 4.40** Venous drainage of the orbit and eyeball. *From R. Drake, A.W. Vogal, A.W.M. Mitchell, Gray's Anatomy for Students, third ed., Churchill Livingstone, London, 2014 with permission.*

Table 4.16 Normal Values for an Adult Human Eye [50]

Physical Parameter	Normal Value
Radius of anterior chamber a (m)	5.5×10^{-3}
Total width of anterior chamber l (m)	11×10^{-3}
Coefficient of linear expansion of aqueous humor α (/K)	3.0×10^{-4}
Height of anterior chamber h_0 (m)	2.75×10^{-3}
Dynamic viscosity μ of aqueous humor (Pa s)	1.0×10^{-3}
Density ρ_0 of aqueous humor (kg/m^3)	1.0×10^3

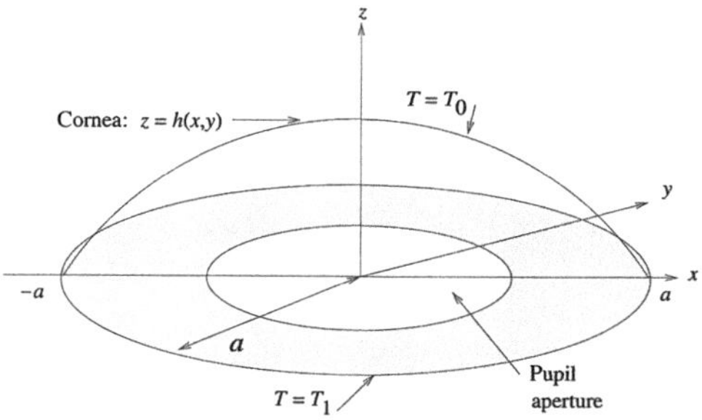

■ **FIGURE 4.41** Schematic illustration of anterior chamber. *From A.D. Fitt, C.R. Canning, M.J. Greaney, J.N. Dewynne, Fluid flow in the anterior chamber of a human eye, IMA J. Math. Appl. Med. Biol., 19 (2002) 31–60 with permission.*

Fig. 4.41 shows a schematic diagram of the model. The model presented the following parameters [51]:

$$p = p_a + (x + a)g\,\rho_0\left[1 - \frac{\alpha(T_1 - T_0)}{2}\right] \tag{4.27}$$

$$u = -\frac{(T_1 - T_0)g\,\alpha\,z}{12\,\nu\,h}(2z - h)(z - h) \tag{4.28}$$

$$v = 0 \tag{4.29}$$

$$w = -\frac{(T_1 - T_0)g\,\alpha\,z^2 h_x}{24\,\nu\,h^2}(z^2 - h^2) \tag{4.30}$$

$$u_{\max} \sim (T_1 - T_0)1.98 \times 10^{-4}\ m/s/K \tag{4.31}$$

$$h \le \frac{k_1}{\sqrt{(T_1 - T_0)}} \tag{4.32}$$

where the p denotes pressure, p_a denotes ambient pressure, t denotes time and ν, k, respectively, denote the fluid kinematic viscosity, thermal conductivity at constant pressure. Also, u, v and w are scalar components of velocity, g is gravity acceleration, ρ_0 is the fluid density at temperature T_0, T_1 is close to body temperature, h is coronial height, α is the coefficient of linear thermal expansion of the fluid (for water at 30°C, $\alpha = 3 \times 10^{-4}$ per K), k_1 is 7.5×10^{-4} mK$^{1/2}$.

Intraocular pressure (IOP), is a measure of the fluid pressure (hydrostatic pressure) within the eye. This pressure plays a key role in maintaining the appropriate distance between the retina and the lens in the eye. Mathematically, the IOP is

$$P_{IO} = \frac{F}{C} + P_{EV} \tag{4.33}$$

where P_{IO} is the intraocular pressure, F is the formation rate of aqueous fluid, C is the outflow rate of the aqueous humor (function of pressure) and P_{EV} is the episcleral venous pressure.

IOP remains constant in the normal eye, usually within ± 2 mmHg of its normal level, which averages about 15 mmHg [1], the normal rang is between 10 and 20 mmHg. The normal episcleral venous pressure range is from 8 to 10 mmHg; variability depends on the technique of measurement [52].

The relation of pressure–volume for human eye is the equation that relates variations in IOP to variations in intraocular volume [53]:

$$\Delta V = -49.8 + 30.2 \ln P + 0.242\, P \tag{4.34}$$

where ΔV is the increment of volume and P is the IOP.

4.8.3 Effect of Eye Diseases on Biofluid Flow

There are more than 55 eye diseases and only a few of them have a connection with biofluids problems. The most important of them are:

- *Conjunctivitis (pink eye)*: The conjunctiva produces mucus to lubricant (moist) the surface of your eye and protected and contains small blood vessels, when the conjunctiva becomes irritated or inflamed, the small blood vessels become larger making the eye appears red.
- *Diabetic retinopathy*: In this disease the blood vessels in the retina are damaged from diabetic retinopathy and can cause vision loss due to biofluid leakage into center of eye. Sometimes these vessels close off completely which causes the lack of blood transportation to the eye tissues.

- *Glaucoma*: This disease is a term for several eye disorders that can damage the optic nerve. The optic nerve transfers visual information from the eyes to the brain. In the most cases, the glaucoma is the effect of abnormal high pressure inside the eyeball (the IOP). The high pressure (more than around 22 mmHg) can lead to vision loss or blindness by damaging the optic nerve cells.

4.9 REPRODUCTIVE SYSTEM

4.9.1 Male Reproductive System

Fig. 4.42A illustrates the several parts of the male reproductive system, and Fig. 4.42B represents the structure of the testis and epididymis. The testis is composed of up to 900 coiled seminiferous tubules, the average length of each tubule is more than 1.5 m. The sperm are formed inside

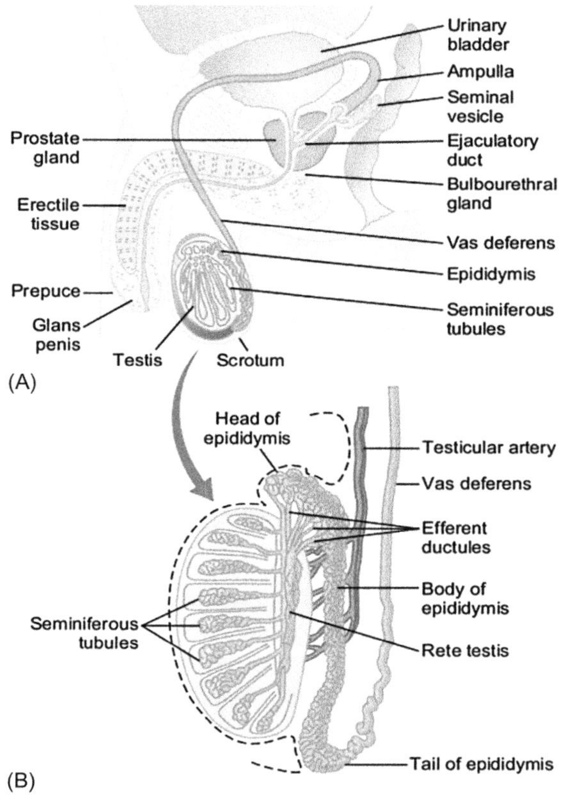

■ **FIGURE 4.42** Schematic of (A) male reproduction system and (B) internal structure of the testis and relation of the testis to the epididymis.

testis then empty into the epididymis, another coiled tube with a length of about 6 m. Then the sperm continues its way into the vas deferens, the ampulla and to the prostate gland. Two seminal vesicles on the prostate gland secret prostate fluid into the ampulla, then the contents pass into an ejaculatory duct, and then into the internal urethra. The normal sperm are capable of moving inside the fluid medium at velocities of 1–4 mm/minutes. Normally the movement of sperm is enhanced in a neutral and slightly alkaline medium, as exists in the ejaculated semen. The sperm will die in strong acidic medium [1].

Fig. 4.43 shows schematic and dimensional information of human spermatozoon. As the figure shows, the human sperm has four elements: the head ($L = 4-5\ \mu m$, $D_{max} \sim 2-3\ \mu m$); the middle piece ($L = 5-6\ \mu m$, $D \sim 0.5-1\ \mu m$); the principle piece ($L = 45-50\ \mu m$) and the end or tail ($L = 4-6\ \mu m$).

Semen, has an average pH of about 7.5 which is slightly alkaline to neutralize the other acidic portion of semen [1]. The average volume of the semen per ejaculate is between 2.5 and 3.5 mL after several days of abstinence. Each milliliter of semen normally contains about 100 million sperms, men with sperm number less than 20 million/mL are sterile [54].

A semen analysis is the usual test to measure the amount of semen a man produces and explains the number and quality of sperm and other elements in the semen sample. Table 4.17 shows the standard ranges for a complete semen analysis in a normal man without vasectomy.

4.9.1.1 *Penis Structure and Blood Circulation*

Structure: The penis is composed of three cylindrical erectile tissues, the two dorsolateral corpora cavernosa which are bonded together and single corpus spongiosum lies in the ventral groove. The deep fascia penis (or Buck's fascia) binds the three parts together in the pendulous portion of the penis. Dartos fascia (without fat tissue) surrounds the deep fascia

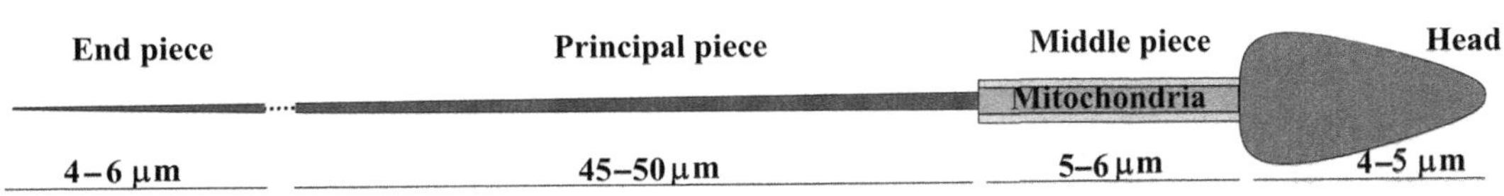

■ **FIGURE 4.43** Dimensional schematic of the human spermatozoon.

Table 4.17 Normal Semen Analysis Ranges [45]

Component	Normal Range
Liquefaction	Liquefied after 30 min
Color	White or grayish white
Viscosity	Viscous
Volume	2.0–5.0 mL
pH	7.2–7.8
Concentration	20–250 million/mL
Agglutination	None
Motility (sperm movement)	>60% progressively motile
Penetration	>30 mm
Viability	>75%
Normal morphology	Laboratory dependent (>75%)
Immature forms	<2%
Leukocytes	None to occasional
Red blood cells	None
Epithelial cells	None to few
Bacteria	None
Fructose	1 + to 4 +

penis and includes the superficial blood vessels, lymphatics and nerves. The skin is the external layer of the penis. It is thin, and comprises smooth muscle fibers, the dartos muscle, to provide the wide difference range in size for the penis between the flaccid and erect conditions [55]. The urethra is a tube that connects the bladder to the external urethral orifice. In men, the urethra is located down the middle of the penis.

4.9.1.2 *Penile Blood Circulation*

The penis blood supply is derived from the internal iliac artery, the pudendal artery and the penile artery, respectively. The penile artery branches to the following main arteries:

- Dorsal artery: This artery supplies different areas, including the head of the penis, the scrotum and the anus.
- Cavernosal artery: This artery provides the blood flow to the corpora cavernosa (corporal bodies).
- Urethral artery: This artery supply the corpus spongiosum.

Other penile arteries are the circumflex artery, superficial artery and helicine artery.

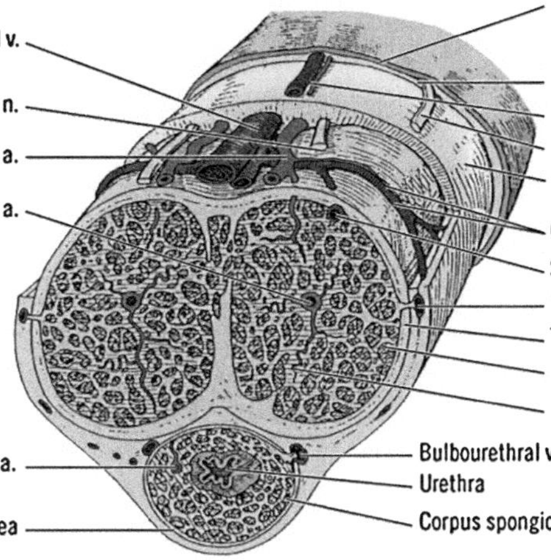

■ **FIGURE 4.44** Cutaway view of penile shaft and penile blood vessels. *From G.T. MacLennan, Penis and male urethra, Hinman's Atlas of UroSurgical Anatomy, second ed., Saunders Publishing, Philadelphia, Pennsylvania, 2012 [64].*

The penile venous system can be classified into:

- superficial dorsal vein,
- deep dorsal vein,
- circumflex vein,
- periurethral vein and
- emissary vein.

Fig. 4.44 shows a cutaway view of the blood vessel distribution in the penile shaft.

Penile erection is determined by pressure variations in the cavernosal arterioles and sinuses tissues. The vasculature mechanism of erectile tissue is composed of arterioles and hollow blood-filled sinuses (sinusoid), see Fig. 4.44. In the flaccid state, this part is contracted, allowing only a small volume of arterial flow for oxygen supply and nutritional purposes. The partial pressure of oxygen (PO_2) in the blood is about 35 mmHg. Dilation of the penile arteries is the first stage in the erection development and it increases the blood flow and pressure into the lacunar space. The incoming blood will be blocked by the expansion of sinus and, venous outflow is decreased by compression of venular spaces. This reduces the venous outflow to a lowest rate, leading to an increase in intracavernosal pressure, which is around 100 mmHg. As a result, the erection mechanism contains sinusoidal relaxation, arterial dilation and venous compression [56]. There are several reports about the perfusion rate necessary to produce and maintain the erection in men, their reported rate is from 20 to 170 mL/minutes [57–59]. Table 4.18 shows the perfusion rate and the blood pressure necessary to produce and maintain an erection in different men from 17 to 45 years old.

4.9.2 Female Reproductive System

Fig. 4.45 illustrates the major parts of the human female reproductive system. The most important parts are the ovaries, fallopian tubes, uterus and vagina. Development of ova in the ovaries is the first step of reproduction. If a single ovum has been fertilized by a sperm, it implants in the uterus and it grows into a fetus, a placenta and fetal membranes, and finally into a baby. During the reproductive years of an adult female (between 13 and 46 years of age) around 400 to 500 of the primordial follicles grow to expel their ova. After that period only a few primordial follicles remain in the ovaries, and they will be degenerated soon thereafter.

Blood circulation of the female reproductive system: The terminal aorta divides into internal iliac arteries that branch into the vaginal and uterine arteries. The vaginal and uterine arteries have been retitled the dorsal and middle uterine arteries, respectively. Subdivisions of the uterine artery anastomose with subdivisions of the vaginal and ovarian arteries to form an arterial manifold that perfuse the internal genital organs [60].

Table 4.18 Hemodynamics Parameters (the Blood Pressure and Flow Rate) During the Penile Passive Erection [59]

	Intracavernous Pressure (mmHg)		Flow Rate (mL/min)	
Age of Patients	**At Rest**	**At Passive Erection**	**To Induce Erection**	**To Maintain Erection**
17	7	90	100	40
22	6	95	90	30
25	8	93	120	35
30	10	90	80	25
35	10	95	110	30
36	9	90	120	40
36	11	92	100	25
42	8	93	80	20
44	7	95	120	30
45	8	91	110	40

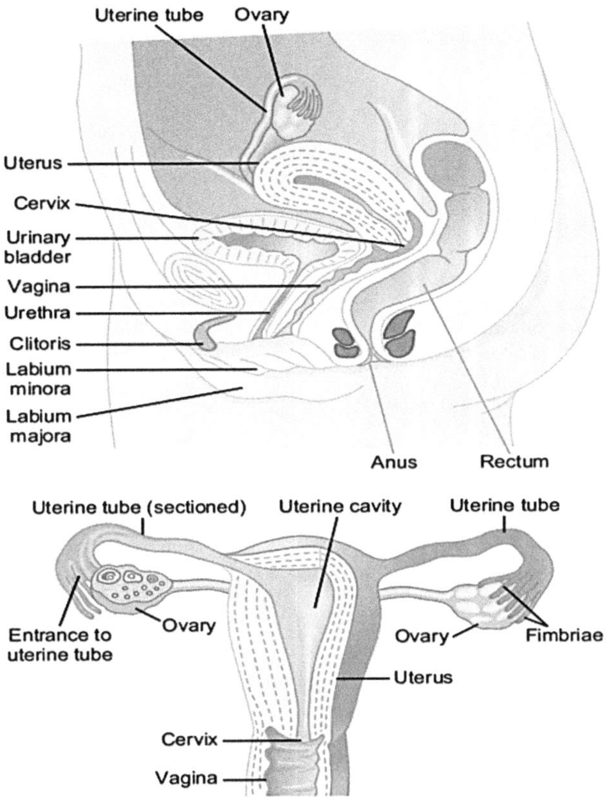

■ **FIGURE 4.45** Schematic of female reproductive organs.

These arteries, uterine, vaginal and ovarian, have central roles in the blood circulation of the female reproductive tract:

- *Uterine artery*, supplies blood to the uterus.
- *Vaginal artery*, supplies blood to the vagina and some portion (fundus) of the bladder.
- *Ovarian artery*, supplies blood to the ovary.

Fig. 4.46 shows composite diagram of left-sided uterine arteries and right-sided uterine veins showing their systematic origins and insertions in a nonpregnant woman.

The blood flow rate of the uterine artery is one of important aspects in female reproduction system during pregnancy. To control the flow rate, the image processing methods (ultrasound or MRI) can be used as the best method to measure the hemodynamics features of the blood through the uterine artery.

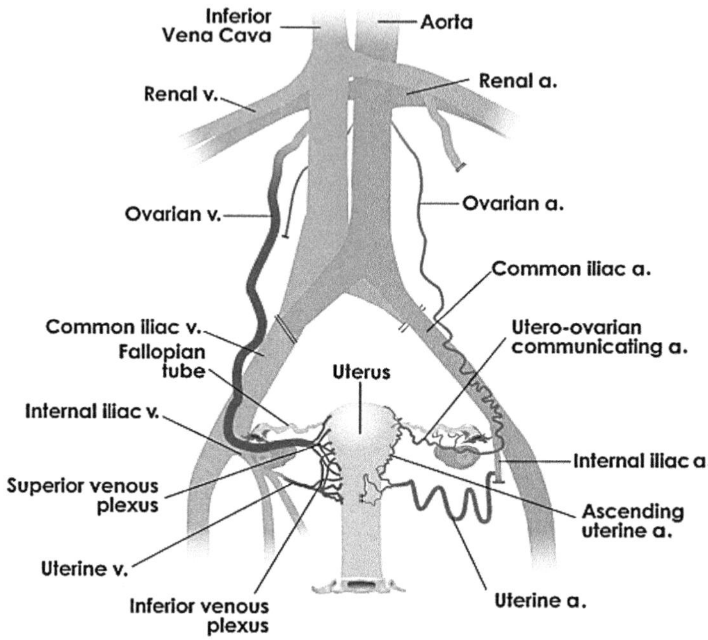

■ **FIGURE 4.46** Composite drawing of the arteries and veins of the female reproductive system in a nonpregnant woman. *From F. Burbank, Fibroids, Menstruation, Childbirth, and Evolution: The Fascinating Story of Uterine Blood Vessels, Wheatmark Publishing, Tucson, Arizona, 2009, [62] with permission.*

Table 4.19 Mean Hemodynamics Measurements of Uterine Vessels[a] [61]

Quantity	Artery	Vein
Velocity (cm/s)	5.33 (1.72)	3.97 (1.49)
Flow rate (mL/min)	203 (159)	274 (218)
Area (cm^2)	0.59 (0.25)	0.96 (0.40)

[a] *The mean and standard deviation of fluid properties over nine volunteers with normal pregnancies.*

The experiments show that the measured blood velocity for the mean velocity are 5.33 and 3.97 cm/seconds and for the unilateral flow rate for the arteries and veins are 203 and 274 mL/minutes, respectively. The power of pulsatile flow through the arterial vessel is more than venous ones and the results showed that the blood flow regime in the uterine vessels is laminar [61]. Table 4.19 shows the averaged mean measurements of uterine vessels over nine volunteers with normal pregnancies.

4.9.3 Effect of Reproductive System Diseases on Biofluid Flow

There are more than fifteen main reproductive diseases for male and female. Some of the major diseases which can effect on biofluid include:

Male diseases:

- *Varicocele* is dilatation and tortuosity of the small veins in the scrotum. This disease can decrease production of the sperm, oxygenated blood supply and nutrient product for living tissues in reproductive system.
- *Hydrocele* is abnormal accumulation of fluid around the testis. The acute large hydrocele causes testis atrophy because of blood vessel blocking or compression.
- *Priapism* is a persistent erection of the penis occurring for several hours, without or with any sexual activation. This disease occurs due to lack of the blood drain function from the tissues of the erect penis and damages the penile tissues as a result of the shortage of oxygenated blood.
- Benign prostatic hypertrophy involves an increase in the number of prostatic cells or enlargement of the prostate gland which decreases the urine flow rate during voiding.
- Cancer in the reproductive organs which can effect on the blood rate in the tissues.

Female diseases:

- *Irregular periods* occur because of fluctuations in levels of hormones during puberty.
- *Menorrhagia* is an abnormal heavy bleeding (period) and it can be an indication of disorders of the uterus such as infection, fibroids, endometriosis, uterine polyps or cancer.
- *Abnormal bleeding* occurs as a result of disorders of the vulva, cervix, vagina, uterus, ovaries and fallopian tubes.
- Cancer in the reproductive organs which can have an effect on the blood rate in the tissues.

4.10 ENDOCRINE SYSTEM BIOFLUIDS

4.10.1 The Thyroid Gland

The thyroid gland is located in the front of trachea, see Fig. 4.47. In adults, its average weight is about 20 g. The thyroid structure is composed of right and left lobes and a connecting branch [3]. The average dimensions of this gland are 45 mm (length) by 15 mm (thickness) by

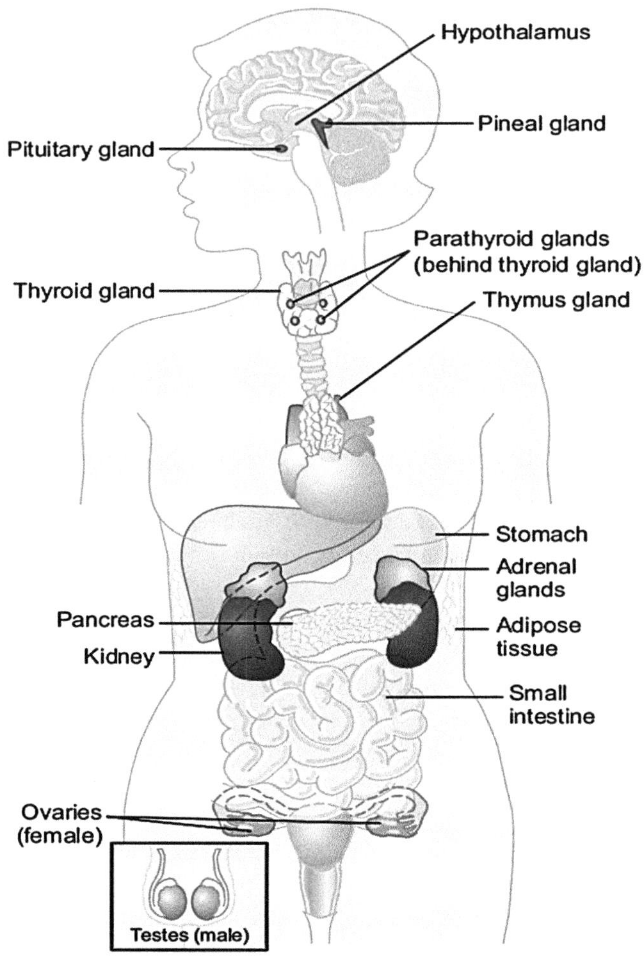

■ **FIGURE 4.47** Anatomical loci of the major endocrine glands. *From A.C. Guyton, J.E. Hall, Textbook of Medical Physiology, eleventh ed., Elsevier, Inc., Philadelphia, PA, 2006 with permission.*

25 mm (width). The gland produces thyroid hormones, the most important hormones are etraiodothyronine (thyroxine or T4) and triiodothyronine (T3) which are peptides containing iodine.

Thyroid hormones increase ions active transport through cell membranes. These hormones cause an increased need for vitamins. Without thyroid hormones, the basal metabolic rate falls almost to one-half normal range. A great increase in thyroid hormone level almost always decreases the body weight [1].

Thyroid hormones can effect on the cardiovascular and other organs [1]:

- increasing blood flow and CO 60% or more above normal,
- increasing heart rate,
- increasing the rate and depth of respiration,
- increasing gastrointestinal motility,
- increasing the rates of secretion of most other endocrine glands and
- without any effect of mean arterial pressure.

Since the thyroid has an organ with the duty of hormone secreting, it is extremely vascularized. The superior and inferior thyroid arteries and thyroidea ima supply the thyroid gland and the superior, middle and inferior thyroid veins drain the venous blood.

The blood flow in the thyroid gland (under basal conditions) is 1% of the blood circulation or 50 mL/minutes (160 mL/minutes/100 g) [1].

4.10.2 **The Pituitary Gland (or Hypophysis)**

In humans, the pituitary gland is about 8 mm in diameter (as large as a pea) and located at the base of the brain, see Fig. 4.47. During puberty and pregnancy the diameter will be between 10 and 12 mm. It is composed of three lobes: anterior; intermediate and posterior.

The hormones secreted by the pituitary gland can have an effect on biofluid aspects such as:

- blood pressure,
- procedure of water reabsorption in kidney,
- regulation of water and osmolality and
- production of breast milk.

Two set of vessels supply the blood of the pituitary gland:

1. The superior hypophyseal artery, from the internal carotid arteries to supply the pars tuberalis.
2. The inferior hypophysial artery, from internal carotid artery to supply pars nervosa.

The posterior and anterior hypophyseal veins drain the deoxygenated blood and hormones from the pituitary gland.

4.10.3 **The Adrenal Glands**

The two adrenal glands lie at the superior poles of the two kidneys and the weight of each one is about 4 g, see Fig. 4.47. These glands are responsible for producing a wide variety of hormones. The adrenal cortex secretes two major types of adrenocortical hormones, the mineralocorticoids and the glucocorticoids. Also it secretes small amounts of sex hormones, especially androgenic hormones. Blood flow in the adrenal glands is about 0.5% of the blood circulation or 25 mL/minutes (300 mL/minutes/100 g) [1].

Three arteries supply oxygenated blood for the adrenal glands:

1. The inferior suprarenal artery, which is originated from the renal artery.
2. The middle suprarenal artery, which is originated from the abdominal aorta.
3. The superior suprarenal artery, which is originated from the inferior phrenic artery.

Two veins transport deoxygenated blood from the adrenal glands:

1. The right suprarenal veins, returns the deoxygenated blood to inferior vena cava.
2. The left suprarenal veins, returns the deoxygenated blood to the left renal vein.

4.10.4 The Endocrine Pancreas

Functionally, the pancreas is two glands in one organ which has both exocrine and endocrine functions. Fig. 4.47 shows anatomical location of the pancreas.

The exocrine gland and the endocrine gland secrete their products into ducts (duct glands) and into the blood, respectively. The pancreas is responsible for controlling:

- the blood sugar level by secreting the hormones such as insulin and amylin (decrease blood sugar) also glucagon (increase blood sugar),
- inhibition of endocrine pancreas by somatostatin and
- inhibition of exocrine pancreas by pancreatic polypeptide.

Regarding the region of the pancreas, several blood arteries supply oxygenated blood to the tissues:

- The head region is supplied by the superior mesenteric artery and the superior pancreaticoduodenal artery.
- Other parts of the pancreas are supplied by the splenic artery pancreatic branches.

In the pancreas venous system [63]:

- The pancreatic veins collect the deoxygenated blood from the neck, body and tail of the pancreas and connect to the splenic vein.
- The pancreaticoduodenal veins are located close to their corresponding arteries and drain into the splenic or portal veins.

The pancreas secretes several enzymes for the digestion of proteins, carbohydrates and fats. It also secretes large quantities of bicarbonate ions, which have an important role in neutralizing the acidity of the stomach chyme. Dailey secretion of the pancreatic secretion is around 1000 mL with a pH around 8.0–8.3 [1].

4.10.5 **The Parathyroid Glands**

Normally, a human has four parathyroid glands, two inferior and two superior, see Fig. 4.47. The lower glands are larger than the upper glands. The primary role of these glands are the regulation of phosphate and calcium metabolism.

The inferior thyroid artery supplies major portion of both the superior and inferior parathyroid glands.

Parathyroid hormone (PTH) is one of the most important hormones for the regulation of calcium in the body. In the kidney this hormone increases tubular reabsorption of calcium and it also decreases the phosphate reabsorption [1].

Fig. 4.48 illustrates the effects on the blood calcium and phosphate concentrations after infusing and continuing PTH into an animal for several

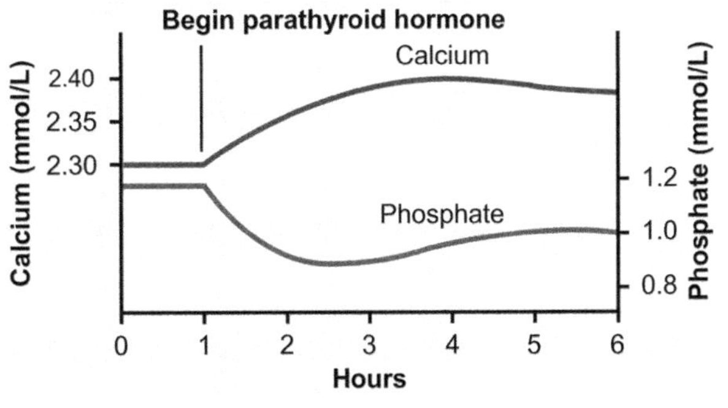

■ **FIGURE 4.48** Effect of the PTH infusion on calcium and phosphate concentrations at a moderate rate. *From A.C. Guyton, J.E. Hall, Textbook of Medical Physiology, eleventh ed., Elsevier, Inc., Philadelphia, PA, 2006 with permission.*

hours. As the figure shows, at the beginning of PHT infusion, the concentration of calcium ion begins to increase and reaches a plateau in around 4 hours after the hormone infusion because of: (1) the effect of PTH in increasing calcium and absorption of phosphate from the bone and (2) a decrease in the excretion of calcium by the kidneys. The phosphate concentration decreases very fast in comparison to the increase of calcium and reaches a depressed level within 1 or 2 hours due to an increase in the renal phosphate excretion [1].

4.10.6 Effect of Endocrine System Diseases on Biofluid Flow

There are more than 14 diseases for the endocrine system. Some of main diseases which can have an effect on biofluids include:

The thyroid:

- Hyperthyroidism (overactive thyroid): The gland secretes too much (5–15 times more) hormone (T4 and T3) and
- Hypothyroidism (underactive thyroid): The gland secretes its hormones by less than the normal amount.

The pituitary gland:

- Hyperpituitarism: This disease occurs due to hypersecretion of pituitary hormones.
- Hypopituitarism: This disease is a reduction in secretion of the pituitary gland hormones. The indications of this disease are varied and depend on the kind of reduced hormones.
- Sheehan syndrome: This may occur in some females because of blood loss and hypovolemia after delivery and around 50% of patients with this disease lose all their pituitary gland hormones.
- Growth hormone deficiency: This is a lack of growth hormone production in the body.
- Cushing's disease: The pituitary gland abnormally increases the secretion of adrenocorticotropic hormone from the anterior pituitary. This disease can increase blood pressure.
- The syndrome of inappropriate antidiuretic hormone (ADH) secretion: In this disorder the gland secretes an extra amount of ADH and the result is a low level of sodium and an increase in the total body fluid.
- Diabetes insipidus: In this disease, the pituitary function is affected by stopping the ADH. Excessive urination, extreme thirst and a high sodium level in the blood are the important signs of this disease.

The adrenal glands:

- Addison's disease (adrenal insufficiency): The glands don't secrete enough cortisol hormones (in some cases, aldosterone hormones as well).
- Cushing's disease: occurs when the level of cortisol is abnormally high in the body and its biofluid symptoms are high blood pressure, high blood sugar and increased dehydration and urination.
- Pheochromocytomas: These tumors develop in the medulla of the adrenal gland and cause overproduction of catecholamines hormones which are involved in controlling blood pressure. Overproduction of these hormones causes high blood pressure in most patients.
- Primary Hyperaldosteronism (Conn's Syndrome): Conn's syndrome is an overproduction of aldosterone by the adrenal glands. This disease causes high blood pressure, low level of potassium and heart problems.

The endocrine pancreas:

Disorders of the pancreas are categorized as:

- acute pancreatitis,
- chronic pancreatitis and
- hereditary pancreatitis.

The parathyroid glands:

- Hyperparathyroidism: It is an overproduction of PTH due to overactivity of the glands. This disease can lead to the abnormal levels (high or low) of the blood calcium.
- Hypoparathyroidism: It is an underproduction of PTH due to low activity of the glands. This syndrome can lead to low levels of blood calcium.

CHAPTER SUMMARY

- The *heart* is composed of *two separate main pumps* which cooperate to flow the blood through the circulatory system. Each pump has two chambers to pump the blood: the atrium (upper chamber) and the ventricle (lower chamber).
- The normal duration of the *cardiac cycle* for a heart rate of *75 beats/minutes is 0.8 seconds.*
- The period of relaxation is called *diastole*, in which the ventricle fills with blood, and the period of ventricle contraction is called *systole*.

- The *ECG* is a general clinical device used to measure the electrical activity of the heart. The shape of *ECG* shows the heart function and it can reveal some heart problems.
- *PCG* records the sound of opening and closing of the heart valves during the cardiac cycle.
- Each mammalian heart has *four valves*: the *mitral* valve; the *tricuspid* valve; the *aortic* valve and the *pulmonary* valve. These valves prevent the blood flowing in the reverse direction, like a regular check valve in a fluid circuit.
- *Cardiac volumes measurements*:

$$\text{SV (stroke volume)} = \text{EDV (end diastolic volume)} - \text{ESV (end systolic volume)},$$
$$\text{SV} = 70\,\text{mL} \quad \text{and}$$
$$\text{CO (cardiac output)} = \text{SV} \times \text{heart rate} \left(\frac{\text{beats}}{\text{minutes}} \right)$$

- The *coronary circulation* is responsible for transporting blood to the heart muscles (myocardium).
- The *resting coronary blood flow* is averages about 225 mL/minutes or 0.8 mL/minutes/g of heart muscle, which represents about 4–5% of the total CO.
- *Heart failure* can be affected by:
  - end phase of coronary artery diseases;
  - myocardial infarction;
  - cardiomyopathies (heart muscle diseases) because of viral, bacterial, alcohol or idiopathic origins;
  - chronic arrhythmias;
  - valvular disease;
  - congenital disease;
  - pressure load, such as hypertension;
  - volume load, such as arterial venous shunt and
  - high output failure, such as thyrotoxicosis or pregnancy.
- The main function of the *respiratory system* is to supply oxygenated blood to living organs and to remove carbon dioxide.
- The *maximum volume of the lungs* and their subsystem is around 5–6 L.
- The lungs can be *inflated and deflated* in two ways: (1) by upward and downward movement of the diaphragm to change the chest cavity size and (2) by movement of the ribs to change the anteroposterior diameter of the chest cavity.
- The *pulmonary artery pressures* are 15–30 mmHg for systolic and 4–12 mmHg for diastolic pressure.

- The *blood volume of the lungs* is about 450 mL, approximately 9% of the total blood volume of the whole circulatory system.
- Hemoglobin (Hb) is the iron (Fe) containing protein molecule in the RBCs and its main role in blood circulation system is the transportation of oxygen to living tissues.
- The *transportation of blood gases* (such as O_2 and CO_2) across the alveolar membrane occurs *by diffusion*. Two important factors in the gas transportation and exchange in alveoli capillaries are the local concentration and the total flux of gas across the capillary wall.
- *Four factors have an effect on gas exchange*: The thickness of the membrane; the surface area of the membrane; the diffusion coefficient of the gas and the pressure difference between the two sides of the alveolar membrane.
- The *kidneys* provide multiple *functions*:
  - excretion of metabolic waste products and foreign chemicals;
  - regulation of water and electrolyte balances;
  - regulation of body fluid osmolality and electrolyte concentrations;
  - regulation of arterial pressure;
  - regulation of acid-base balance;
  - secretion, metabolism and excretion of hormones and
  - gluconeogenesis.
- The *blood flow in kidneys* is around 22% of CO or 1100 mL/minutes.
- Each kidney in a human includes about *1 million nephrons* (the filtration units), which are located on the cortex of the kidney.
- This equation provides a conceptual framework to examine kidney functions:

$$\text{Urinary excretion rate } (1.5\ \text{L/day}) = \text{Filtration rate } (180\ \text{L/day}) - (\text{Reabsorption rate} + \text{Secretion rate})\ (178.5\ \text{L/day})$$

- The renal glomerulus system can filter 180 L of whole blood per day and the average *GFR* is 125 mL/minutes.
- To separate plasma from whole blood, the *glomerular capillary wall* is made from three layers: endothelial cells; GBM and podocytes.
- The tubule can select ions for *reabsorption or secretion from the tubule* wall, depending on body needs.
- The *gastrointestinal (GI)* tract consists of the glands and hollow organs from the mouth to the anus.
- Naturally, *gastrointestinal contractions* occur rhythmically, and this rhythm is determined mainly by the frequency which is called "*slow waves*."
- The blood vessels of the gastrointestinal system are called the *splanchnic circulation*, Fig. 4.32 shows schematic of this circulation.

- The splanchnic circulation contains around *15%* of the total blood volume and has a vital role in the cardiovascular system.
- The *small intestine pushes* the contents toward the anus at a velocity range of 0.5–2.0 cm/seconds.
- *Daily water turnover by the intestines* is around 2000 mL of ingested fluid plus 7000 mL of secretions from the mucosa of the gastrointestinal system and related glands. Around 98% of this fluid is reabsorbed, with a daily fluid loss of around 200 mL in stools.
- The *liver is supplied* by two blood sources, the *portal vein* and the *hepatic arteries* which provide around 75% and 25% of liver blood, respectively.
- The *blood flow in the livers* is 27% of the blood circulation or 1350 mL/minutes (95 mL/minutes/100 g).
- The *blood flow in the brain* is 14–15% of the blood circulation or 700 mL/minutes (50 mL/minutes/100 g).
- The brain capillary only transports water, O_2, CO_2 (or other gases) and glucose by diffusion from the capillary membrane (barrier).
- *Poiseuille's equation* is the best equation to represent the pressure drop and flow rate in *cerebral arteries*.
- In the brain region, the pressure gradient between the arteries and the veins is called the CPP which is important in the brain perfusion. Mathematically CPP is

$$\text{CPP} = \text{MAP} - \text{ICP}$$

  where MAP is mean arterial pressure and the ICP is intracranial pressure. The normal range of the ICP is from 0 to 10 mmHg. Also, the normal range of the MAP is between 60 and 130 mmHg.
- *Synovial joints* which are separated by a cavity are the most common joint in mammals.
- The main role of synovial fluid is to lubricate the cartilage of a synovial joint to reduce friction and wear and it also absorbs movement shocks. *Synovial fluid* is a viscous fluid (liquid), clear and straw colored, an ultrafiltrate or plasma dialysate, and contains glucose, acids and proteins.
- The *ophthalmic artery* supplies the blood flow to the structures in the orbit. This artery is a branch of the internal carotid artery, and passes into the orbit through the optic canal with the optic nerve.
- The orbit has two venous networks: the *superior ophthalmic vein*; and the *inferior ophthalmic vein*.
- *IOP*, is a measure of the fluid pressure (hydrostatic pressure) within the eye:

$$P_{\text{IO}} = \frac{F}{C} + P_{\text{EV}}$$

where P_{IO} is intraocular pressure, F is the formation rate of aqueous fluid, C is the outflow rate of the aqueous humor (function of pressure) and P_{EV} is the episcleral venous pressure.

- The relation of the *pressure–volume for the human eye* is:

$$\Delta V = -49.8 + 30.2 \ln P + 0.242\, P$$

where ΔV is the increment of volume and P is the IOP.

- The *sperm* are formed inside testis and then empty into the epididymis, another coiled tube with a length about 6 m.
- Two seminal vesicles on the *prostate gland* secrete prostate fluid into the ampulla, then the contents pass into an ejaculatory duct and then into the internal urethra.
- The *normal sperm* are capable to move inside the fluid medium at velocities of *1–4 mm/minutes*.
- *Semen*, has an average *pH of about 7.5* which is slightly alkaline to neutralize the other acidic portion of semen.
- *Each milliliter* of semen normally contains about *100 million sperm*, men with a sperm number of less than 20 million/mL are sterile.
- The *penis* is composed of *three* cylindrical erectile tissues, the *two dorsolateral corpora cavernosa*, which are bonded together, and the *single corpus spongiosum* lies in the ventral groove.
- The *penile artery branches* to the following main arteries: the dorsal artery, the cavernosal artery and the urethral artery.
- The *penile venous system* can be classified into: the superficial dorsal vein, the deep dorsal vein, the circumflex vein, the emissary vein and the periurethral vein.
- *Penile erection*, is determined by pressure variations in the cavernosal arterioles and the sinuses tissues. The vasculature mechanism of erectile tissue is composed of arterioles and hollow blood-filled sinuses (sinusoid).
- During the reproductive years of an adult female (between 13 and 46 years of age) around *400 to 500* of the *primordial follicles* grow to expel their ova.
- The uterine, vaginal and ovarian arteries have central roles in the *blood circulation of the female reproductive* tract.
- The *blood flow rate of the uterine artery* is one of the most important aspects of the female reproduction system during pregnancy.
- The thyroid gland produces thyroid hormones, the most important hormones are etraiodothyronine (thyroxine or T4) and triiodothyronine (T3) are peptides containing iodine.
- Since the thyroid has an organ with the duty of hormone secreting, it is extremely vascularized.

- The *hormones* secreted by the *pituitary gland* can have an effect on biofluid aspects such as:
  - blood pressure,
  - procedure of water reabsorption in kidney,
  - regulation of water and osmolality and
  - production of breast milk.
- The *adrenal cortex* secretes two major types of adrenocortical hormones, the mineralocorticoids and the glucocorticoids, also it secretes small amounts of sex hormones.
- The *pancreas is responsible* for controlling:
  - the blood sugar level by secreting the hormones such as insulin and amylin (decrease blood sugar) and also glucagon (increase blood sugar);
  - inhibition of endocrine pancreas by somatostatin and
  - inhibition of exocrine pancreas by pancreatic polypeptide.
- The *primary role* of the *parathyroid glands* is the regulation of phosphate and calcium metabolism.

PROBLEMS

1. Explain the heart function phases.
2. How many filaments are composed in cardiac myofilaments, determine their function and geometry.
3. The normal beat and duration of cardiac cycle in an adult is . . . beats/minutes and . . . seconds.
4. Explain the differences between Phase 2 and 4 in the cardiac cycles?
5. Diastole and systole occur in the period of . . .. and . . ., respectively.
6. What are the normal blood pressures (systolic and diastolic) in an adult?
7. Draw a normal ECG wave and explain each wave.
8. What is CO in an adult normal runner with 100 beats/minutes?
9. What is role of mitral, aortic, tricuspid and pulmonary valves in cardiac cycles?
10. Which circulation have supply duty of the heart muscles, explain the blood stream of this circulation?
11. What is valve regurgitation?
12. How many generations of airways are defined in the lung? Also explain the geometry of the alveoli in the lungs.
13. What are the systolic and diastolic pressure ranges in the lungs' arteries?
14. What is the name of the artery in the blood circulation which is responsible for the transfer of deoxygenated blood?

15. What are the alveolar and pleural pressures ranges in a normal adult?
16. Explain the mechanism of oxygen transport in a RBC in the lung.
17. What would happen to the lungs and breathing of a human who is located in the center of a tornado (there is a great pressure drop)?
18. Explain the role of stenosis in the blood velocity and pressure gradient in the heart valves with stenosis problems.
19. Explain the glomerular filtration in human kidney and its rate, also numerically, describe the role of pressures in bowman capsule and GFR.
20. What is the electrical charge of GBM and why?
21. Considering Table 4.9, which substance is the most reliable material to control performance of the kidney function and why?
22. Which organ has the maximum blood flow rate among human organs?
23. How much water (mL) will be reabsorbed by the gastrointestinal system per day?
24. A kid with diarrhea needs low ORS, calculate the amount of substances (ingredients) to make 3800 mL of ORS.
25. Describe the best equation to calculate and simulate the biofluid flow in the brain, the liver, the lung alveoli and glomerular capillaries.
26. Explain and characterize the synovial fluid properties.
27. What is IOP and describe its normal condition?
28. Calculate the variation in intraocular volume for an eye with a hypertension around 20% more than average pressure.
29. Why is semen pH more than 7 (7.2–7.8)? What is your opinion on the pH of the female reproductive system, is it an acid, base, or neutral medium?
30. What is the average flow rate to induce and maintain normal penile erection?
31. Describe the blood circulation of the female reproductive system.

REFERENCES

[1] A.C. Guyton, J.E. Hall, Textbook of Medical Physiology, eleventh ed., Elsevier, Inc., Philadelphia, PA, 2006.

[2] R.E. Klabunde, Cardiovascular Physiology Concepts, Wolters Kluwer and Lippincot Williams & Wilkins, Baltimore, MD, 2011.

[3] W.F. Boron, E.L. Boulpaep, Medical Physiology: A Cellular and Molecular Approach, second ed., Saunders Elsevier, Philadelphia, PA, 2009.

[4] D.R. Rubenstien, W. Yin, M.D. Frame, Biofluid Mechanics, An Introduction to Fluid Mechanics, Macrocirculation and Microcirculation, Elsevier, Inc., Philadelphia, PA, 2012.

[5] L. Waite, J. Fine, Applied Biofluid Mechanics, McGraw-Hill, New York, 2007.

[6] S.J. Melby, M.E. Klingensmith, L.E. Chen, S. Glasgow, T.A. Goers, The Washington Manual of Surgery, Wolters Kluwer Health, Lippincut Williams & Wilkins, Philadelphia, PA, 2008.

[7] L. Waite, Biofluid Mechanics in Cardiovascular Systems, Mc Graw-Hill, New York, 2006.

[8] K.S. Kunzelman, R. Cochran, S. Murphree, W. Ring, E. Verrier, R.C. Eberhart, Differential collagen distribution in the mitral valve and its influence on biomechanical behaviour, J. Heart Valve Dis. 2 (2) (1993) 236–244.

[9] D. Bouchner, K. Lim, Mechanical properties of human mitral valve chordae tendineae, Can. J. Physiol. Pharmacol. 53 (1975) 330–339.

[10] D.N. Ghista, A. Rao, Structural mechanics of the mitral valve: stresses sustained by the valve, J. Biomech. 5 (1972) 295–307.

[11] D.J. Schneck, J.D. Bronzino (Eds.), Biomechanics, Principles and Applications, CRC Press LLC, Boca Raton, FL, 2003.

[12] T. Ramanathan, H. Skinner, Coronary blood flow, Contin. Educ. Anaesth. Crit. Care Pain 5 (2) (2005) 61–64.

[13] L. Wijayasiri, K. McCombe, A. Patel, The Primary FRCA Structured Oral Examination Study Guide 1 (Masterpass Series), Radcliffe Publishing Ltd, Oxon, 2010.

[14] World Health Organization, The top 10 causes of death. [Online]. <http://www.who.int/mediacentre/factsheets/fs310/en/index2.html>, 2015.

[15] J.G. Kegel, M.B.D. Schalet, W.J. Corin, A.S. Iskandrian, Simplified method for calculating aortic valve resistance: correlation with valve area and standard formula, Cathet. Cardiovasc. Diagn. 30 (1993) 15–21.

[16] B.P. Griffin, S.R. Kapadia, C.M. Rimmerman, The Cleveland Clinic Cardiology Board Review, second ed., Lippincott Williams & Wilkins, Philadelphia, PA, 2012.

[17] H. Baumgartner, J. Hung, J. Bermejo, J.B. Chambers, A. Evangelista, B.P. Griffin, et al., Echocardiographic assessment of valve stenosis: EAE/ASE recommendations for clinical practice, Eur. J. Echocardiogr. 10 (2009) 1–25.

[18] P. Lancellotti, C. Tribouilloy, A. Hagendorff, L. Moura, B. Popescu, E. Agricola, et al., European Association of Echocardiography recommendations for the assessment of valvular regurgitation. Part 1: aortic and pulmonary, Eur. J. Echocardiogr. 11 (3) (2010) 223–244.

[19] P. Lancellotti, L. Moura, L.A. Pierard, E. Agricola, B.A. Popescu, C. Tribouilloy, et al., European Association of Echocardiography recommendations for the assessment of valvular regurgitation. Part 2: mitral and tricuspid regurgitation (native valve), Eur. J. Echocardiogr. 11 (2010) 307–332.

[20] E. L. LLC, <www.edwards.com>, 2009. [Online]. Available: <http://ht.edwards.com/scin/edwards/sitecollectionimages/edwards/products/presep/ar04313hemodynpocketcard.pdf>.

[21] D. Chemla, Castelain V, M. Humbert, J.L. Hébert, G. Simonneau, Y. Lecarpentier, et al., New formula for predicting mean pulmonary artery pressure using systolic pulmonary artery pressure, Chest J. 126 (4) (2004) 1313–1317.

[22] Y.C. Fung, Biomechanics Circulation, Springer, New York, 1997.

[23] J. Keener, J. Sneyd, Mathematical Physiology, Springer-Verlag, New York, 1998.

[24] J.B. West, Pulmonary Pathophysiology, the Essentials, Lippincott Williams & Wilkins, a Wolters Kluwer Business, Baltimore, MD, 2013.

[25] Deleted in review.

[26] J. Oliver, Nephrons and Kidneys, Harper & Row, Hoeber Medical, New York, 1968.

[27] K. Peter, Untersuchungen über Bau und Entwicklung der Niere, Gustav Fischer, Jena, 1909.

[28] H. Smith, The Kidney: Structure and Function in Health and Disease, Oxford University Press, New York, 1951.

[29] R. Osterby, Morphometric studies of the peripheral glomerular basement membrane in early juvenile diabetes, I: Development of initial basement membrane thickening, Morphomet. Stud. Peripher. Glomer. Basement Membra Diabetol. 8 (1972) 84–92.

[30] B. Stanton, B. Kaissling, Regulation of renal ion transport and cell growth by sodium, Am. J. Physiol. Renal. Physiol. 257 (1989) F1–10.

[31] B. Trump, R. Bulger, Morphology of the kidney, in: B. Becker (Ed.), Structural Basis of Renal Disease, Harper & Row, New York, 1968, pp. 1–92.

[32] P. Mundel, W. Kriz, Structure and function of podocytes: an update, Anat. Embryol. (Berl) 192 (1995) 385–397.

[33] R.J. Alpern, S.C. Hebert, The Kidney: Physiology and Pathophysiology, Academic Press, Amsterdam, 2007.

[34] R.W. Schrier, Atlas of Diseases of the Kidney, vol. 1, Wiley-Blackwell, Philadelphia, PA, 1999.

[35] M.A. Perazella, R.F. Reilly, Chronic kidney disease: a new classification and staging system, Hos. Physic. 39 (3) (2003) 18–22 and 45.

[36] P.R. Kvietys, The Gastrointestinal Circulation, Morgan & Claypool Life Sciences, San Rafael, CA, 2010.

[37] K.E. Barrett, S. Boitano, S.M. Barman, H.L. Brooks, Ganong's Review of Medical Physiology, McGraw-Hill, New York, 2010.

[38] World Health Organization, The Treatment of Diarrhoea—A Manual for Physicians and Other Senior Health Workers, World Health Organization, Geneva, 2005.

[39] W. Blonski, T. Siropaides, K. Reddy, Coagulopathy in liver disease, Curr. Treat. Options Gastroenterol. 10 (6) (2007) 464–473.

[40] H.G. Bogren, M.H. Buonocore, W.-Z. Gu, Carotid and vertebral artery blood flow in left-and rightlhanded healthy subjects measured with MR velocity mapping, J. Magn. Reson. Imaging. 4 (1994) 37–42.

[41] Deleted in review.

[42] R. Drake, A.W. Vogal, A.W.M. Mitchell, Gray's Anatomy for Students, third ed., Churchill Livingstone, London, 2014.

[43] J.D. Bronzino, D. Schneck, Biomechanics: Principles and Applications, CRC Press LLC, Boca Raton, FL, 2003.

[44] M.W. Ropes, E.C. Rossmeisl, W. Bauer, The origin and nature of normal human synovial fluid, J. Clin. Invest. 19 (1940) 795–799.

[45] A. Mundt, K.S. Lillian, Graff's Textbook of Routine Urinalysis and Body Fluids, Lippincott Williams & Wilkins, a Wolters Kluwer business, Philadelphia, PA, 2011.

[46] D. Dowson, Bio-tribology, Faraday Discuss. 156 (2012) 9–30.
[47] B. Troy, P.B. David, Remington: The Science and Practice of Pharmacy, Wolters Kluwer, Philadelphia, PA, 2005.
[48] M. Ramachandran, Basic Orthopaedic Sciences: The Stanmore Guide, CRC Press, Boca Raton, FL, 2006.
[49] M.E. Blewis, G.E. Nugent-Derfus, T.A. Schmidt, B.L. Schumacher, R.L. Sah, A model of synovial fluid lubricant composition in normal and injured joints, Eur. Cells Mater. 13 (2007) 26–39.
[50] A.D. Fitt, G. Gonzalez, Fluid mechanics of the human eye: aqueous humour flow in the anterior chamber, Bull. Math. Biol. 68 (1) (2006) 53–71.
[51] A.D. Fitt, C.R. Canning, M.J. Greaney, J.N. Dewynne, Fluid flow in the anterior chamber of a human eye, IMA J. Math. Appl. Med. Biol. 19 (2002) 31–60.
[52] M. Moster, P. Ichhpujani, Episcleral venous pressure and glaucoma, J. Curr. Glaucoma Pract. 3 (1) (2009) 5–8.
[53] O. Geyer, D.M. Silver, Pressure-volume relation for the living human eye, Curr. Eye Res. 20 (2) (2000) 115–120.
[54] K.E. Barrett, S. Boitano, S.M. Barman, H.L. Brooks, Ganong's Review of Medical Physiology, McGraw-Hill Companies, New York, 2010.
[55] G. Jordan F. Schreiter, Reconstructive Urethral Surgery, Heidelberg: Springer, 2006.
[56] R.C. Webb, K.P. Nunes, Mechanisms in erectile function and dysfunction: an overview, Erectile Dysfunction—Disease-Associated Mechanisms and Novel Insights Into Therapy, InTech, Rijeka, 2012.
[57] J. Devlin, H.F. Newman, J.D. Northup, Mechanism of human penile erection, Invest. Urol. 1 (1964) 350–353.
[58] V. Michal, R. Kramar, J. Pospichal, L. Hejhal, Arterial epigastricocavernous anastomosis for the treatment of sexual impotence, World J. Surg. 1 (1977) 515–520.
[59] C.C. Schulman, E. Wespes, Parameters of erection, Brrfrsh J. Urol. 56 (1984) 416–417.
[60] G. Meschia, Circulation to female reproductive organs, Supplement 8: Handbook of Physiology, the Cardiovascular System, Peripheral Circulation and Organ Blood Flow, John Wiley & Sons, Inc., New York, 2011.
[61] B. Issa, R.J. Moore, R.W. Bowtell, P.N. Baker, I.R. Johnson, B.S. Worthington, et al., Quantification of blood velocity and flow rates in the uterine vessels using echo planar imaging at 0.5 Tesla, J. Magn. Reson. Imaging 31 (2010) 921–927.
[62] F. Burbank, Fibroids, Menstruation, Childbirth, and Evolution: The Fascinating Story of Uterine Blood Vessels, Wheatmark Publishing, Tucson, Arizona, 2009.
[63] M. Feldman, L.S. Friedman, L.J. Brandt, Sleisenger and Fordtran's Gastrointestinal and Liver Disease—Pathophysiology, Diagnosis, Management, Saunders-Elsevier Health Sciences, Philadelphia, PA, 2015.
[64] G.T. MacLennan, Penis and male urethra, Hinman's Atlas of UroSurgical Anatomy, second ed., Saunders Publishing, Philadelphia, Pennsylvania, 2012.

Chapter 5

Biofluid Flow in Artificial, Assistive and Implantable Devices

LEARNING OBJECTIVES

- Explanation and categorization of the blood pumps.
- Total artificial heart and its types, functions and implantation procedures.
- Classification of prosthetic heart valves, advantages and disadvantages of mechanical and bioprosthetic valves and comparing heart valve fluidic aspects, such as velocity and shear stress.
- Definition and classification of artificial lungs, mechanical ventilators and assistive devices related to the lung diseases.
- Explanation of artificial kidney and hemodialysis machines and their elements.
- Classification and formulation of artificial vessels and grafts.
- Definition of stent.

5.1 BLOOD PUMPS AND ARTIFICIAL HEART

Blood transportation failure in living organs results in a lack of function in the transportation of oxygen and nutrients to the organ. If the heart is failing or due to other medical issues, blood pumps can help to maintain blood circulation in normal flow to provide enough oxygen and nutrients to the living tissues.

In general mechanical engineering, pumps can be categorized into two main types: positive displacement and centrifugal (rotary) pumps. In biofluid mechanics, the heart is responsible for pumping the blood through the circulation system. The heart may fail due to medical problems, such as coronary heart diseases, heart attack, diseased muscles (cardiomyopathy), severe viral infection of the heart or other related diseases.

Biofluid Mechanics. DOI: http://dx.doi.org/10.1016/B978-0-12-802408-9.00005-3

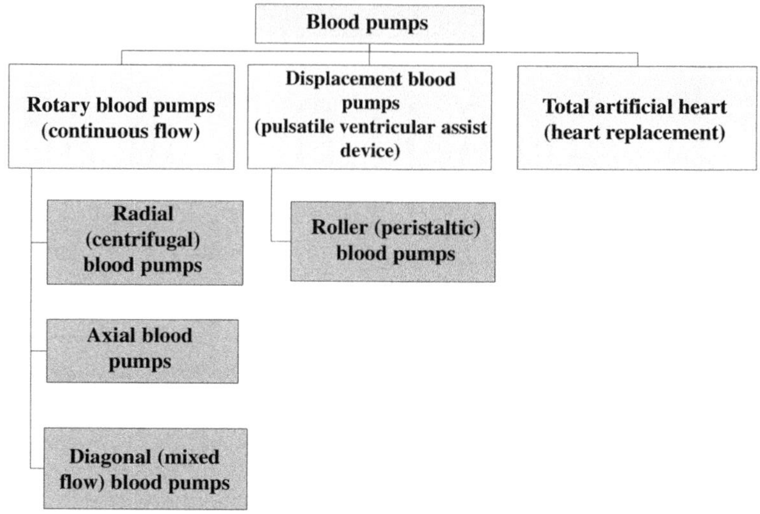

■ **FIGURE 5.1** Blood pumps categorization and their applications.

Blood pumps can be classified into three major types: displacement; rotary and total artificial heart (TAH) pumps. The rotary pumps can be classified into three subsystems: radial; axial and diagonal pumps. Fig. 5.1 shows the categorization of the blood pumps and their applications.

5.1.1 Rotary Blood Pumps

Rotary blood pumps are known as one of the best therapies for end-stage heart failure. In rotary pumps the mechanical energy transfers to the fluid by velocity changes within the impeller vanes. Depending on the geometry of impeller, these pumps can be categorized into three main groups: radial (centrifugal); axial and diagonal (mixed flow) pumps. Rotary pumps are the best option for high flow rates at differential pressure lower than 500 mmHg. Radial pumps are ideal for producing high pressures and low flows, whereas axial pumps produce high flow rates at low pressure difference. The diagonal pumps are known as mixed flow systems, this design has the ability to generate high flow rates at high pressures. The rotary blood pumps are categorized by low trauma of blood and lower levels of anticoagulation which mean less hemorrhages.

The radial (centrifugal) blood pumps are made from several parts, commonly including the inlet–outlet ports, pump housing, shaft, vanes (impellers), bearing, magnet and magnet housing. The body parts of the axial pumps include the flow straightener, impeller, diffuser, motor stator

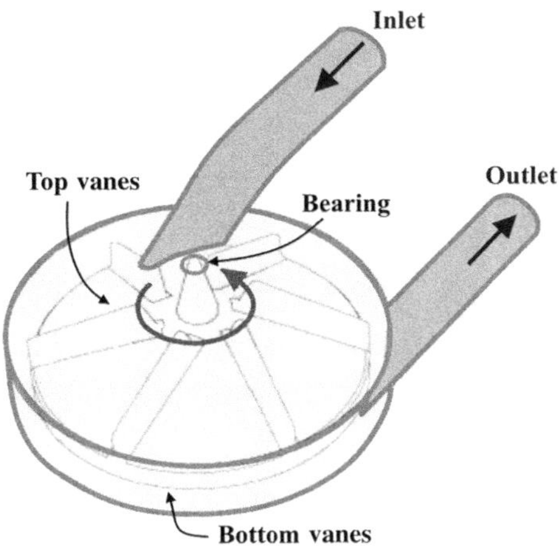

■ **FIGURE 5.2** Schematic of a radial (centrifugal) blood pump. *Modified from M. Behr, D. Arora, Y. Nose, T. Motomura, Performance analysis of ventricular assist devices using finite element flow simulation, Int. J. Numer. Methods Fluids 46 (2004) 1201—1210 with permission.*

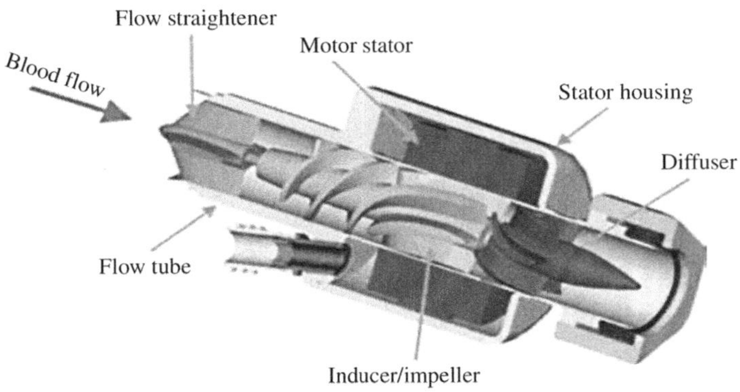

■ **FIGURE 5.3** Cutaway view of an axial blood pump. *From M. Akdis, H. Reul, Mechanical blood pumps for cardiac assistance, Appl. Bionics Biomech. 2 (2) (2005) 73—80 with permission.*

and stator housing. Figs. 5.2 and 5.3 shows the schematic of one kind of these blood pumps.

For cardiopulmonary bypass (CPB), an ideal blood pump should have the following aspects [3]:

- minimum allowable hemolysis and cell damages,
- blood flow rate of 7 L/minutes against a pressure of 500 mmHg,
- manual and digital control and
- high accuracy to control blood flow rate.

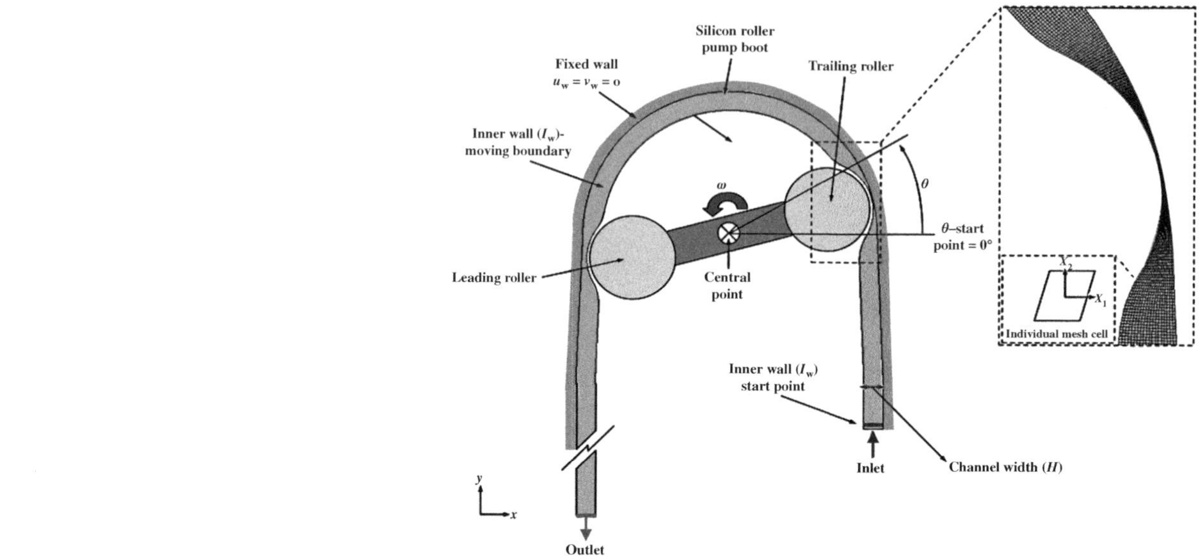

■ **FIGURE 5.4** Schematic of a roller pump operation. *Modified from J.W. Mulholland, J. Shelton, X. Luo, Blood flow and damage by the roller pumps during cardiopulmonary bypass, J. Fluids Struct. 20 (2005) 129–140 with permission.*

5.1.2 Displacement Blood Pump

The roller pump comes under the category of positive displacement pumps and has been the most commonly used pump for CPB for the past 50 years. The golden era of this pump is passing due to developments in systems using radial (centrifugal) pumps. The roller pump contains a tubing placed inside a curved wall. This wall is located at the travel perimeter of rollers installed on the ends of rotating arms. One of these arms always compresses the tubing at all times. The blood is pushed ahead by this compression and the moving roller, which produces continuous blood flow. The output of the rotary pump is determined by the frequency of rotation or revolutions per minute (RPM) of the pump. The volume of pumped blood depends on the size of the tubing, the length of the track and the RPM of motor [3].

The roller pumps are categorized as single, double and multiple roller pumps. The single roller pumps were used for CPB in the middle of the 20th century. The flow in this type was more pulsatile. The most commonly used pump for CPB is the double-roller pump. This pump consists of a 210° semicircular support plate and two rollers with the rotating arms set 180° apart. When one roller ends its operational phase, the second roller has already begun its phase. Due to hemolysis issues, the third type (multiple roller pump) is not clinically available [3]. Fig. 5.4 shows a schematic of a roller pump (pumping chamber). As the figure shows, there are two rollers (leading and trailing). The central point of rotation is connected to an electric motor which controls the speed of rotation and the blood flow rate of the blood pump. Compressing of the tube by the rollers leads to an increasing hemolysis (blood cell damage) which is an important issue for patients (see Fig. 5.4).

There are three materials currently used for tubing in the medical device industry: silicone; latex and polyvinyl chloride (PVC). Regarding microparticle releasing, the PVC is the best material for tubing because the latex generates more hemolysis and the silicon tube releases more microparticles than the PVC. One of the common problems in blood pumps is the tube occlusion. Occlusion can be controlled by either increasing or decreasing the compression of the tube by the rollers and the optimization between increasing and decreasing the compression is very vital in pumping the blood, because more compression increases hemolysis and less generates occlusion, although the priority is with forward output of the blood [3].

Selection of these pumps depends on several factors such as flow rate, pressure, rate of hemolysis, operation time, etc. The blood flow rate and pressure (pump head) are the most important aspects among pump features. Fig. 5.5 shows an approximate comparison between a centrifugal pump and a positive displacement pump. As the figure illustrates, the performance graph of these pumps is completely different. The centrifugal pump has a wide range of flow rate depending on pressure, whereas the positive displacement pump has a limited range of flow rate regardless of pressure. Another issue in blood transportation mechanisms is hemolysis or blood cell damage (see chapter: Physical Injury to Blood). Estimation of the hemolytic performance among the blood pumps requires experimental tests and it is related to design aspects such as the contact surface (roughness), pumping chamber shape, vanes shape, values of stress (compress or turbulent) in pumping zones and rotating speeds, which are different for each pump. There are several experimental and numerical studies on hemolytic performance among the blood pumps, however most of the research did not consider all those design aspects together in their analysis.

5.1.3 **Total Artificial Heart**

A TAH is a device that replaces the two ventricles of the heart. Generally, patients with end-stage heart failure use this device to keep them alive while they are on the waiting list for a heart transplant or are not eligible for a heart transplant.

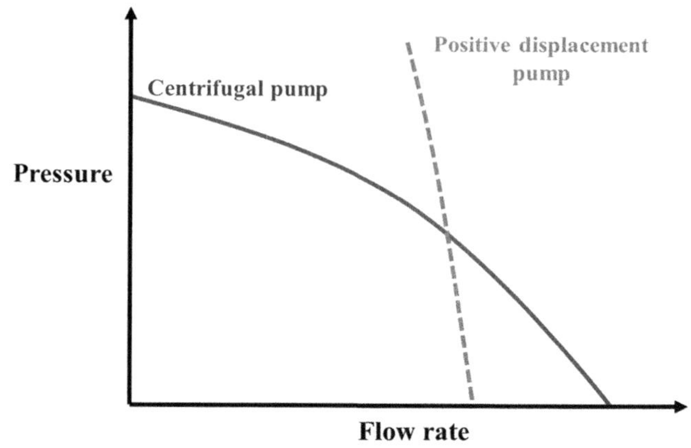

■ **FIGURE 5.5** An approximate comparison between the total pressure (head) and the flow rate for a centrifugal pump and a positive displacement pump.

There are several TAHs used around the world but this section describes only two of them, which are approved by Food and Drug Administration (FDA) of United States of America: the SynCardia TAH, with an implant rate of more than 90% of all implanted TAH around the world and AbioCor TAH which is made by Abiomed company.

SynCardia: The SynCardia TAH is the only implantable device that eliminates the indications and source of end-stage heart failure due to biventricular damage. Similar to existing methods of heart transplant, the SynCardia TAH replaces the four heart valves and both failing heart ventricles.

The procedure of the SynCardia TAH implantation is started by removal of the native ventricles, removal of the native heart valves, installation of the inflow and the outflow connectors, placement of the left and right artificial ventricles (AVs) and eventually installation of the right and the left drive lines. Fig. 5.6 shows a schematic of the SynCardia TAH implantation procedure.

The SynCardia TAH consists of two separate blood pumps that are pneumatically driven and generate pulsatile blood flow in the TAH. Each pump is made from a rigid plastic chamber and contains two mechanical valves. A diaphragm separates the blood side from the air side inside AVs. The blood output is equal to maximum stroke volume, 70 mL. This TAH is controlled by an external console. The console contains a control panel, pressurized air tanks, vacuum pump and battery for transport.

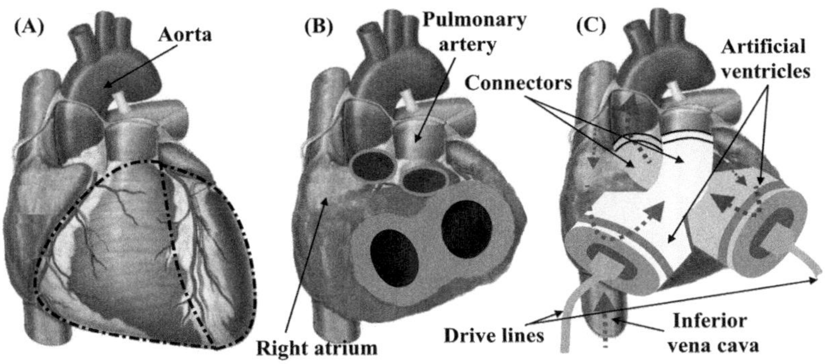

■ **FIGURE 5.6** Schematic of the SynCardia TAH implantation procedure. (A) The whole heart; (B) removing ventricles and heart valves and (C) installation of connectors, AVs and pneumatic drive lines. *Modified from T. SynCardia, Image Library, The SynCardia, [Online]. http://www.syncardia.com/media/multimedia-library-dp1.html, 2014.*

During systole, a pulse of compressed air is sent into the air sac, which displaces blood from the blood sac, and the two mechanical valves (tilting disc) in each pump ensure unidirectional discharge. Individually, the pressure of air on the diaphragm is controlled for each pump. The drive pressure on the left side must be at least 60 mmHg above systemic pressure and normally is set between 150 and 200 mmHg. The right drive pressure must be at least 30 mmHg above pulmonary pressure and is normally set between 55 and 90 mmHg. During diastole action, the air sac is opened to atmospheric pressure since the patient's venous pressure is enough to fill the blood sac. The console can produce a heartbeat between 90 and 130 beats/minutes [6].

For the CardioWest model of TAH, three in vitro reliability tests were conducted by SynCardia [7]:

- First test: Four TAH units were run for a period of 6 months. During this period there were no failures or abnormalities detected.
- Second test: Four TAH units were tested in a "run to failure" study design and are ongoing. After 35 months of testing, there were no failures or defects detected.
- Third test: Three TAH units were under assessment, the TAHs had expired their 3-year sterilization expiration date. This test method helped to provide information about the effects of long-term storage on the fatigue resistance properties of the TAH. After 24 months of testing, there were no failures or defects.

The result of reliability tests revealed that the mean time between failures for CardioWest model of TAH is 2916 days (7.99 years) [7]. The cardiac

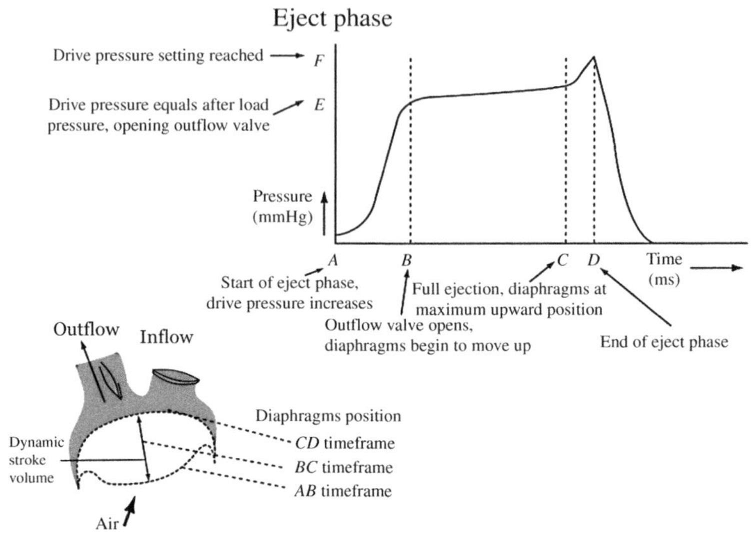

■ **FIGURE 5.7** Diaphragm position and pressure–time waveform, TAH (CardioWest TAH, SynCardia). *From I. Syncardia Systems, SynCardia CardioWest TAH instructions for use, [Online]. www.fda.gov/ohrms/dockets/ac/04/briefing/4029b1_final.pdf with permission.*

output is related to the volume of airflow out of drive lines and the plots of left and right sided cardiac output are continuously displayed for each AV. Fig. 5.7 shows fully the eject blood plot or TAH pressure–time waveform from each AV with each TAH beat. As the figure shows, there are clear differences between normal heart waveform and the TAH especially in the BC timeframe. During systole, the waveform of the normal human heart has a parabola shape (see Section 4.1) but the waveform is a flat line with low gradient in the CardioWest TAH (Fig. 5.7, *BC* timeframe).

Abiomed: The AbioCor TAH is the world's first completely self-contained TAH. The AbioCor is designed for a patient who wants to be mobile with normal activity. This TAH is planned for use with patients who have biventricular heart failure or patients who cannot have a heart transplant. The complete AbioCor system contains of an internal thoracic unit, an internal rechargeable battery, an internal electronics system, an external battery, handheld alarm monitor and a computer console. Fig. 5.8 illustrates the connection of the implanted components and the external components of the complete AbioCor system. The implanted elements receive power and command signals from the AbioCor console [8].

The clinical trial for the AbioCor took more than 3 years (between Jul. 2001 and Nov. 2004) and 14 patients with end-stage heart failure were implanted with this TAH. The results proved the advantage of this machine [8].

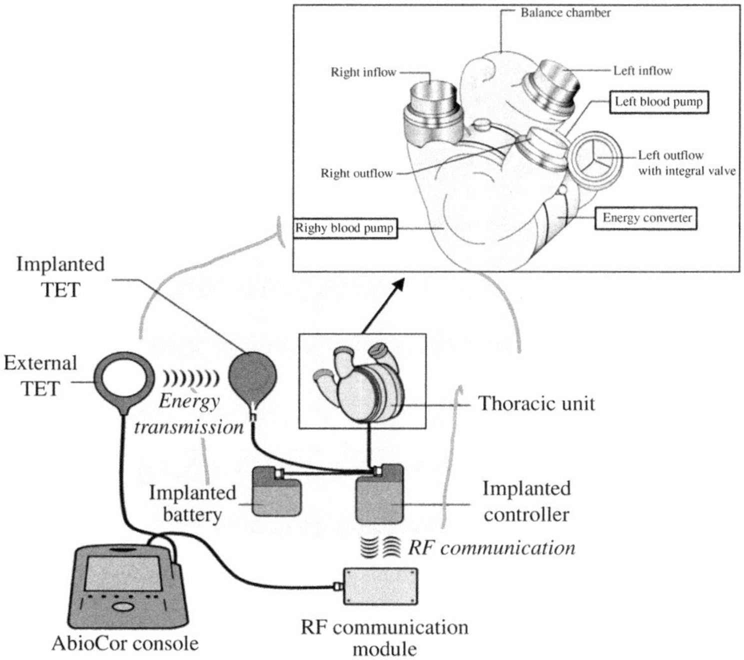

■ **FIGURE 5.8** Relationship of implantable elements to external elements of the AbioCor system. *Modified from I. ABIOMED, AbioCors implantable replacement heart, instructions for use, [Online]. http://www.fda.gov/ohrms/dockets/ac/05/briefing/2005-4149b2_01_ABIOMED%20INSTRUCTIONS%20FOR%20USE.pdf with permission.*

The thoracic unit (Fig. 5.8) consists of two blood pumps which are sealed to and separated by the energy converter. The blood pumps are a hard shelled chamber containing a sac filled with blood. The gap between the energy converter and the sac is filled with a hydraulic fluid. The duty of the energy converter is to move the hydraulic fluid from one side to the other, squeezing the sac in one pump and forcing blood out of it. The converter pumps hydraulic fluid in each direction alternately, thereby, the left and the right blood pump alternately fills and ejects blood into blood circulation system. This mechanism can produce a blood flow rate up to 12 L/minute [8].

5.2 PROSTHETIC HEART VALVES

Over a single lifetime, normal heart valves will open and close at least 3×10^9 times [9]. As was explained in the chapter "Tissue Engineering of Cardiovascular System," heart valve failure affects hemodynamic performance in two ways; stenosis and insufficiency (incompetence). To restore the normal function of a heart valve, a repair surgery or replacement of the diseased valve with a prosthetic heart valve is necessary. Presently, approximately 300,000 prosthetic heart valves are implanted each year throughout the world.

Classification of prosthetic heart valve: The ideal prosthetic heart valve should mimic the features of a normal heart valve, including being implantable, great biocompatibility (also thromboresistance), similar hemodynamics, long durability and high reliability. The prosthetic heart valves are categorized into two main types: mechanical and bioprosthetic valves.

Table 5.1 explains the two main types of prosthetic heart valves and their subsystems.

Table 5.1 Classification of Prosthetic Heart Valves

Mechanical Prosthetic Valves	Bioprosthetic Valves
Caged ball valves, which contains a silastic ball with a circular sewing ring and a metal cage, this type is no longer implanted. 	*Stented bioprostheses* have two types: Porcine bioprosthetic valves, consisting of three porcine aortic valve leaflets crosslinked with glutaraldehyde and attached on a metallic or polymer supporting stent. 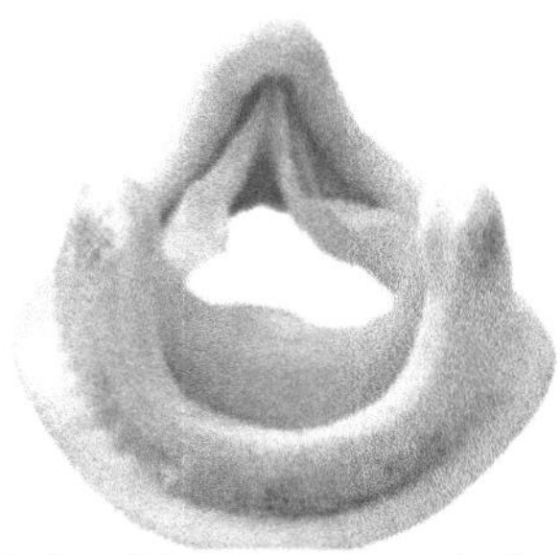 Pericardial valves are made from sheets of bovine pericardium mounted a supporting stent. 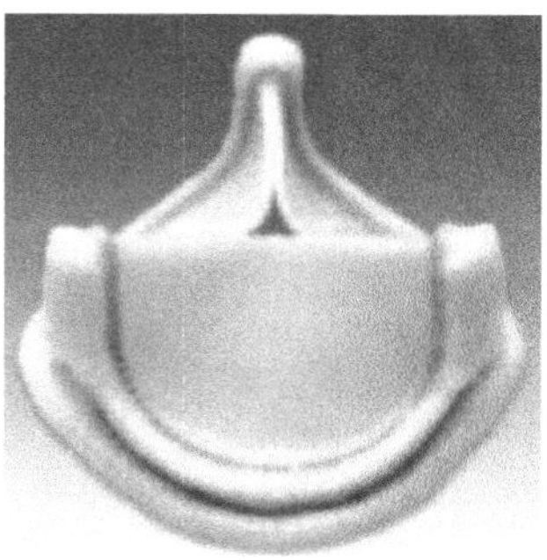

(Continued)

Table 5.1 Classification of Prosthetic Heart Valves *Continued*

Mechanical Prosthetic Valves	Bioprosthetic Valves
Monoleaflet valves are created from a single disk protected by metal struts (central and lateral). The opening angle ranges from 60° to 80°. 	*Stentless bioprostheses* are produces from whole porcine aortic valves or bovine pericardium. 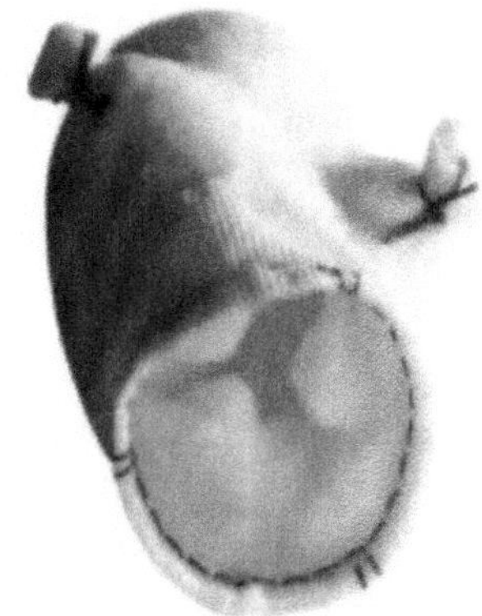
Bileaflet valves are created from two semilunar disks attached to a rigid valve ring by small hinges. Open valve contains three orifices. The opening angle of the leaflets ranges from 75° to 90°. 	*Percutaneous bioprostheses*: Percutaneous aortic valve implantation is developing as an alternative to standard aortic valve replacement (AVR) in patients who considered to be at high operative risk. 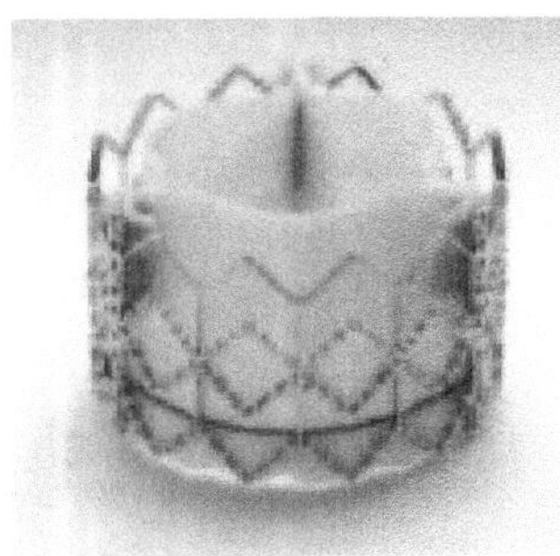 (The valve is expanded over a balloon)

From [10] with permission.

Advantage and disadvantage of mechanical and bioprosthetic valves [11]:

Mechanical heart valves: The best advantage of mechanical valves is more reliability or long durability and the main disadvantage of them is the continuous use of anticoagulation therapy to decrease the risk of thrombosis and thromboembolic problems.

Bioprosthetic valve: The main advantage of tissue valves is a lower rate of thromboembolic problems and its main disadvantage is large pressure drops.

From a stability point of view, the artificial heart valve must be durable to around 3.2 billion cycles in a lifetime (or around 100,000 cycles/day) under a cyclical blood pressure of about 120 mmHg ($\sim$200 g/cm^2) [12] in human blood, which is an aggressive and corrosive fluid to metals due to chloride solutions and ionic composition and protein concentration.

Pressure drop (blood flow potential energy loss) has a main role in prosthetic heart valve design and it occurs when viscous blood flows through the heart valve. This pressure gradient is an indication of heart valve efficiency. A large pressure drop across a prosthetic aortic valve requires a higher systolic pressure in the left ventricle. For better efficiency of prosthetic valves, the pressure drop should be minimized (optimized) to support normal flow rate in the blood circulation.

With regard to stenosis issues, the effective orifice area, EOA, is the typical factor for assessment of stenosis severity. For prosthetic heart valves, the same factor can be used to show the quality and efficiency of the valve. For in vitro conditions, the EOA can be written as [13]:

$$\text{EOA (cm}^2) = \frac{A_V Q}{\sqrt{Q^2 + KA_V^2 \Delta P}} \tag{5.1}$$

when $Q^2 \ll KA_V^2 \Delta P$ this equation is reduced to

$$\text{EOA (cm}^2) = \frac{Q}{(C\sqrt{\Delta P})} \tag{5.2}$$

where $K = 2540$, A_V is the cross-section area of left ventricular outflow tract (LVOT), ΔP is the mean systolic/diastolic pressure drop, Q is the blood flow rate and C is a constant and considered equal to 51.4 by Ref. [13] and 51.6 by Ref. [14].

A larger EOA shows a smaller pressure drop or a smaller energy loss. Also, another formula has been written for prosthetic valve as [15]:

$$\text{EOA (cm}^2) = 0.785 \times (\text{prosthesis size in cm})^2 \times (\text{VTI}_{\text{LVOT}}/\text{VTI}_{\text{Aortic prosthesis}}) \tag{5.3}$$

where VTI is velocity time integral of blood flow.

An in vitro investigation was done to evaluate the fluid performance characteristics of prosthesis valves such as the blood flow velocity and turbulent shear stress. The results were obtained at systolic peak at a cardiac output of 6.0 L/minute and a heartbeat rate of 70 beats/minute. Shear stress is an important feature in hemodynamics due to the membrane damage of red blood cells or hemolysis [11]. All downstream distances are measured from the valve sewing ring.

Governing equations: To derive the governing equations, the best assumption is to consider blood flow as a Newtonian fluid regardless of the flow characteristics, such as the laminar or turbulent conditions. Considering the blood fluid characteristics (incompressible, motion viscous fluid, etc.), the Navier–Stokes equations can be introduced as the best governing equations for the biofluid mechanics of heart valves. The general equations can be written as (from chapter: Fluid Mechanics and Biofluids Principles):

$$\rho\frac{\partial \mathbf{V}}{\partial t} + \rho\mathbf{V}\cdot\nabla\mathbf{V} = -\nabla P + \mu\nabla^2\mathbf{V} + \rho g$$

$$\tau = \mu\frac{\partial \mathbf{V}}{\partial y}$$

These equations can be generalized to three-dimensional flow (see chapter: Fluid Mechanics and Biofluids Principles).

With consideration of some characteristics of blood flow such as, incompressible fluid, inviscid flow (rare and ideal condition!), laminar and steady, the Bernoulli equation can be used to calculate and analyze the conservation of kinematic and energy (from chapter: Fluid Mechanics and Biofluids Principles):

$$\rho\mathbf{V}\nabla\cdot\mathbf{V} = -\nabla P + \rho g$$

or

$$\frac{P_2 - P_1}{\rho} + \frac{1}{2}(\mathbf{V}_2^2 - \mathbf{V}_1^2) + g(z_2 - z_1) = 0$$

Estimation is one of the engineering techniques to provide the approximate result when there are some limitations to the measurement of engineering items, such as pressure, velocity, etc. Measurement limitations in in vivo conditions lead to the use of estimation solutions.

The average velocity, V (m/seconds), in the ascending aortic cross section is measured noninvasively and the pressure drop, ΔP (mmHg), can be calculated using the equation [16]:

$$\Delta P \cong 4V^2 \tag{5.4}$$

Using this equation and noninvasive measurement of the aortic root velocity using the Doppler method, the pressure drop across the aortic valve can be calculated.

In the following, three mechanical heart valves and two tissue heart valves are described to compare the blood flow velocities and the shear stresses in order to understand performance of each valve.

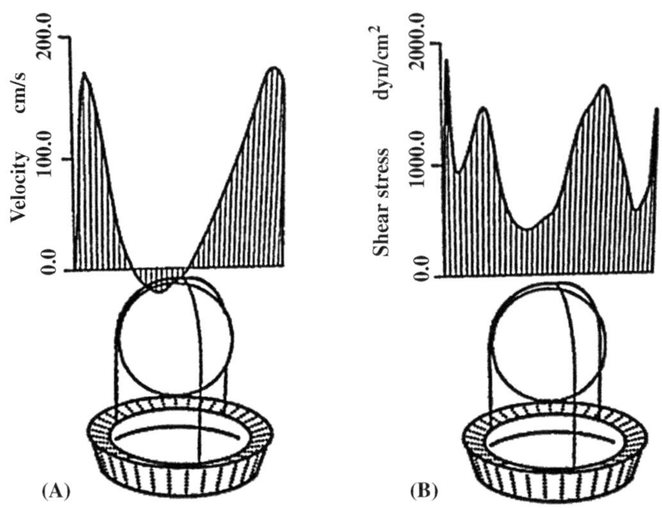

■ **FIGURE 5.9** (A) Velocity profile and (B) shear stress for the Starr Edward ball valve 30 mm downstream at peak systole (1 dyn/cm^2 = 0.1 Pa). *From A.P. Yoganathan, Cardiac valve prostheses, Biomedical Engineering Handbook, CRC Press LLC, Boca Raton, FL, 2000 with permission.*

Cage-ball valves: The Starr Edward (Model 1260, in use clinically) is studied for this category of the prosthetic valves. The flow developing from the valve shaped a circumferential jet that separated from the ball first hit and then flowed along the wall. The results show high flow velocities in the annular region. The maximum velocity was 220 cm/seconds at 12 mm downstream of the valve during peak systole. The peak systolic velocity was 180 cm/seconds at 30 mm downstream of the valve, furthermore the maximum reverse velocity measured was −25 cm/seconds at the same point. The maximum turbulent shear stress was 1850 (mean 1100) dyn/cm^2 (1 dyn/cm^2 = 0.1 Pa), which was experienced at the edges of the jet [11,17]. Fig. 5.9 shows the velocity profile, flow velocity quantities and shear stress and its profile at 30 mm downstream.

Tilting disc valve: The Medtronic-Hall tilting disc valve is investigated for this group of prosthetic valves. High velocity flows (jetlike) were detected from both the minor and major opening discharge parts. The highest velocities occurred at 7 mm downstream of the valve. The measured velocities were 200 and 210 cm/seconds in the minor and the major outlet regions, respectively.

The reverse flow region was detected 2 mm from the wall with a maximum reverse velocity of −25 cm/seconds at peak systole. During the deceleration phase, the location of this region moved to 8 mm from the

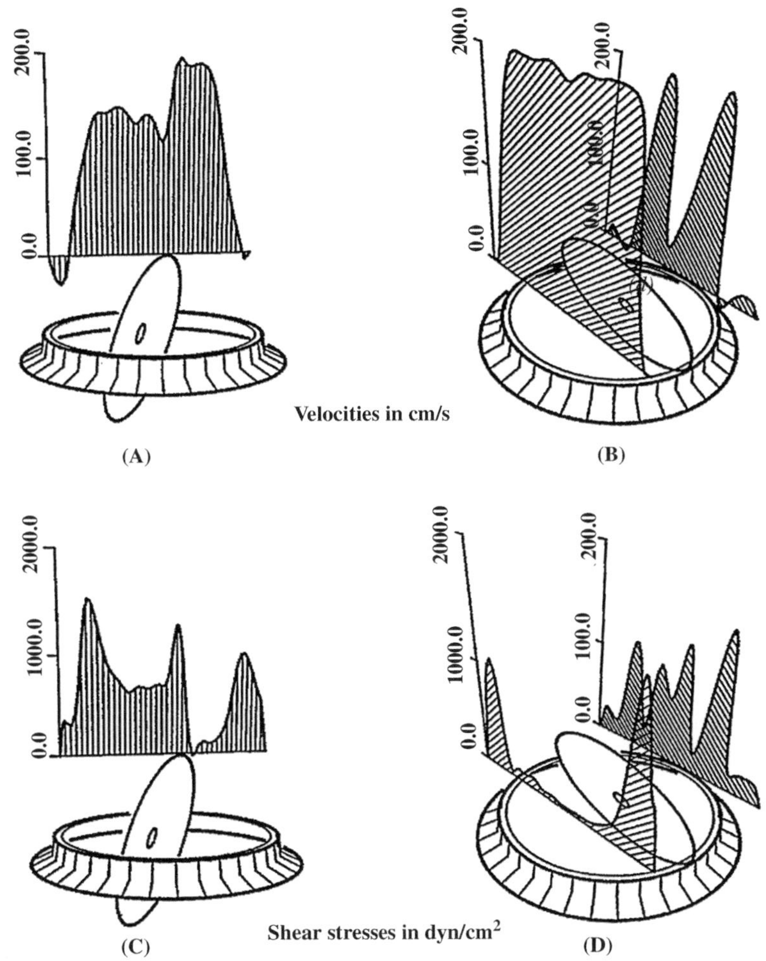

FIGURE 5.10 Velocity and turbulent shear stress profile at peak systole for a Medtronic-Hall tilting disc valve. (A) Velocity profile 15 mm downstream on the centerline across the major and minor outlets of (major outlet to the right); (B) velocity profiles 13 mm downstream in the major and minor outlet; (C) turbulent shear stress profile 15 mm downstream on the centerline across the major and minor outlets (major outlet to the right) and (D) turbulent shear stress profiles 13 mm downstream in the major and minor outlets (1 dyn/cm^2 = 0.1 Pa). *From A.P. Yoganathan, Cardiac valve prostheses, Biomedical Engineering Handbook, CRC Press LLC, Boca Raton, FL, 2000 with permission.*

wall. Also a small region of the flow separation was experienced close to the wall in the major outlet region. The major outlet region produced high turbulent shear stresses which occurred at the edges of the major outlet jet, see Fig. 5.10. The highest turbulent shear stresses measured at maximum systole were 1200 and 1500 dyn/cm^2, 7 and 13 mm downstream of the

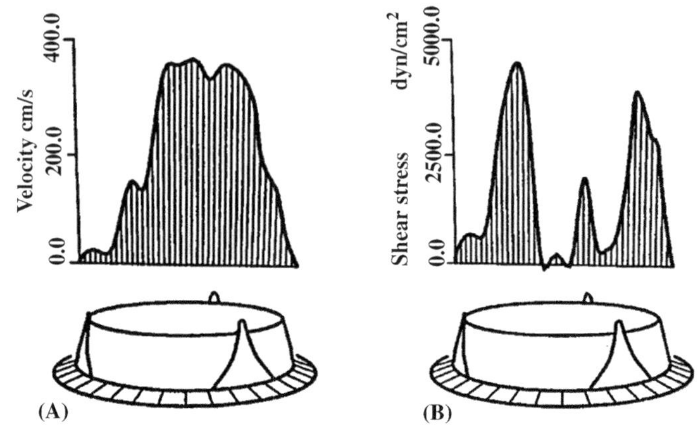

■ **FIGURE 5.11** (A) Velocity profile and (B) turbulent shear stress profile, 15 mm downstream on the centerline for the Carpentier—Edwards 2625 porcine valve, at peak systole. *From A.P. Yoganathan, Cardiac valve prostheses, Biomedical Engineering Handbook, CRC Press LLC, Boca Raton, FL, 2000.*

valve, respectively. As Fig. 5.10 shows, the high turbulent shear stresses occurred more in the minor than in the major outlet region. The figure illustrates the profile of shear stresses across the major and minor outlets 15 mm downstream of the tilting disk valve and it shows a stress around 1450 dyn/cm^2 as the maximum turbulent shear stress at the lower edge of the minor outlet jet [17].

Bileaflet valve: The St Jude Medical valve is investigated for this group of prosthetic valves. The valve has two semicircular leaflets which form three regions for forward blood flow: two lateral outlets and a central outlet. The two lateral outlets create the main part of the blood flow rate. The maximum velocity of 220 and 200 cm/seconds occurs at the centerline plane 8 mm downstream of the valve for the lateral and central outlet jets. As Fig. 5.11 shows, at the location of the two leaflets, the velocity profiles show two defects. The measurements show that the maximum velocity was 220 cm/seconds in the central outlet and also the low velocity reverse flows were experienced adjacent to the hinges mechanism of the valve. Maximal turbulent shear stresses occurred at locations of high velocity gradients and adjacent to the valve leaflets (see Fig. 5.11). At peak systole, the maximum turbulent shear stresses were 1150 and 1500 dyn/cm^2 at 8 and 13 mm downstream, respectively, and it occurred on the centerline plane of the valve. The figure and the profiles across the central outlet confirm that the flow is very disturbed in this area. At peak systole, the maximum turbulent shear stress measured 11 mm downstream in the central outlet at around 1700 dyn/cm^2. Probably, the turbulent shear stress will be more than this amount for the points closer to the valve [17].

Bioprostheses valves: Bioprostheses valves are produced with tissues from porcine heart valves or bovine cardiac tissue because their functions are similar to human heart valves. In the following, two kinds of tissue heart valves will be studied to compare their fluid aspects.

Carpentier–Edwards porcine valve (model 2625): The velocity profiles were investigated along the centerline plane at 10 mm downstream of the valve. For the downstream of 10 mm, the profiles at peak systole show that the maximum velocity of the jetlike flow emerging from the tissue valve is around 220 cm/seconds. When the flow traveled from 10 to 15 mm downstream of the valve, the flow velocity of the jet increased to around 370 cm/seconds at peak systole. As Fig. 5.11 shows, no areas of flow separation were observed during the systolic period. An increase in turbulent shear stresses was observed at the edge of the jet. The maximum measured turbulent shear stress was 2750 dyn/cm^2 at 10 mm downstream of the valve at peak systole. Normally, with an increase downstream of the valve, the turbulent shear stresses at the edge of the jet will increase. During a peak systole, the mean and maximum turbulent shear stress increased to 2000 and 4500 dyn/cm^2, respectively [17].

Carpentier–Edwards pericardial valve (model 2900): During peak systole, a maximum velocity of 180 cm/seconds is observed along the centerline plane 17 mm downstream of the valve. The maximum velocity of the jet is approximately constant as the flow field moved from 17 to 33 mm downstream of the valve at peak systole, see Fig. 5.12. During peak systole and the deceleration phase, a flow separation area is detected at 6 mm from the wall. The increasing and decreasing range of this area

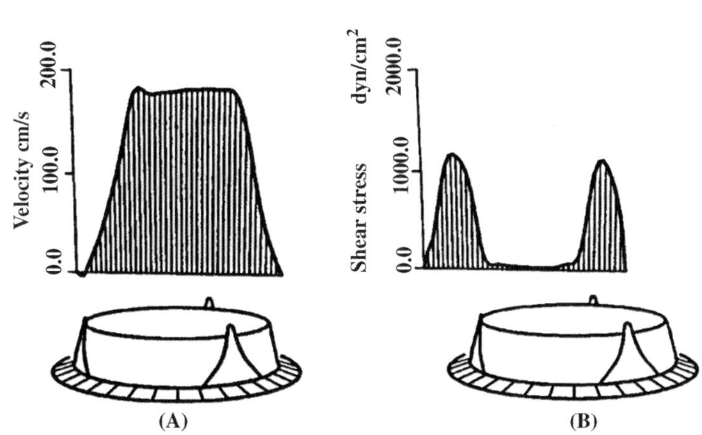

■ **FIGURE 5.12** (A) Velocity profile and (B) turbulent shear stress profile, 17 mm downstream on the centerline for the Carpentier–Edwards pericardial valve, at peak systole. *From A.P. Yoganathan, Cardiac valve prostheses, Biomedical Engineering Handbook, CRC Press LLC, Boca Raton, FL, 2000.*

from the wall is only about 1 mm during the acceleration phase. A shear stress value of 100 dyn/cm^2 is observed along the centerline plane 17 mm downstream of the valve. The maximum turbulent shear stresses was 850 dyn/cm^2 which occurred for a narrow area during peak systole. The amount of turbulence increased as the flow moved from 17 to 33 mm downstream of the valve, during peak systole [17].

5.3 ARTIFICIAL LUNG

The number of patients requiring a lung transplant is increasing. The recorded data showed an 11% increase in the number of applicants on the lung transplant list from 1997 to 2007 [18]. Each year several hundred thousand patients die of some form of lung disease around the world. The artificial lung (AL) can be considered as an alternative solution for patients with end-stage lung diseases. An AL is a device that delivers enough oxygen to the blood and removes carbon dioxide from the blood. The ALs can be used as an assistive or supportive device.

Generally, the ALs are classified as [19]:

- Extracorporeal: Mainly, it is used for surgeries requiring the CPB method, but also is used less often to support patients with respiratory failures. The common extracorporeal technologies are CPB or extracorporeal membrane oxygenation (ECMO).
- Paracorporeal: This type is a wearable device that will be attached directly to patients.
- Intravascular: In this technology, respiratory catheters are located within the vena cava through a peripheral vein.
- Intrathoracic: This method approaches the devices that will be located internally within the abdominal or thoracic cavities.

ECMO circuit: ECMO equipment include: biocompatible tubing, AL or membrane oxygenator, heat exchanger (to adjust the blood temperature), roller pump, oxygen cylinder, drug infusion (heparin or other liquids) ports. Fig. 5.13 illustrates a simple schematic of the ECMO circuit and the elements.

ECMO configurations: ECMO was developed by a technical revolution in the membrane oxygenators in last century (in the 1960s). Biocompatibility is one of the best features of these devices during gas exchange. The new ECMOs are less harmful as blood is exposed to oxygen and also they enhance gas transfer. Three configurations of ECMO tubing are defined to connect the AL to the body: (1) Veno-Arterial or VA; (2) single-site double lumen Veno-Venous or VV and (3) double lumen VV. Both VV and VA methods require a blood pump to provide

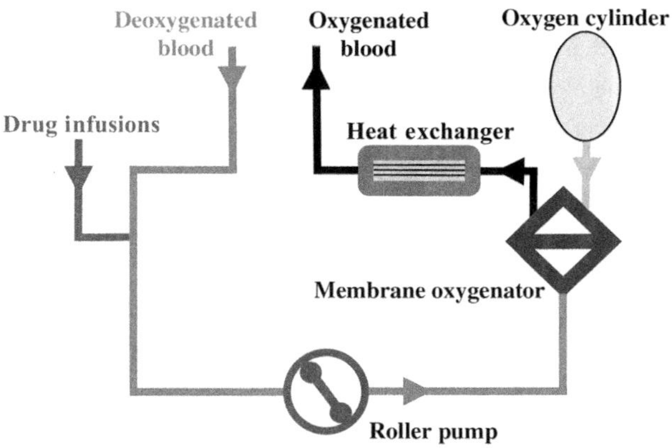

■ **FIGURE 5.13** Schematic of ECMO circuit and the elements.

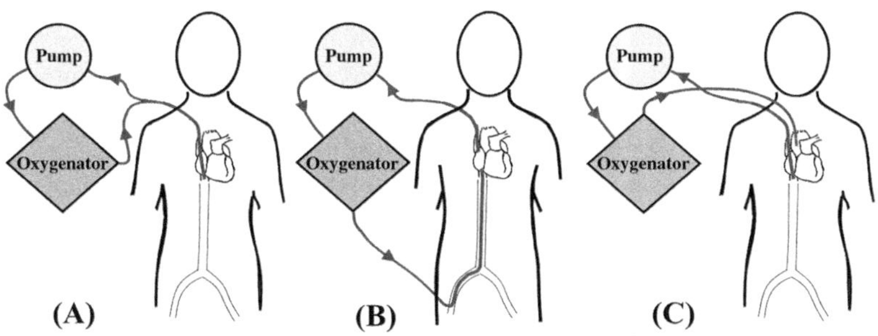

■ **FIGURE 5.14** The configurations of extracorporeal blood flow (A) single-site double lumen VV; (B) two site VV and (C) VA. *Modified from D. Hayes Jr., J.D. Tobias, J. Kukreja, T.J. Preston, R. Yates, S. Kikby, et al., Extracorporeal life support for acute respiratory distress syndrome, Ann. Thorac. Med. 8 (3) (2013) 134—141.*

enough flow rates (around 3—5 L/minutes) to assure sufficient organ perfusion and oxygenation in the patient. Also an AV method can be used without a blood pump. A pumpless arrangement depends on flow resistance in the AL blood circulation. These pumpless extracorporeal lung assist (PECLA) devices generate flow rates of 0.8—1.5 L/minutes which is sufficient for removing CO_2 [20]. Fig. 5.14 shows the variable forms of ECMO support currently used in patients.

There are several alternative extracorporeal devices to support the patients with lung diseases [20]:

- *Arteriovenous CO_2 removal (AVCO_2R)*: The AVCO_2R was developed using a low resistance hollow fiber gas exchanger to provide lung rest in the severe lung failures. This technique only removes CO_2 from the blood without any O_2 transfer. The experiments showed an approximate removal of 70% CO_2 without any hemodynamic compromise or instability [21,22].

- *PECLA, interventional lung assist (ILA) and Novalung*: The fluid circuit of these devices uses a hollow fiber gas exchanger without a blood pump (pumpless devices). The PECLA benefits, such as lower rate of bleeding, hemolysis and mechanical trauma to the blood system, create a superior role for these devices versus lung assist devices that require a pump. The ILA removed around 50% of total CO_2 created with rapid normalization of respiratory acidosis. The ILA has several disadvantages including arterial damage, immobilization and cardiovascular steal [20].
- *Intravenacaval (intravascular) oxygenator to remove CO_2*: IVOX is a CO_2 removal device. The IVOX components contain several biocompatible hollow fibers with silicone coating and heparin bonding, to create a thin membrane, which will be located in the vena cava to provide gas exchanges (blood oxygenation and CO_2 removal) without the need for extracorporeal circulation or blood transfusion. The experimental results show a limited capacity for gas exchange in IVOX and researchers do not recommend this device as total support for patients with acute lung failures [20].
- *Intravascular lung assist device (ILAD)*: ILAD provides sufficient gas exchange without enormous pressure drop. Geometry, configuration and surface area of membrane fibers have an effect on gas transfer efficiency. The in vitro experiments proved that the ILADs with cross flow fibers transfer at least 10 times more O_2 than parallel flow ILADs of the same surface area, and six to eight times more CO_2. The experiments proved that the ILAD with 0.4–0.6 m^2 of surface area can transfer 100 mL/minutes of both CO_2 and O_2 [23].
- *Hattler respiratory assist catheter*: The Hattler's invention was a respiratory assist catheter which includes a small pulsating balloon to allow the blood mixing and gas exchange. The balloon is located in the middle of a hollow fiber bundle. In an in vitro model, large balloon volume and high pulsation rate lead to an increase of both O_2 loading and CO_2 removal [24].
- *Decap and hemolung respiratory assist system*: In this method, a combination of Decap (CO_2 removal) and hemodialysis are performed by a system. To have an intravascular access, a single double lumen cannula is inserted into the femoral vein and a blood pump creates enough pressure to flow the blood through a membrane oxygenator and then a hemofilter. This mechanism does not allow for total gas exchange [20].

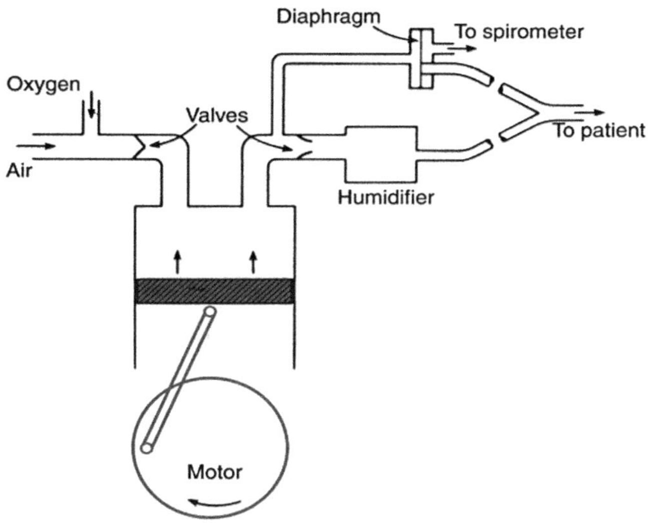

■ **FIGURE 5.15** Schematic of a CV ventilator. *From J.B. West, Pulmonary Pathophysiology: The Essentials, Lippincott Williams & Wilkins, a Wolters Kluwer Business, Baltimore, MD, Philadelphia, PA, 2013 with permission.*

Mechanical ventilators: Generally, ventilators provide support to patients who have respiratory failures. These failures have an effect on the air flow into the lungs which occurs because of pressure gradient (negative) between the atmosphere and the alveoli. Mechanical ventilators create a pressure gradient to move the air flow between the ventilator machine and the lungs.

The basic functions of the mechanical ventilators are under the subsystem of volume control or pressure control. The ventilators types can be categorized as [25]:

- *Constant volume (CV) ventilators*: The CV ventilators supply a preset volume of fluid (gas) to the patient. The function of these machines is based on cylinder-piston mechanism. The preset volume of ventilation can be adjusted by the stroke and frequency. The CV ventilators are known as robust and reliable machines for long-term ventilation. These machines are used extensively in anesthesia. Fig. 5.15 illustrates the CV mechanism. The advantage and disadvantage of this kind of ventilator are a known volume and developing high pressure, respectively.
- *Constant pressure (CP) ventilators*: The CP ventilators supply fluid (gas) at a preset pressure. They are small, inexpensive devices and do not need electrical power but instead work off a source of compressed

gas with a minimum pressure of 50 lb/in^2. These machines have some disadvantages such as delivery relation of gas volume to the compliance of the lung or chest wall and also more airway resistance can reduce the ventilation due to equilibration of pressure between the machine and the alveoli.

- *Tank ventilators*: The tank ventilators transport negative pressure (less than atmospheric) to the body of patients, excluding the head. A rigid box (iron lung) is connected to a pump (large-volume and low-pressure) that controls the respiratory cycle. Due to access problems to the patient and the large dimensions, these kinds of ventilators are no longer used in the cure of acute respiratory failure. Frequently, it can be considered as a system reserved for patients who have partially recovered from neuromuscular respiratory failure.
- *Patient-cycled (PC) ventilators*: In these ventilators, the patient can make an inspiratory effort by applying the inspiratory phase. The term "assisted ventilation" is sometimes given to this mode of operation. Many CP ventilators have this ability for their patients.

Because of the compressible aspect of gases and the compliance of ventilator system tubing, as much as 3–5 mL/cm H_2O can be compressed in the ventilator system. For patients with a small tidal volume, this amount can have an effect on ventilator performance [26].

The pressure–time graph illustrates the changes in the airway pressure in unit of cmH_2O, or mbar. For a volume-controlled ventilation and constant flow rate (flow), the airway pressure can be affected by resistance and compliance amounts. Fig. 5.16 shows a pressure–time graph for volume-controlled ventilation. The pressure increases between points A and B at the beginning of inspiration. The pressure is product of resistance and flow rate. After point B the pressure rises in a straight line, until point C or the peak pressure. At point C the ventilator applies the tidal volume and no further flow is delivered ($\dot{V} = 0$). The pressure gradient is dependent on the inspiratory flow rate and the overall compliance C [27],

$$\frac{\Delta p}{\Delta t} = \frac{\dot{V}}{C} \tag{5.5}$$

The pressure "p" drops to plateau pressure. This pressure drop is equal to the increase in pressure amounts caused by the resistance at the beginning of the inspiration period. The connecting line between points A and D runs parallel to the line B–C, the plateau pressure is related to the compliance and the tidal volume. Due to lung recruitment and leaks in the system, there is a slight decrease in pressure (points D to E). During the plateau period no volume is delivered to the lung, and inspiratory flow

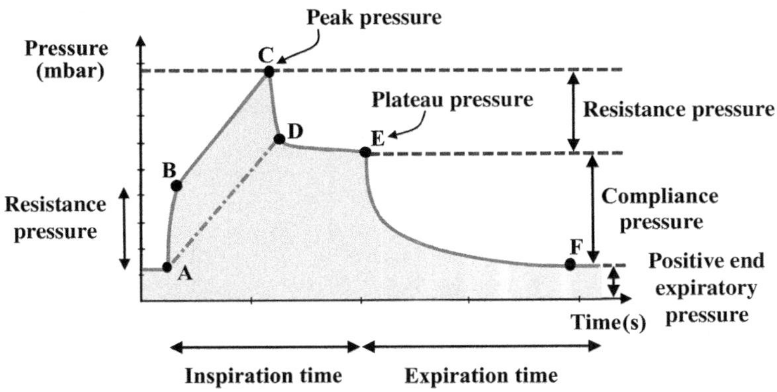

■ **FIGURE 5.16** Diagram of pressure—time for volume-controlled constant flow ventilation. *From F. Rittner, M. Doring, Curves and loops in mechanical ventilation, [Online]. http://www.draeger.com/sites/assets/PublishingImages/Products/savina-300/UK/9097421-Fibel-Curves-Loops-en.pdf, 2012.*

rate is zero. In relation to the elastic recoil forces of the thorax, expiration begins at point E as a passive process and air is forced against atmospheric pressure out of the lung. The compliance pressure or variance between plateau pressure, E, and positive end expiratory pressure (PEEP), F, is achieved by dividing tidal volume (V_T) by compliance C. After finishing the expiration phase, the pressure again reaches the positive end expiratory level (PEEP) at F point [27].

Fig. 5.17 shows a comparison between pressure, flow and volume for volume- and pressure-oriented ventilation (eg, PCV/BIPAP). These variables define the performance of a mechanical ventilator. In pressure-oriented ventilation, the pressure graphs are quite different. There are four phases to a mechanical ventilation: flow phase of inspiration; pause phase of inspiration; flow phase of expiration and pause phase of expiration. During the flow phase of inspiration, the volume increases but it remains constant during the flow pause (plateau line), which means there is no further volume entering the lung. The value of the maximum volume means the relocated tidal volume to the lung and does not denote the whole volume in the lung. During the expiration period, the transferred volume decreases due to the passive form of exhalation. The flow (flow rate) is the volume of gas which passes per unit time. During inspiration the flow amount rises to achieve the amount set on the ventilator and remains constant until the tidal volume VT, similarly set on the ventilator, has been delivered (the square part under the curve). At the beginning of the pause phase (plateau line) the flow drops down to zero. The expiratory flow begins at the end of the pause phase, the progress of which depends on the resistances of the ventilation system and on features of the lung and its airways. Normally,

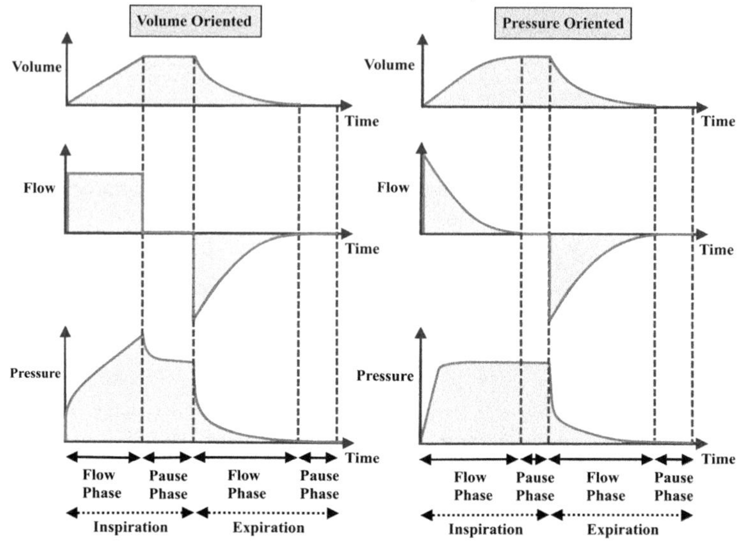

■ **FIGURE 5.17** Diagrams of pressure, flow and volume for volume-oriented and pressure-oriented ventilation. *From F. Rittner, M. Doring, Curves and loops in mechanical ventilation, [Online]. http://www.draeger.com/sites/assets/PublishingImages/Products/savina-300/UK/9097421-Fibel-Curves-Loops-en.pdf, 2012.*

constant flow is an aspect of a classic volume-oriented mode of ventilation and decelerating flow is an aspect of a pressure-oriented ventilation mode. The amounts of pressure have a relation to the volume in the lung. For instance, as the filling volume in the lung increases the pressure in the lung also rises. But this relationship is not governed between pressure and flow during inspiration, see Fig. 5.17. It is proved that changes in the value of compliance have a reverse effect on the peak and plateau pressures values, but the resistance has a direct relationship on the pressures in the mechanical ventilation [27].

There are several variables which collaborate to define the governing equations of the ventilation, such as tidal volume, inspiratory flow (flow rate), pressure gradient, peak pressure, PEEP, etc. Table 5.2 represents the mathematical relation among all operational variables in mechanical ventilation.

5.4 ARTIFICIAL KIDNEY AND HEMODIALYSIS MACHINE

History: In 1943, Willem Kolff constructed the rotating-drum hemodialysis system, originally made from a wooden core. He used cellophane membranes and an immersion bath in his design. The first recovery of an acute renal failure patient treated with hemodialysis was reported in that

Table 5.2 The Govern Equations Relating the Variables in the Mechanical Ventilation

Variables			
f	Frequency (breath/min)	T_E	Expiratory time (s)
P	Pressure	T_I	Inspiratory time (s)
V	Volume	$e \cong 2.72$	Euler's number (Natural logarithm base)
t	Time (s)	V_D	Physiologic dead space volume
τ	Time constant (s)	PIP	Peak inspiratory pressure
$\dot{V}$	Flow rate (L/s)	PEEP	Positive end expiratory pressure

Mode	Parameter	Symbol	Equation
Pressure control	Tidal volume (L)	V_T	$V_T = \Delta P \times C \times (1 - e^{-t/\tau})$
	Pressure gradient (cmH_2O)	ΔP	$\Delta P = PIP - PEEP$
	Inspiratory flow rate (L/s)	$\dot{V}_I$	$\dot{V}_I = \left(\frac{\Delta P}{R}\right)(e^{-t/\tau})$
Volume control	Tidal volume (L)	V_T	$V_T = \frac{\dot{V}_E}{f} = \overline{\dot{V}_I} \times T_I$
	Mean inspiratory flow rate (L/min)	$\overline{\dot{V}_I}$	$\overline{\dot{V}_I} = 60 \times \frac{V_T}{T_I}$
Pressure and volume controls	Minute alveolar ventilation (L/min)	$\dot{V}_A$	$\dot{V}_A = f \times (V_T - V_D)$
	Minute ventilation exhaled (L/min)	$\dot{V}_E$	$\dot{V}_E = f \times V_T$
	Expiratory volume (L)	V	$V = C\Delta P(e^{-t/RC})$
	Ventilator frequency (cycle/min)	f	$f = \frac{60}{(T_I + T_E)} = \left(\frac{60}{I}\right)\left(\frac{I}{I+E}\right)$
	Time constant (s)	τ	$\tau = R \times C$
	Total cycle time or ventilator period (s)	TCT	$TCT = T_I + T_E$ $TCT = \frac{60}{f}$
	Expiratory flow (L/s)	$\dot{V}_E$	$V = -\left(\frac{\Delta P}{R}\right)(e^{-t/\tau})$
	Clastance (cmH_2O/L)	E	$E = \frac{1}{C}$
	Compliance (L/cmH_2O)	C	$C = \frac{\Delta V}{\Delta P}$
	Resistance (cmH_2O/L/s)	R	$R = \frac{\Delta P}{\Delta \dot{V}}$
	Ratio of inspiratory time to expiratory time	$I:E$	$I:E = \frac{T_I}{T_E}$
	Mean airway pressure (cmH_2O)	$MAWP$	$MAWP = \frac{f \times T_I}{60} \times (PIP - PEEP) + PEEP$

Modified from [28].

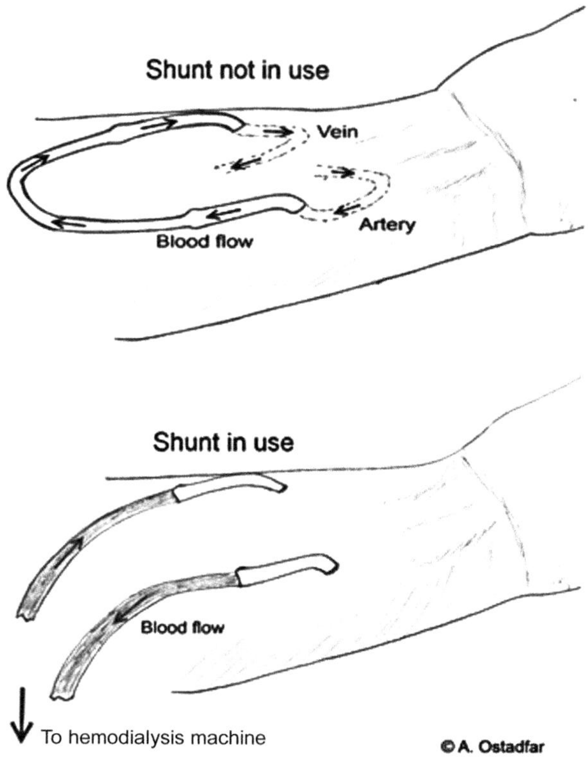

■ **FIGURE 5.18** Schematic of Quinton and Scribner shunt method. *From A. Ostadfar, Design and experimental proof of selected functions in implantable artificial kidney, [Online]. http://summit.sfu.ca/system/files/iritems1/13866/etd8164_AOstadfar.pdf, 2013 with permission.*

year [29–31]. Kolff is considered to be the Father of Artificial Organs, and he is regarded as one of the most important physicians of the 20th century [29,32,33]. In 1946, Nils Alwall introduced a new method with a vertical stationary drum kidney and circulating dialysate around the membrane [29,34]. He also was the inventor of the arteriovenous shunt for dialysis. Dr Alwall collaborated with Swedish businessman Holger Crafoord to establish one of the famous companies that would manufacture hemodialysis equipment over the past 50 years, Gambro. The compact lightweight AK 96 self-care machine is one of the new generations of hemodialysis machines for in-house treatment [29,35].

In 1960, a major breakthrough occurred with the creation of an exterior Teflon bypass by Quinton, Scribner and Dillard [29,36]. This arteriovenous shunt provided for the continuous circulation of the blood when the patients were not connected to the dialysis machine. It gives ready access for long-term hemodialysis. This invention opened the new horizon to chronic renal disease therapy, see Fig. 5.18.

Fig. 5.19 illustrates the essential elements and a blood flow diagram of a hemodialysis machine. Whole blood enters the machine from the vascular entrance via the arterial blood line; it passes through the dialyzer and the blood is returned to the patient via the venous blood line. Dialysate is prepared and fed into the dialyzer, where water and substance exchange take place. After passing through the dialyzer, the dialysate is dumped into a drain. The blood circuit is separated from the dialysate circuit; they are only in indirect contact in the dialyzer. Fig. 5.19 shows the basic components in a typical hemodialysis machine. Usually, blood flow in hemodialysis machines is at the rate of 200–600 mL/minutes [29].

5.4.1 **Hemodialysis Membranes and the Dialyzer**

The performance of the dialyzer membrane depends on the biomaterial used, its pore size and density, thickness, biocompatibility, hydraulic permeability and hydrophilic/hydrophobic properties [29,37].

To use membranes for applications such as hemodialysis, the membranes need to meet the requirements discussed below.

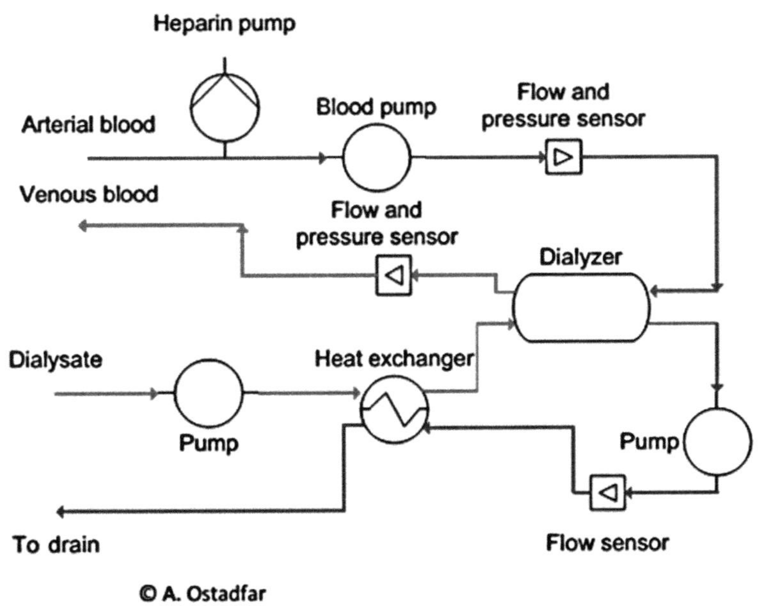

■ **FIGURE 5.19** Schematic flow diagram and basic components in typical hemodialysis machine. *From A. Ostadfar, Design and experimental proof of selected functions in implantable artificial kidney, [Online]. http://summit.sfu.ca/system/files/iritems1/13866/etd8164_AOstadfar.pdf, 2013 with permission.*

5.4.2 Membrane Biocompatibility

One of the most important aspects of a hemodialysis membrane is biocompatibility; that is, the compatibility of the dialyzer membrane in contact with blood. Three major concerns are vital in hemodialysis biocompatibility: clotting, complement activation and adsorption of endotoxins and cytokines [29].

- *Clotting*: Clotting is usually controlled through the use of systemic heparinization, although a small percentage (<5%) of dialysis treatments are performed heparin-free due to allergic reactions to heparin [29,37]. Heparin-coating is an alternative method to prevent clotting, and Heparin-coated dialyzers were introduced several years ago to protect the membrane from clotting [29,38].
- *Complement activation*: Proteins of complement are so-named because they complement antibody activity to remove pathogens. Complement activation during hemodialysis was first recognized by the rapid drop in white blood cell counts (neutropenia) in the first 30 minutes of dialysis. Regenerated cellulose membranes activate complement through the alternate pathway [25,29,39]. Modified cellulose membranes approach the biocompatibility profile of synthetic materials in terms of complement activation and neutropenia (an abnormally low number of neutrophils—a type of white blood cells) [29,37].
- *Adsorption of endotoxins and cytokines*: Endotoxins are bacterial products released from Gram-negative bacteria upon death. Because endotoxin fragments fall in the middle molecule range, they may be inadvertently transported from the dialysate to blood during high-flux (flow) dialysis, leading to cytokine generation that adds to the chronic inflammatory state of dialysis patients [29,37].

The open pore in high-flux membranes affords more adsorptive potential than do low-flux counterparts. In addition, synthetic membranes, many of which are fundamentally hydrophobic, normally are much more adsorptive than hydrophilic cellulosic membranes [29,40].

The adsorption features, distribution of hydrophobic and hydrophilic areas, and charge distribution on the surface and in the pores are significant factors that determine membrane biocompatibility [29].

5.4.3 Membrane Materials

Dialyzer membranes can be categorized based on their chemical compositions as either cellulosic, modified cellulosic or synthetic [29,37].

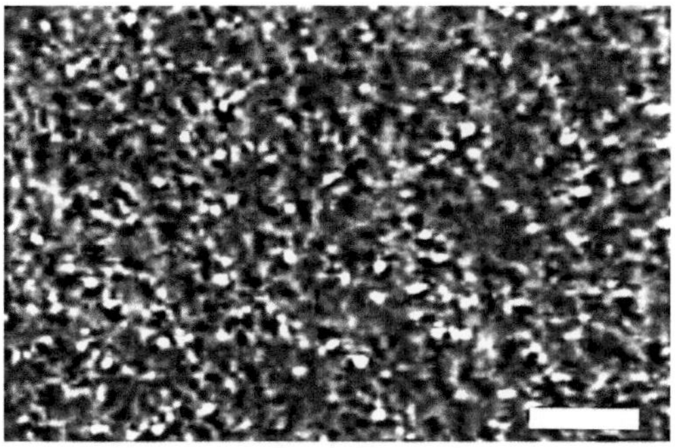

■ **FIGURE 5.20** Cross-section of a synthetic hemodialysis membrane, scale bar is 10 μm. *From A. Ostadfar, Design and experimental proof of selected functions in implantable artificial kidney, [Online]. http://summit.sfu.ca/system/files/iritems1/13866/etd8164_AOstadfar.pdf, 2013 with permission.*

- *Cellulosic*: These membranes were exclusively used in the 1940s through the 1960s. Cellulosic membranes are polysaccharide membranes derived from cotton linters. Activating the complement cascade makes these membranes bio-incompatible [29,41]; however, because of their lower cost in some parts of world, medical systems still use them.
- *Modified cellulosic*: Modified cellulosic membranes are made more biocompatible by substituting the hydroxyl groups with other moieties or by coating the membrane with a biocompatible coating [29].
- *Noncellulosic synthetic membranes*: Currently, these membranes are very common in the dialysis field; the term "synthetic membrane" is used to denote all polymeric membranes that are not cellulose based. Fig. 5.20 illustrates a synthetic membrane. The most used membranes in this category are polysulfone, polycarbonate and polyurethane [29].

5.4.4 **Hollow Fibers**

The first membrane materials were saponified cellulose acetate and Cuprophan [29,42]. The idea of the Cuprophan hollow fiber was visualized in 1969 by Heinz Ruck [29,43], which revolutionized dialysis technology. Currently, the hollow fiber module is a hollow fiber dialyzer which consists of semipermeable polymeric capillaries sealed inside a tubular cartridge. Fig. 5.21 shows cross-sections of a hollow fiber. The hollow fiber has three regions: lumen of the feeding line, membrane and extra-capillary space (ESC) for permeate (filtrated) fluid [29].

The schematic of a hollow fiber dialyzer is illustrated in Fig. 5.22. This figure also illustrates a cutaway view of a regular dialyzer which is used in hemodialysis machines.

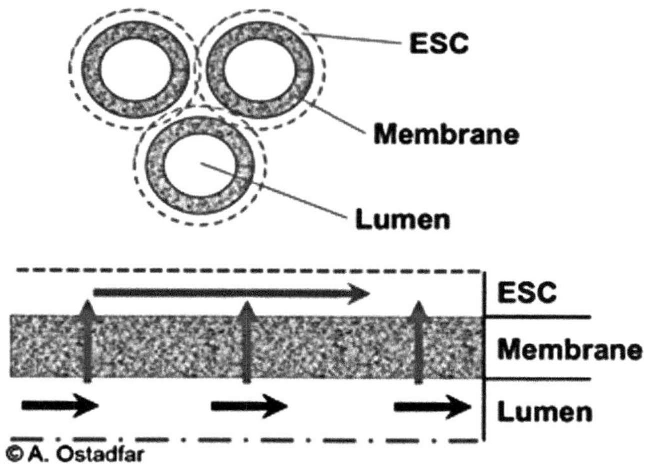

■ **FIGURE 5.21** Cross-sections of the hollow fiber. Top: Radial cross-section. Bottom: Longitudinal cross-section. *From A. Ostadfar, Design and experimental proof of selected functions in implantable artificial kidney, [Online]. http://summit.sfu.ca/system/files/iritems1/13866/etd8164_AOstadfar.pdf, 2013 with permission.*

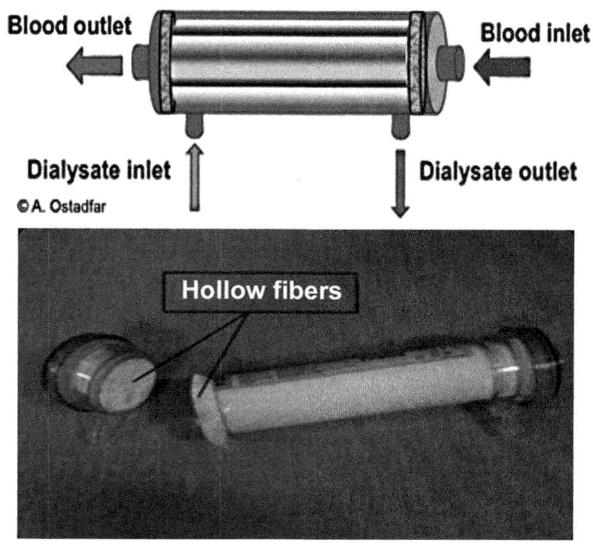

■ **FIGURE 5.22** Hollow fiber dialyzer. Top: Schematic of dialyzer. Bottom: Cutaway view of polysulfone dialyzer module. *From A. Ostadfar, Design and experimental proof of selected functions in implantable artificial kidney, [Online]. http://summit.sfu.ca/system/files/iritems1/13866/etd8164_AOstadfar.pdf, 2013 with permission.*

The thickness of the membrane wall is in the range of 50–100 μm. Double-layer asymmetric hollow fibers [44] are also commercially available. Typical internal diameters for hollow fibers are from approximately 100 μm to 1.5 mm [29].

The ideal membrane design for hemodialysis is the hollow fiber, which has the benefit of a very high membrane area for a given volume. A large fraction of microporous membranes are still made of cellulose derivatives, but several other synthetic polymers are being used, including polysulfone, polycarbonate, polyamide and polyacrylonitrile [45]. The mechanism of filtration in a dialyzer and hemodialysis machine is ultrafiltration to separate waste products from whole blood.

5.5 ARTIFICIAL VESSELS AND GRAFTS

Vascular grafting refers to using transplanted or prosthetic blood vessels to redirect blood flow in a failure region of the body. Grafting procedures can be applied to stenosed arteries by surgery. There are several procedures for grafting [46]:

- Replacement (end to end anastomosis) is normally done by sutures between ends of both host and graft vessels.
- End to side anastomosis or side to end anastomosis.
- Side to side anastomosis offers an alternate method for the blood to supply tissues by bypassing the narrowed or blocked arterial vessels.

Fig. 5.23 shows the blood vessel grafts for distal aortic anastomosis.

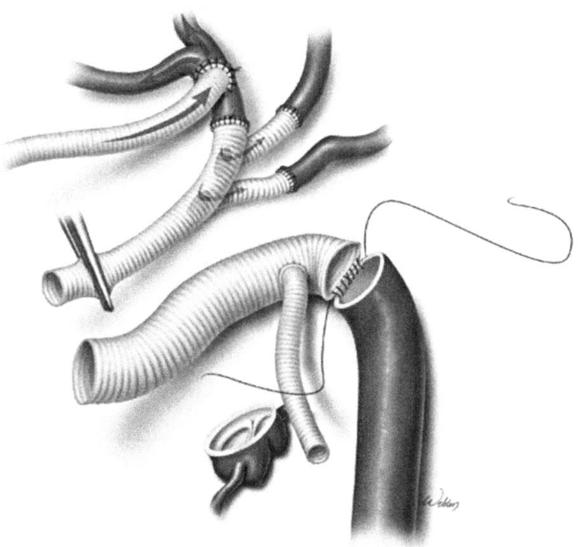

■ FIGURE 5.23 Schematic of distal aortic anastomosis. *From S.A. LeMaire, S.A. Weldon, J.S. Coselli, Total aortic arch replacement: current approach using the trifurcated graft technique, Ann. Cardiothorac. Surg. 2 (3) (2013) 347–352 with permission.*

Vascular grafts can be manufactured from: (1) a synthetic tube, polyethyleneterphtalate (dacron) from woven or knitted fabric material, or expanded polytetrafluoroethylene; (2) homografts (autografts) using homologous (autologous) veins or arteries (eg, saphenous vein for femoral artery) or (3) heterologous vessel (eg, bovine carotid). The blood flow models in artificial vessels and grafts depend on anastomosis methods (parallel side to side, diamond side to side and end to side) and graft/host diameter ratio. Numerical modeling of the blood flow has a significant role to minimize flow turbulences and treatment failures (thrombosis or restenosis) for surgical reconstruction (endarterectomy or bypass) of arterial stenosis [46].

Pressure drop of the graft depends on the blood flow rate and geometry of graft such as diameter and length. Generally, Navier–Stokes equations are the best method to model the blood flow through the tubes, but the simplest approach for calculating pressure drop is Poiseuille's formula as a derived equation for fully developed laminar flow:

$$\Delta P_{\mathrm{G}} = \frac{8\,\mu\,L_{\mathrm{G}}Q}{\pi\,R_{\mathrm{G}}^{4}} \tag{5.6}$$

where ΔP_{G} is graft pressure drop, μ is viscosity, L_{G} is graft length, Q is the blood flow rate and R_{G} is graft radius.

As an alternative, Blasius' equation for fully developed turbulent flow can be considered [48]:

$$\Delta P_{\mathrm{G}} = \frac{0.0665\,\rho\,L_{\mathrm{G}}\nu^{1/4}}{\pi^{7/4}\,R_{\mathrm{G}}^{4.75}}Q^{7/4} \tag{5.7}$$

where ν is the kinematic viscosity and ρ is the blood density, approximately 1060 kg/m^3.

5.6 STENT

Stent is a support element (a mesh tube) to create an open passage through the blood vessels which are narrow or weak. The stent can be used for the heart and coronary vessels, kidney and urinary system.

Stent procedure: A balloon catheter over a guide wire transfers the stent into the narrowed artery. Once in place, the balloon is inflated and the stent expands and holds the artery open permanently, then the balloon is deflated and removed from the artery. Fig. 5.24 illustrates a stent, catheter and guide wire in a coronary artery.

To analyze the flow aspects, Navier–Stokes equations are one of the best approaches to model the blood flow through the stent blood vessel.

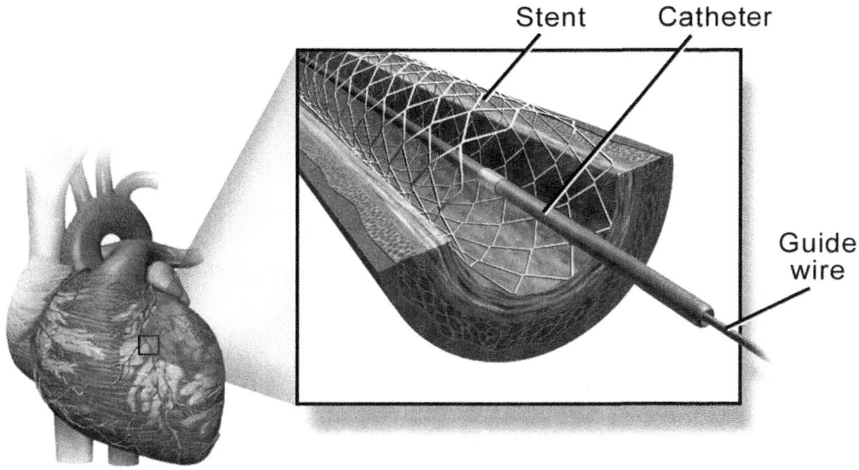

■ **FIGURE 5.24** Schematic of stent after installation and deflation of balloon. *From B. Staff, Blausen 0034 Angioplasty Stent 01.png, Blausen.com, [Online]. https://commons.wikimedia.org/wiki/File:Blausen_0034_Angioplasty_Stent_01.png, 2004 with permission.*

The design and material aspects of the stent have an effect on the blood flow and blood components. The fluid–solid interaction between stent wires and blood flow changes wall shear stress (WSS) amounts, particularly between stent struts. The studies revealed that the stent geometry alters the phase shift between pressure and flow waveforms; additionally the Womersley number has effects on the phase shift. The results showed the variation of the phase shift (φ) between pressure and flow waveform created by different types of stent in four major sections: inlet of artery; proximal; middle and distal portions of the stents [50]. Analytically, there are two methods to calculate phase shift between the blood flow rate and pressure for pulsatile flow in the rigid tubes: the Womersley solution and the Fry solution [51].

Other studies showed that the presence of stent struts increases the WSS also leads to the formation of a recirculation region at the immediate downstream of strut intersection [52]. Generally, the study's results and discussions were done for selected types of stent and the conclusions are related to the geometry of experimented stents and also the biological characters of the blood vessel.

CHAPTER SUMMARY

- *Blood pumps* can be classified into three major types: displacement; rotary and TAH pumps.

- *Rotary pumps* can be classified into three subsystems: radial (centrifugal); axial and diagonal (mixed flow) pumps.
- *Radial pumps* are ideal for producing high pressures and low flows, whereas the *axial pumps* produce high flow rates at low pressure difference.
- The *roller pump* falls under the category of positive displacement pumps and has been the *most commonly used pump* for CPB for the past 50 years.
- The *volume of pumped blood* depends on the size of tubing, length of the track and the RPM of motor.
- The *blood flow rate* and *pressure (pump head)* are most important aspects among pump features (flow rate, pressure, rate of hemolysis, operation time).
- A *TAH* is a device that replaces the two ventricles of the heart.
- To restore normal function of heart valve, a repair surgery or replacement of the diseased valve with a prosthetic heart valve is necessary.
- The *prosthetic heart valves* are categorized into two main types: *mechanical and bioprosthetic (tissue) valves.*
- The best *advantage* of *mechanical valves* is more reliability or long durability and the main *disadvantage* of them is the continuous use anticoagulation therapy to decrease the risk of thrombosis and thromboembolic problems.
- The main *advantage* of *tissue valves* is the lower rate of thromboembolic problems and its main *disadvantage* is large pressure drops.
- *Pressure drop* has a main role in prosthetic heart valve design and it occurs when viscous blood flows through the heart valve and it shows *heart valve efficiency.*
- For prosthetic heart valves, the *EOA* can be used to show the quality and efficiency of the valve.
- The *Navier–Stokes equations* can be introduced as the *best governing equations* for the biofluid mechanics of heart valves.
- An *AL* is a device that delivers enough oxygen to blood and removing carbon dioxide from blood.
- The ALs are *classified* as extracorporeal, paracorporeal, intravascular and intrathoracic.
- *ECMO* equipment include: biocompatible tubing, AL or membrane oxygenator, heat exchanger (to adjust the blood temperature), roller pump, oxygen cylinder and drug infusion (heparin or fluid) ports.
- For a volume-controlled ventilation and constant flow rate (flow), the airway pressure can be affected by resistance and compliance amounts.

- The performance of the *dialyzer membrane* depends on the biomaterial used, its pore size and density, thickness, biocompatibility, hydraulic permeability and hydrophilic/hydrophobic properties.
- In an *artificial kidney* (hemodialysis machine), whole blood enters the machine from the vascular entrance via the arterial blood line; it passes through the dialyzer and the blood is returned to the patient via the venous blood line. Dialysate is prepared and fed into the dialyzer, where water and substance exchange take place. After passing through the dialyzer, the dialysate is dumped into a drain.
- In hemodialysis machines, the *blood circuit is separated from the dialysate circuit*; they are only in indirect contact in the dialyzer.
- The *hollow fiber has three regions*: lumen of the feeding line; membrane and ESC for permeate (filtrated) fluid.
- *Vascular grafting* refers to using transplanted or prosthetic blood vessels to redirect blood flow in a failure region of body.
- Generally for vascular grafting, *Navier–Stokes equations* are the best method to model the blood flow through the graft, but the simplest approach for calculating pressure drop is using *Poiseuille's formula* (Eq. (5.6)) as a derived equation for fully developed *laminar flow*. As an alternative, *Blasius' equation* (Eq. (5.7)) for fully developed *turbulent flow* can be considered.
- *Stent* is a support element (a mesh tube) to create an open passage through blood vessels which are narrow or weak.
- The design and material aspects of stent effect on the blood flow and blood components.
- The studies revealed that the *stent geometry* alters *the phase shift* between pressure and flow waveforms; additionally the *Womersley number* has effects on *the phase shift*.

PROBLEMS

1. Classify the blood pumps and explain which one has the main role in blood transportation?
2. Which characteristics are important in selecting a blood pump?
3. Explain the function and compare both the SynCardia and the Abiomed TAH.
4. Classify the prosthetic heart valves and compare their advantage and disadvantage.
5. Describe three configurations of ECMO tubing.
6. Discuss about Fig. 5.16 and *AB*, *BC*, *CD*, *DE*, and *EF* paths.
7. What is the duty of a dialyzer and explain its construction.

REFERENCES

[1] Deleted in review.

[2] Deleted in review.

[3] S.A. Raskin, Y. Nose, E. Tayama, Blood pumps, Cardiopulmonary Bypass: Principles and Practice, second ed., Lippincott Williams & Wilkins, Philadelphia, PA, 2007, pp. 35–46.

[4] Deleted in review.

[5] Deleted in review.

[6] B.D. Gaitan, C.A. Thunberg, L.G. Stansbury, D.E. Jaroszewski, F.A. Arabia, B.P. Griffith, et al., Development, current status, and anesthetic management of the implanted artificial heart, J. Cardiothorac. Vasc. Anesth. 25 (6) (2011) 1179–1192.

[7] The Syncardia™ total artificial heart: in vivo, in vitro, and computational modeling studies. J Biomech. 2013 January 18; 46(2): 266–275. Available from: http://dx.doi.org/10.1016/j.jbiomech.2012.11.032.

[8] I. ABIOMED, AbioCor® implantable replacement heart, instructions for use, [Online]. <http://www.fda.gov/ohrms/dockets/ac/05/briefing/2005-4149b2_01_ABIOMED%20INSTRUCTIONS%20FOR%20USE.pdf>. (accessed 06.06.15).

[9] M.S. Sacks, A.P. Yoganathan, Heart valve function: a biomechanical perspective, Phil. Trans. R. Soc. B 362 (2007) 1369–1391.

[10] P. Pibarot, J.G. Dumesnil, Prosthetic heart valves, selection of the optimal prosthesis and long-term management, Circulation 119 (2009) 1034–1048.

[11] A.P. Yoganathan, Z. He, S.C. Jones, Fluid mechanics of heart valves, Annu. Rev. Biomed. Eng. 4 (2004) 331–362.

[12] H. Mohammadi, K. Mequanint, Prosthetic aortic heart valves: modeling and design, Med. Eng. Phys. 33 (2011) 131–1447.

[13] L.-G. Durand, D. Garcia, F. Sakr, H. Sava, R. Cimon, P. Pibarot, et al., A new flow model for Doppler ultrasound study of prosthetic heart valves, J. Heart Valve Dis. 8 (1) (1999) 86–95.

[14] L.P. Dasi, H.A. Simon, P. Sucosky, A.P. Yoganathan, Fluid mechanics of artificial heart valves, Clin. Exp. Pharmacol. Physiol. 36 (2009) 225–237.

[15] R.B. Van den Brink, Evaluation of prosthetic heart valves by transesophageal echocardiography: problems, pitfalls, and timing of echocardiography, Semin. Cardiothorac. Vasc. Anesth. 10 (1) (2006) 89–100.

[16] K.B. Chandran, Heart valve prostheses, Encyclopedia of Medical Devices and Instrumentation, John Wiley & Sons, Inc., New York, 2006, pp. 407–426.

[17] A.P. Yoganathan, Cardiac valve prostheses, Biomedical Engineering Handbook, CRC Press LLC, Boca Raton, FL, 2000.

[18] Health Resources and Services Administration, Healthcare Systems Bureau, Division of Transplantation, Annual report of the US Organ Procurement and Transplantation Network and the Scientific Registry of Transplant Recipients: transplant data 1997–2006, Health Resources and Services Administration, Healthcare Systems Bureau, Division of Transplantation, Rockville, MD, 2007.

[19] W.J. Federspiel, R.G. Svitek, Lung, artificial: current research and future directions, Encyclopedia of Biomaterials and Biomedical Engineering, second ed., Marcel Dekker, Inc., New York, 2004, pp. 922–931.

[20] D. Hayes Jr., J.D. Tobias, J. Kukreja, T.J. Preston, R. Yates, S. Kikby, et al., Extracorporeal life support for acute respiratory distress syndrome, Ann. Thorac. Med. 8 (3) (2013) 134–141.

[21] R.L. Brunston Jr., J. Zwischenberger, W. Tao, V.J. Cardenas, D. Traber, A. Bidani, Total arteriovenous CO2 removal: simplifying extracorporeal support for respiratory failure, Ann. Thorac. Surg. 64 (1997) 1599–1604.

[22] J.B. Zwischenberger, S. Conrad, L. Grier, A. Bidani, Percutaneous extracorporeal arteriovenous CO2 removal for severe respiratory failure, Ann. Thorac. Surg. 68 (1999) 181–187.

[23] S.N. Vaslef, L.F. Mockros, R.W. Anderson, Development of an intravascular lung assist device, Am. Soc. Artif. Intern. Organs 35 (3) (1989) 660–664.

[24] B.G. Hattler, L.W. Lund, J. Golob, H. Russian, F. Lann, T.L. Merrill, et al., A respiratory gas exchange catheter: in vitro and in vivo tests in large animals, J. Thorac. Cardiovasc. Surg. 124 (2002) 520–530.

[25] J.B. West, Pulmonary Pathophysiology: The Essentials, Lippincott Williams & Wilkins, a Wolters Kluwer Business, Baltimore, MD, Philadelphia, PA, 2013.

[26] D.R. Hess, N.R. MacIntyre, Mechanical ventilation, Respiratory Care: Principles and Practice Hardcover, Jones & Bartlett Learning LLC, Sudbury, MA, 2011, pp. 462–500.

[27] F. Rittner, M. Doring, Curves and loops in mechanical ventilation, [Online]. <http://www.draeger.com/sites/assets/PublishingImages/Products/savina-300/UK/9097421-Fibel-Curves-Loops-en.pdf>, 2012. (accessed 06.08.15).

[28] R.L. Chatburn, Fundamental of Mechanical Vntilation—A Short Course on the Theory and Application of Mechanical Ventilators, Mandu Press Ltd., Cleveland Heights, OH, 2003.

[29] A. Ostadfar, Design and experimental proof of selected functions in implantable artificial kidney, [Online]. <http://summit.sfu.ca/system/files/iritems1/13866/etd8164_AOstadfar.pdf>, 2013. (accessed 06.09.15).

[30] W. Kolff, H. Berk, De kunstmatige nier: een dialysator met groot oppervlak, Tijdschr. Geneeskd. 87 (1943) 1684.

[31] W. Kolff, H. Berk, The artificial kidney: a dialyzer with a great area, Acta Med. Scand. 117 (1944) 121–134.

[32] T.S. Ing, M. Rahman, M. Kjellstrand, Dialysis: History, Development and Promise, World Scientific Publishing Co. Pte. Ltd., Singapore, 2012.

[33] C.A. Moore, Kolff, 'father of artificial organs,' dies at 97, Deseret. News (2009).

[34] N. Alwall, On the artificial kidney I: apparatus for dialysis of blood in vivo, Acta Med. Scand. 28 (1947) 317–325.

[35] G. Company, Monitors-AK-96, Gambro, 2011, [Online]. <http://www.gambro.com/en/global/Products/Hemodialysis/Monitors/AK-96/>. (accessed 06.09.15).

[36] W. Quinton, D. Dillard, B. Scribner, Cannulation of blood vessels for prolonged hemodialysis, Trans. ASAIO 6 (1960) 104–107.

[37] N.N. Li, A.G. Fane, W.S. Winston Ho, T. Matsuura, Advanced Membrane Technology and Applications, John Wiley & Sons, Inc., New York, 2008.

[38] S. Lavaud, E. Canivet, A. Wuillai, H. Maheut, C. Randoux, J.-M. Bonnet, et al., Optimal anticoagulation strategy in haemodialysis with heparin-coated poly-acrylonitrile membrane, Nephrol. Dial. Transplant. 18 (2003) 2097–2104.

[39] D.E. Chenoweth, A. Cheung, L. Henderson, Anaphylatoxin formation during hemodialysis: effects of different dialyzer membranes, Kidney Int. 24 (6) (1983) 764–769.

[40] W.R. Clark, W. Macias, B. Molitoris, N. Wang, Plasma protein adsorption to highly permeable hemodialysis membranes, Kidney Int. 48 (2) (1995) 481–488.

[41] R.M. Hakim, D. Fearon, J. Lazarus, Biocompatibility of dialysis membranes: effects of chronic complement activation, Kidney Int. 26 (1984) 194–200.

[42] A.R. Nissenson, R.N. Fine, Clinical Dialysis, fourth ed., McGraw-Hill Professional, New York, 2005.

[43] H. Ruck, Application of cellulose products in medicine, pharmaceutics and hygiene. Part 1, das papier 40 (30) (1986) 93–102.

[44] D. Wang, K. Li, W. Teo, Preparation of annular hollow fibre membranes, J. Membr. Sci. 166 (1) (2000) 31–39.

[45] F. Schiith, S.K.W. Sing, J. Weitkamp, Handbook of Porous Solids, Wiley-VCH, New York, 2002.

[46] M. Thiriet, Biology and Mechanics of Blood Flows, Part II: Mechanics and Medical Aspects, Springer Science + Business Media, LLC, New York, 2008.

[47] Deleted in review.

[48] S.A. Jones, S. Jin, A. Kantak, D.A. Bell, W.D. Paulson, Mathematical model for pressure losses in the hemodialysis graft vascular circuit, Biomech. Eng. 127 (1) (2005) 60–66.

[49] Deleted in review.

[50] V. Dehlaghi, S. Najarian, M.T. Shadpour, Effect of stent geometry on phase shift between pressure and flow waveforms in stented human coronary artery, Am. J. Appl. Sci. 5 (4) (2008) 340–346.

[51] L. Waite, J. Fine, Applied Biofluid Mechanics, McGraw-Hill Companies, New York, 2007.

[52] D. Rajamohan, R.K. Banerjee, L.H. Back, Developing pulsatile flow in a deployed coronary stent, J. Biomech. Eng. 128 (2006) 347–359.

Chapter 6

Physical Injury to Blood

LEARNING OBJECTIVES

- Description of the geometry and membrane of RBCs.
- Characterization of deformation and mechanical properties of RBCs.
- Definition, formulation and classification of hemolysis.
- Explanation of hemolysis in the needles, grafts, hemodialysis machine, oxygenation (lung) machines, total artificial heart (TAH) and artificial heart valves.
- Definition of collection, storage and transportation of whole blood and its elements.
- Explanation of anticoagulant-preservatives.
- Classification of storage temperatures and storage equipment for cold blood chain.
- Transport container (box) for blood products.

6.1 GEOMETRY AND CELL MEMBRANE OF RED BLOOD CELL

Blood cells are categorized into three types in mammals: erythrocytes or red blood cells (RBCs); leukocytes or white blood cells (WBC); and thrombocytes or platelets.

The membrane of RBCs is created by a phospholipid bilayer with protein inclusions, bicarbonate HCO^{-3}, ion channels for chloride Cl^-, sialoglycopeptides (blood groups), glucids, cytoskeleton meshing and mooring proteins. Moreover, a mesh made of spectrin, actin and others is placed at the inner surface of the plasmalemma. It binds to the erythrocyte membrane and creates ankyrin–spectrin bonds [1,2]. These elements form the membrane and shape of the RBC. Fig. 6.1 illustrates the RBC and a cutaway view of its membrane. This figure indicates

Biofluid Mechanics. DOI: http://dx.doi.org/10.1016/B978-0-12-802408-9.00006-5

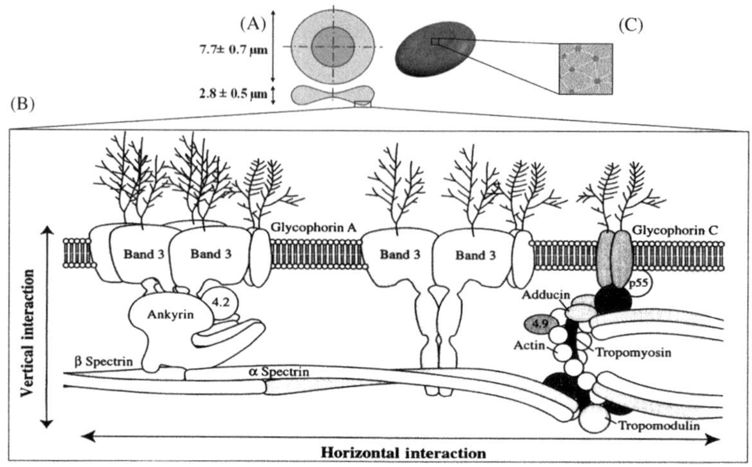

■ **FIGURE 6.1** Cutaway view of a red blood cell membrane. (A) Red blood cell views. (B) RBC membrane. (C) Spectrin network. The bands of 3, 4.2, 4.9, etc. denote the protein elements of membrane. The relative position of the various proteins is correct, but the components are not drawn to scale. *Figure (B) from W.T. Tse, S.E. Lux, Red blood cell membrane disorders, Br. J. Haematol. 104 (1999) 2–13 with permission.*

the vertical and horizontal interaction of the membrane components. There are different membrane proteins in Hereditary Spherocytosis (HS) or Hereditary Elliptocytosis and Hereditary Pyropoikilocytosis (HE/HPP) [3]:

- Vertical interaction, HS: band 3, 20%; protein 4.2, 5%; ankyrin, 45%; b spectrin, 30%.
- Horizontal interaction, HE/HPP: b spectrin, 5%; a spectrin, 80%; protein 4.1, 15%.

Geometry of red cell: The oval of Cassini is the most popular and precise method to describe erythrocytes (RBC).

The Cassini profile depends on the parameters a and c and is generally characterized by evaluating the dimensionless ratio $\varepsilon = a/c$. Fig. 6.2 shows Cassini ovals for several ranges of a and c. In this figure:

$$d_1 d_2 = c^2 \tag{6.1}$$

$$F_1 F_2 = 2a \tag{6.2}$$

$$\varepsilon = \frac{a}{c} \tag{6.3}$$

For RBC application, the range of ε is between $\frac{\sqrt{2}}{2}$ and 1.

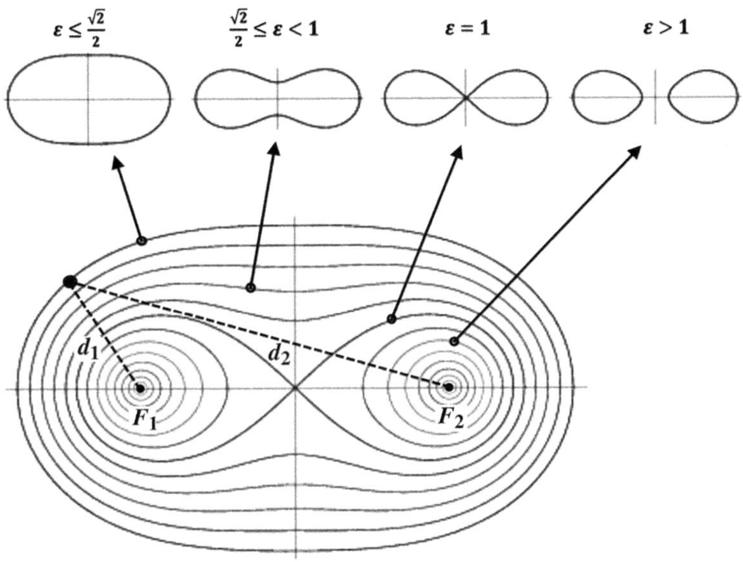

■ **FIGURE 6.2** Cassini ovals with respect to the ratio of ε. *Modified from E.W. Weisstein, Cassini Ovals. From MathWorld—A Wolfram Web Resource, Wolfram Research, Inc., 2015. [Online]. Available: <http://mathworld.wolfram.com/CassiniOvals.html>. with permission.*

The diameter (d) and thickness (t), volume (V), surface area (A) and cross-section area (A_C) of RBC can be denoted by [5];

$$d = 2\sqrt{(a^2+c^2)} = 2c\sqrt{(1+\varepsilon^2)} \tag{6.4}$$

$$t = \frac{c}{\varepsilon} \tag{6.5}$$

$$V = \frac{4}{3}\pi c^3 V(\varepsilon) \tag{6.6}$$

where the factor of $V(\varepsilon)$ is

$$V(\varepsilon) = \frac{\sqrt{(1-\varepsilon^2)}(1+2\varepsilon^2)}{4} + \frac{3\arccos(1-2\varepsilon^2)}{8\varepsilon} \tag{6.7}$$

$$A = 4\pi c^2 A(\varepsilon) \tag{6.8}$$

where the factor of $A(\varepsilon)$ is

$$A(\varepsilon) = \frac{\sqrt{2}}{\varepsilon}\left[E\left(\arccos\sqrt{\frac{(1-\varepsilon^2)}{(1+\varepsilon^2)}}, \sqrt{\frac{(1+\varepsilon^2)}{2}}\right) - \frac{(1-\varepsilon^2)}{2}F\left(\arccos\sqrt{\frac{(1-\varepsilon^2)}{(1+\varepsilon^2)}}, \sqrt{\frac{(1+\varepsilon^2)}{2}}\right)\right] \tag{6.9}$$

Here $F(\psi,k)$ and $E(\psi,k)$ denote the first and the second kind of incomplete elliptic integrals.

$$A_C = \pi c^2 A_C(\varepsilon) \tag{6.10}$$

where the factor of $A_C(\varepsilon)$ is

$$A_C(\varepsilon) = \frac{2E(\varepsilon^2)}{\pi} \tag{6.11}$$

where $E(k) = E\left(\frac{\pi}{2}, k\right)$ is the complete integral of the second kind.

There are some approximate formulas to simplify and calculate RBC geometry parameters [5]. Table 6.1 shows these formulas. Also circumference (length of cross-section) and elastic free energy of the RBC outer membrane are represented in this table too. The table formulation is valid for $\frac{\sqrt{2}}{2} \leq \varepsilon \leq 0.998$.

A recall from chapter "Fluid Mechanics and Biofluids Principles" (Eq. (1.88)) denotes that the average unstressed shape of RBC in Cartesian coordinates (x, y, z) is

$$y = R\sqrt{\left(1 - \frac{x^2+z^2}{R^2}\right)}\left[c_0 + c_1\frac{x^2+z^2}{R^2} + c_2\frac{(x^2+z^2)^2}{R^4}\right]$$

where, $R = 3.91\ \mu m$, $c_0 = 0.0135805$, $c_1 = 1.001279$, $c_2 = -0.561381$.

It is proved that the geometrical shape of Erythrocytes (RBC) optimizes (minimize) the elastic energy of deformation, also it is the best shape to

Table 6.1 Approximate Formulas of RBC Geometry Parameters with Respect to c and ε

Quantity	Expression	Approximation
Volume	$\frac{4}{3}\pi c^3 V(\varepsilon)$	$V(\varepsilon) = (2.88 - 1.7\varepsilon)\varepsilon$
Surface area	$4\pi c^2 A(\varepsilon)$	$A(\varepsilon) = \frac{2.88 - 1.7\varepsilon}{2.82 - 1.98\varepsilon}$
Area of cross-section	$\pi c^2 A_C(\varepsilon)$	$A_C(\varepsilon) = \frac{1.05 - 0.94\varepsilon}{1 - 0.83\varepsilon}$
Circumference	$2\pi c L(\varepsilon)$	$L(\varepsilon) = \frac{2.5 - 2.3\varepsilon}{10 - 9.31\varepsilon}$
Elastic free energy	$\frac{D}{2}U(\varepsilon)$	$U(\varepsilon) = \frac{151.3 - 148.23\varepsilon}{10 - 17.21\varepsilon - 7.25\varepsilon^2}$

From B. Angelov, I.M. Mladenov, On the geometry of red blood cell, Geom. Integr. Quant. 300 (1999) 27–46, doi:10.7546/giq-1-2000-27-46 with permission.
"D" is the bending rigidity constant.

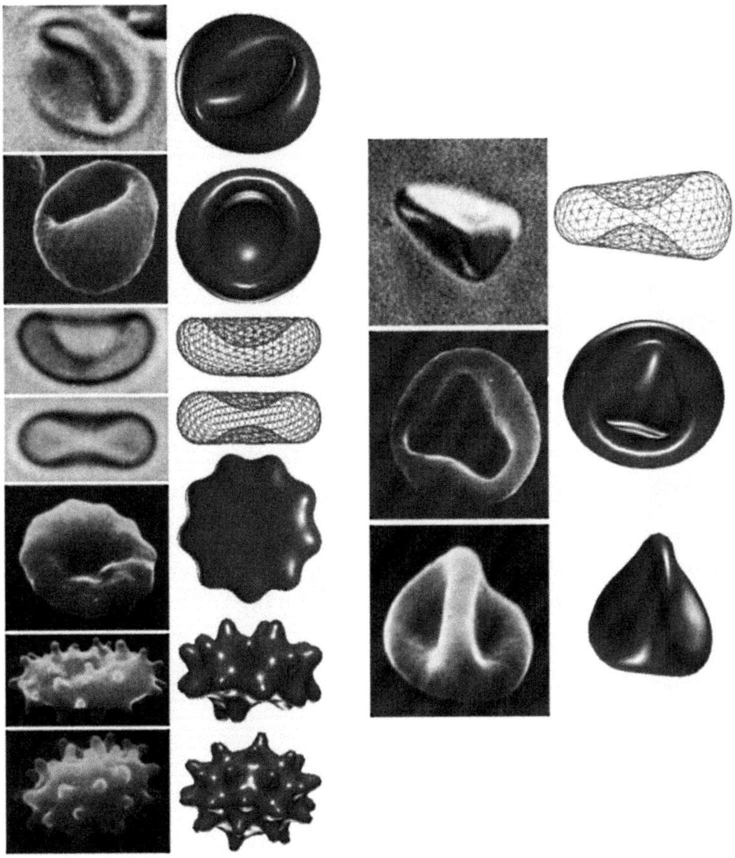

■ **FIGURE 6.3** Deformed shapes of RBC. *From H.W. Gerald Lim, M. Wortis, R. Mukhopadhyay, Stomatocyte—discocyte—echinocyte sequence of the human red blood cell: evidence for the bilayer—couple hypothesis from membrane mechanics, Proc. Natl. Acad. Sci. (PNAS) 99 (26) (2002) 16766—16769 with permission.*

maximize gas transportation and to move through the micro vessels and capillaries. Several geometrical parameters, such as volume, area, diameter and thickness, are associated to define the RBC shape. Fig. 6.3 illustrates the deformed shapes of RBCs.

6.2 MECHANICAL PROPERTIES AND DEFORMATION OF RBC

Mechanical aspects of RBCs are governed by elastic and viscous properties of the cell membrane. These aspects occur due to the viscose role of the lipid bilayer, the elastic role of the spectrin network, the bending resistance factor, and the viscosity of the fluids outside and inside of the RBC. One of the realistic models in analyzing the RBC membrane is a network of springs with viscoelastic character combined with bending energy and limitations for conservation of external area and cell volume. In this model each spring in the network is linked to a single spectrin tetramer between two adjacent actin connections. The multiscale RBC model will be useful to reproduce the realistic mechanical and rheological properties [7].

During oxygen transporting, a healthy RBC must pass through small capillaries with inner diameters smaller than the RBC diameter. Due to this function, the RBC needs to change its shape from its normal biconcave form to a parachute or bullet form. This means a maximum allowable strain on RBC membrane under the local biofluid pressure gradient. Therefore, the RBC mechanical characteristics help us to understand the blood cell deformation and their lysis in vivo and in vitro.

There is a derived formula for the membrane shear modulus in a regular hexagonal network. The formula offers an expression for the shear modulus of the RBC membrane (spectrin network) [8]:

$$\mu_0 \equiv \frac{d\tau_{xy}}{d\gamma} = \frac{\sqrt{3}k_B T}{4pL_{max}x_0}\left\{\frac{3}{4(1-x_0)^2} - \frac{3}{4} + 4x_0 + \frac{x_0}{2(1-x_0)^3}\right\} \tag{6.12}$$

$$x_0 = \frac{L_0}{L_{max}} \tag{6.13}$$

where μ_0 is the initial membrane shear modulus (in units of force per unit length), τ_{xy} is in-plane stress, γ is the shear strain, p is the persistence length, k_B is the Boltzmann constant, T is the temperature, L_{max} is the maximum extension length of the spectrin link. Considering, $p = 8.5$ nm, the equilibrium spring length or $L_0 = 87$ nm, and $L_{max} = 238$ nm, gives linear shear modulus $\mu_0 = 7.98\ \mu N/m$ [8]. This amount is in the normal range of shear module which was previously estimated between 5 and 8 μN/m [9] and experimented between 4 and −9 μN/m [10].

The linear elastic Young's modulus (E) and Poisson's ratio (ν) of the RBC membrane (Spectrin network) is denoted as [8];

$$E = \frac{4K\mu_0}{K + \mu_0} \tag{6.14}$$

$$\nu = \frac{K - \mu_0}{K + \mu_0} \tag{6.15}$$

Table 6.2 The Simulation Results for the Mechanical Parameters of Red Blood Cell Membrane (Spectrin Network)

Shear Modulus (μ_0)	Elastic Area Compression Modulus (*K*)	Young's Modulus (*E*)	Poisson's Ratio (*v*)
~8 μN/m	~16 μN/m	~22 μN/m	1/3 − 1/2

where K is the linear elastic area compression modulus;

$$K = \frac{\sqrt{3}\,k_B T}{4pL_{max}(1-x_0)^2}\left\{\left(q+\frac{1}{2}\right)(6-9x_0+4x_0{}^2)+\frac{1+2(1-x_0)^3}{(1-x_0)}\right\} \tag{6.16}$$

where q is a constant factor which can be considered equal to one ($q=1$) [8].

A coarse grained molecular dynamics simulation is done [8] by using the following parameter set:

$$L_0 = 75\text{ nm}, L_{max} = 3.17\,L_0 = 238\text{ nm}, p = 0.1\,L_0 = 7.5\text{ nm}, q = 1,$$

These input parameters resulted following linear properties, Table 6.2 shows the mechanical parameters of RBC membrane (Spectrin network).

Laser tweezers is one of the valuable experimental methods to examine the RBC deformation. In this optical method, the laser tweezers use a focused laser beam to trap micro- or nanoparticles in the focus of the beam and drag the particles around by moving the beam. Generally, the experiment setup of the laser tweezers system includes a single bead gradient optical trap, an inverted microscope and a laser module. Silica beads, (around 4 μm in diameter) are optically trapped by the laser beam to reveal the mechanical aspects of RBC. The range of stretching force for the experiment was from 0 to 200 pN.

Fig. 6.4 illustrates a comparison for deformation of RBC between the experimental results [11] and finite element analysis (FEA) results for variety of the shear modulus (μ_0). The figure reveals the computational results by using the shear modulus $\mu_0 \sim 8$ μN/m match well with the experimental results, while $\mu_0 = 5-12$ μN/m covers most of the experimental scatter. As a function of the external forces, the figure shows an increase and a decrease in the axial and the transverse diameters of the RBC, respectively [8]. Considering stretching forces in both FEA and experiment methods, it was cleared that the deformation induced folding for variety of shapes.

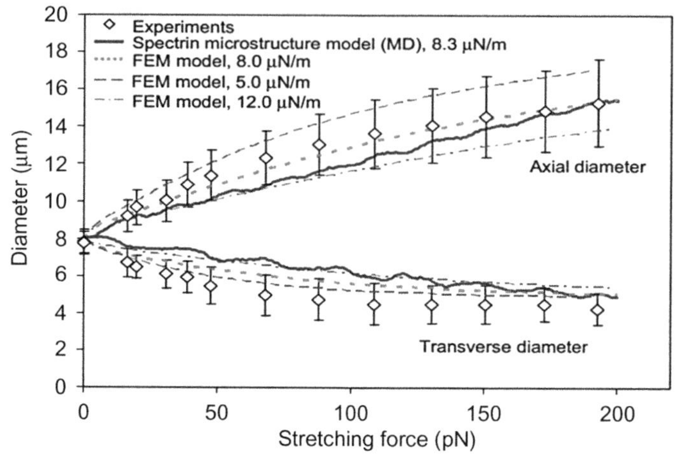

■ **FIGURE 6.4** Diameter-force function (deformation) of RBC for finite element model (FEM) simulation and experimental results for stretching force range from 0 to 200 pN. *From M. Dao, J. Li, S. Suresh, Molecularly based analysis of deformation of spectrin network and human erythrocyte, Mater. Sci. Eng. C 26 (2006) 1232—1244 with permission.*

6.3 HEMOLYSIS CHARACTERIZATION

Hemolysis occurs due to breakage of the RBC membrane, and then the RBC hemoglobin will be released into the blood plasma. Several factors play key roles in hemolysis;

- physiological diseases of the RBC membrane
- shear stress
- turbulence flow
- pressure oscillations
- flow features or geometry of the device
- interaction with bio materials.

Recently, various types of implantable assistive devices have been developed and used for treating patients with chronic diseases. In an implantable artificial device, the roughness of the surfaces and geometry of parts introduce the shear stress. This shear stress is well known as a factor causing hemolysis [12,13]. The estimation of hemolysis within an implantable artificial device is important for the development of the blood-compatibility of these devices. As explained before, hemolysis is the damage of the RBCs' membrane, and it is assessed by the amount of hemoglobin released from inside the RBCs to the plasma. Hemolysis or damage of the RBCs has been modeled with shear stress and exposure time (or accumulated number of events) in previous studies [13—16]. Predicting or estimating hemolysis is one of the best methods to analyze the problem without using physical experiments, and this method is a very valuable tool for some devices, such as the implantable artificial kidney [13].

An important factor in the development of an implantable artificial device is the decrease of hemolysis and thrombus creation so that durable use of these devices is feasible. Many studies have investigated hemolysis in an implantable pump, and, while shear stress [17–19], solid friction [20], and heat generation [21,22] have been listed as the main causes of hemolysis, it has also been suggested that the roughness of the surface has an influence [23–25]. It is reported by Yamazaki et al. [25] that hemolysis levels increase with a raise in the arithmetic mean level of roughness (*Ra*), and they evaluated the detailed flow in the vicinity of the roughened surfaces [13].

Blood cells are exposed to positive pressures up to 200 kPa (~2 atm, ~1500 mmHg) and negative pressures down to −70 kPa (~ −0.7 atm, −500 mmHg). Outside the body, when RBCs are exposed to pressures up to 150 atm, neither hemolysis nor a measurable change of osmotic fragility result [13,26]. Recently, the hemolysis threshold was recognized at 130 MPa (~1300 atm) [27,28]. Negative pressure also does not cause hemolysis [29]. Only the dynamic element is related to blood damage caused by shear. No general pressure limits for the dynamic pressure can be prescribed for avoiding blood damage because, first, pressure drop is not directly related to shear and, second, blood damage is a nonlinear function of shear and exposure time [30]. Shear originates in the difference of fluid velocity between the blood layers in the channel; this difference creates a force which is called shear. The velocity difference between the two flow layers divided by the distance between the two flow layers is defined as the shear rate, measured in units per second. The wall shear rate and shear stress for cylindrical channels can be calculated by Eqs. (6.17) and (6.18), which show the shear rate for general fluids and Newtonian fluids, respectively [13]:

$$\gamma = \frac{8v}{d} \tag{6.17}$$

$$\gamma = \frac{32Q}{d^3\pi} \tag{6.18}$$

$$\tau = \gamma \cdot \mu \tag{6.19}$$

where γ is the wall shear rate, v is fluid velocity, Q is the volumetric flow, d is the channel diameter, τ is the wall shear stress in a cylindrical channel and μ is the viscosity.

Eq. (6.20) defines the pressure drop in a cylindrical channel, and Eq. (6.21) describes wall shear rate in a similar channel as a function of the pressure gradient [13]:

$$\Delta P = \frac{128 l \mu Q}{d^4 \pi} \tag{6.20}$$

$$\gamma = \frac{\Delta P d}{4\ \mu l} \tag{6.21}$$

As explained before, hemolysis is a nonlinear function of shear and exposure time (t), and, regarding these relations, most researchers [12–17] based their hemolysis prediction upon the following power-law equation [12,13,18,31–34]:

$$\frac{\Delta Hb}{Hb} = C \cdot t^a \cdot \tau^b \tag{6.22}$$

where ΔHb is the increase of the plasma free (released) hemoglobin concentration and Hb is hemoglobin concentration, τ (N/m^2) is the shear stress and $t(s)$ is the exposure time.

Two sets of numbers are defined for the constants a, b and C. The first equation (Eq. (6.23)) was introduced by Giersiepen et al. [13,35]:

$$\frac{\Delta Hb}{Hb} = 3.62 \cdot 10^{-7} \cdot t^{0.785} \cdot \tau^{2.416} \tag{6.23}$$

The second equation (Eq. (6.24)) was proposed by Heuser et al. [13,36]:

$$\frac{\Delta Hb}{Hb} = 1.8 \cdot 10^{-6} \cdot t^{0.765} \cdot \tau^{1.991} \tag{6.24}$$

The threshold of shear stress for exposure times above 100 seconds is around 150 Pa, which approaches the shear rate of $\sim 45{,}000$ seconds^{-1} for a blood viscosity of ~ 3.3 mPa·second. For specified short times that approach the residence time inside the cannulas, the threshold was about 400–500 Pa [37]. Allowing for relaxation after shear stress reduces hemolysis remarkably [13,38].

The Normalized Index of Hemolysis (NIH) is a standard clinical (experimental) indicator used to describe hemolysis in flow loop tests. The NIH is proportional to the increase of plasma free hemoglobin concentration, the volume of circulated blood, the volumetric flow rate of blood flow, and the hematocrit and time of the circulations [13,39].

$$NIH\left(\frac{g}{100\ \text{L}}\right) = \Delta Hb \cdot V \cdot \frac{(100 - Hct)}{100} \cdot \frac{100}{Q \cdot t} \tag{6.25}$$

where ΔHb is the increase of plasma free hemoglobin concentration (mg/L), V is the volume of circulated blood (L), Hct is the blood hematocrit, Q is volumetric flow rate of the blood flow (L/minutes) and t is the time of the test (minutes).

Surface roughness and the chemistry of the surface have major roles in hemolysis, and they may even govern capillary flow, especially at shear stress below the critical limit of ~ 150 Pa [13,40]. The influence of surface roughness on hemolysis has been investigated for medical devices, especially for centrifugal and impeller pumps used for heart assistive systems or heart–lung machines. Roughness exceeding 0.1 μm appear to increase hemolysis rates significantly [13,41].

Siliconizing (silicon coating) the internal surface of stainless steel tubing decreases hemolysis [42]. Currently, this method is used for blood access cannulas in hemodialysis systems. Decreasing the rate of hemolysis is possibly a combination effect of reduced roughness and modified chemical interface [13].

Table 6.3 [37,43] summarizes research regarding hemolysis and the condition under which it was studied. This table shows all the results together which makes it simple to observe why there have been many disagreements among various researchers as to the mechanism and levels of threshold for RBC damage [13].

In field of bioparticle separation, the cross-flow filtration method used to minimize cell deposition on the membrane surface also affects the hemolysis rate. The transmembrane pressure drop, wall shear rate and pore size are major factors that result in hemolysis in filtration membranes. These factors produce shear stress and membrane tension on RBC membranes which cause lysis.

Fig. 6.5 shows a model of the cell deformation in a micropipette. Regarding the figure and Laplace's law for wall tension, two equations were derived [13]:

$$P_3 - P_1 = \frac{2\sigma}{R_p} \tag{6.26}$$

$$P_3 - P_2 = \frac{2\sigma}{R_c} \tag{6.27}$$

where P is pressure, σ is membrane tension and R_p and R_c are the pore and cell radii, respectively.

Table 6.3 Summary of the Effect of Shear Stress on Hemolysis [13]

Type of Exposure	Order of Magnitude of Exposure Time (s)	Threshold Level of Damage (Pa)	References and Comments
Turbulent jet	10^{-5}	4000	Forstrom and Blackshear
Oscillating wire	10^{-4}	560	William Stal. (human, canine)
Oscillating bubble	10^{-3}	450	Rooney (human, canine)
Capillary flow	10^{-2}	500	Bacher and Williams (bovine blood)
Capillary flow	10^{-2}	450–700	Keshaviah and Blackshear (canine blood)
Concentric cylinder	10^{2}	150	Leverett et al.
Concentric cylinder maximum stress, 60 Pa	10^{2}–10^{3}	Relatively little hemolysis per unit time	Shapiro and Williams (surface effects dominate)
Concentric cylinder maximum stress, 25 Pa	10^{3}	Relatively little hemolysis per unit time	Knapp and Yarborough (surface effects dominate)
Concentric cylinder maximum stress, 60 Pa	10^{3}	Relatively little hemolysis per unit time	Steinbach and Blackshear (surface effects dominate)

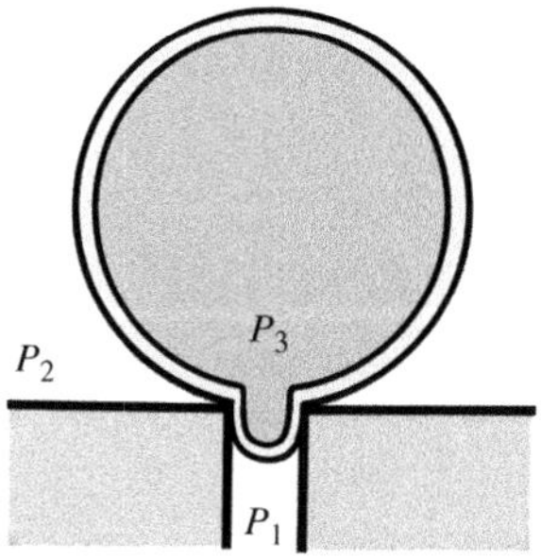

■ **FIGURE 6.5** Deformation of a red blood cell in a micropore. *Modified from A. Ostadfar, Design and experimental proof of selected functions in implantable artificial kidney. [Online]. Available: <summit.sfu.ca/system/files/iritems1/13866/etd8164_AOstadfar.pdf>, 2013 with permission.*

Two equations (Eqs. (6.28) and (6.29)) are introduced for membrane tension as a function of time [13,44]:

$$\sigma_L = \frac{33}{1+0.2\,t}, \text{ for } t \leq 4 \text{ seconds} \tag{6.28}$$

$$\sigma_L = 2 + \frac{60}{t}, \text{ for } t > 4 \text{ seconds} \tag{6.29}$$

where σ_L is lysis membrane tension (dyne/cm) and t is time (seconds).

Three hemolysis experiments conducted by researcher that all had similar tension for lysis: Blachshear and Anderson (33 dyne/cm), Rand (28.6 dyne/cm) and Zydney (25.6 dyne/cm) [13,45].

Considering Table 6.3, the recorded stress range for hemolysis is between 150 and 4000 Pa. Obviously, the stress will be affected by the pore edge, the shape of the pore inlet and the flow regime of the fluid in biofluid separation [13].

Fig. 6.6 is a 3D plot of Eq. (6.23). Designers can use this figure as an assistive tool for predicting when hemolysis will occur. In this graph, the exposure time was considered between 0 and 1 second with a shear stress between 0 and 250 Pa. As the figure shows, increasing the time and stress will cause the blood damage, $\frac{\Delta Hb}{Hb}$, or D to increases to around 0.22 [13].

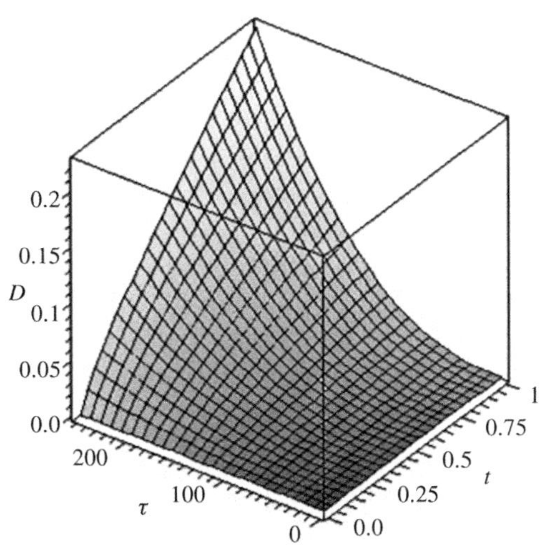

■ **FIGURE 6.6** Hemolysis prediction, $\frac{\Delta Hb}{Hb}$, (or D) using Eq. (6.23), regarding τ (Pa) and t (s). *From A. Ostadfar, Design and experimental proof of selected functions in implantable artificial kidney. [Online]. Available: <summit.sfu.ca/system/files/iritems1/13866/etd8164_AOstadfar.pdf>, 2013 with permission.*

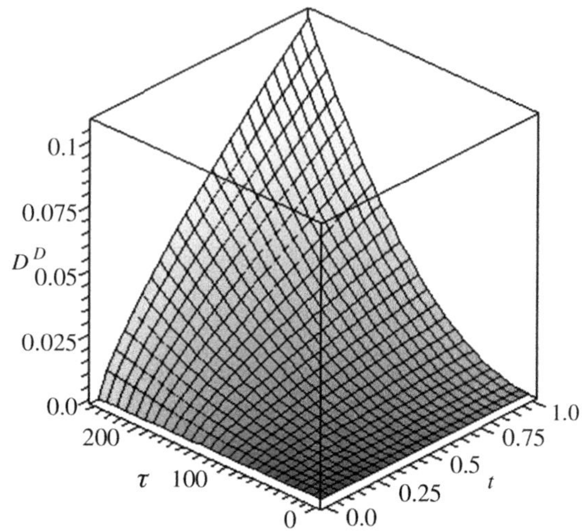

■ **FIGURE 6.7** Cell damage (Hemolysis) prediction, $\frac{\Delta Hb}{Hb}$, (or D) using Eq. (6.24), regarding τ (Pa) and t (s). *From A. Ostadfar, Design and experimental proof of selected functions in implantable artificial kidney. [Online]. Available: <summit.sfu.ca/system/files/iritems1/13866/etd8164_AOstadfar.pdf>, 2013 with permission.*

Similar to Fig. 6.6, Fig. 6.7 illustrates a 3D plot for Eq. (6.24). The range of time and stress were similar to Eq. (6.23). As the figure shows, increasing the time and stress will cause the blood damage, $\frac{\Delta Hb}{Hb}$, or D to increase to around 0.1 [13].

As the figures show, these two models differ even in a similar situation. This difference is probably caused because of an unknown mechanical characteristic of blood cell hemolysis. Most researchers prefer to use Eq. (6.23) in their research and manuscripts as the main equation in hemolysis field, but, for the realistic purposes, they need to have a fully functional product and then they can measure the hemolysis rate of the product, such as in Fig. 6.8. Fig. 6.8 illustrates an example for NIH during a hemolysis test for a pediatric centrifugal pump [13,39].

Most foci on hemolysis were on external assistive devices; recently, focus has turned to implantable heart pumps because these machines are operational in medical equipment, and researchers can make experiments using these machines. In some cases, such as an implantable artificial kidney, we have to use numerical or simulation methods to understand hemolysis aspects in these systems. During device design, the prediction of when hemolysis will occur can be calculated using Eq. (6.23) to evaluate

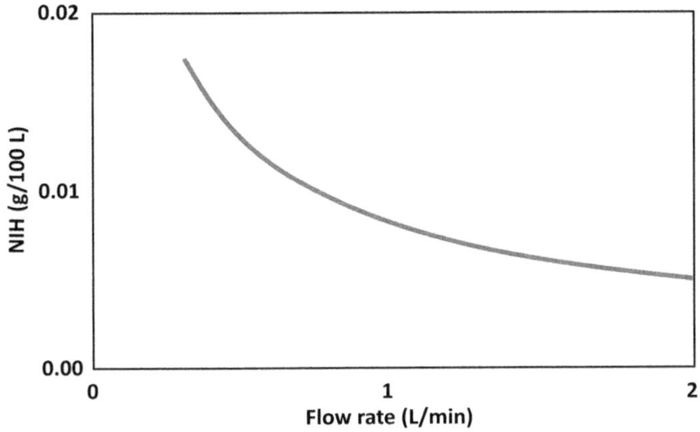

■ **FIGURE 6.8** Normalized index of hemolysis as a function of flow rate. *From O.K. Baskurt, Handbook of Hemorheology and Hemodynamics, IOS Press, Amsterdam and Washington, DC, 2007.*

or predict cell damage in the system. The designers can change their design and compare shear stress and other design parameters to the threshold of damage. This redesign and comparison is a useful tool in engineering design. Some software are very helpful for engineering design purposes. FEA methods and its software are a very useful method to understand the mechanical damage in blood cell membranes, and it helps to reduce the problem of hemolysis on artificial surfaces [13].

6.4 HEMOLYSIS IN NEEDLES AND GRAFTS

Generally in health care systems, needles are used with syringes, catheters, evacuated tubes and butterfly sets. Needles are composed of various materials; titanium, stainless steel, iron, alloys, etc. Generally, a needle has an extended sharp end to make a hole in the skin, blood vessels or other living tissues. Needles are standardized by gauge (#6 to #34), which is inversely related to needle size [46] (bigger gauge denotes smaller diameter).

Hemolysis can be caused by shear stresses of the blood flowing through needles (or grafts), on the RBC membrane during blood sampling (venipuncture blood draws), blood transfusion or hemodialysis (phlebotomy needle). Hemolysis in needles depends on several factors including the blood flow pressure, the diameter and material of the needle (or graft). Furthermore, formulation and calculation of the blood flow depend on Newtonian and non-Newtonian properties of the blood. For Newtonian blood flow which is common in the calculations and simulations, the shear stress (τ) at the wall is

$$\tau = \frac{\Delta P\, R}{2L} \tag{6.30}$$

where ΔP is the blood pressure difference between upstream and downstream of the flow in the tube (needle or graft), R and L are the radius and length of the tube, respectively.

The exposure time grows with length of the needle (or graft) and is inversely related to blood flow velocity. The RBC exposure time varies from the center line to the tube wall. The exposure time for fully developed Newtonian flow at the average velocity can be written as [47];

$$t = \frac{8\mu}{\Delta P}\frac{L^2}{R^2} \tag{6.31}$$

where μ is the viscosity for the Newtonian blood flow.

Table 6.3 is a good reference to estimate or predict the RBC damage or hemolysis in respect to shear stress and time.

Considering Eqs. (6.22)–(6.25) also (6.30) and (6.31); Needle hemolysis is related to the blood viscosity, the blood pressure (inside the needle or graft) and geometrical aspects of the needle or graft, such as length and radius.

6.5 HEMOLYSIS AND HEMODIALYSIS PROCESS

Hemodialysis machines are made of several parts, including blood pump, cannula (or graft), membrane (dialyzer), heat exchanger, etc. One of the most important parts in the hemodialysis machine is the blood pump to produce enough pressure for circulating the blood. Any misalignment or maladjustment in the blood pump generating more pressure or high shear stress on formed elements of the blood, such as RBCs and results in mechanical hemolysis. Sections 6.3 and 6.4 describe hemolysis in grafts and filter pores which are used in hemodialysis machines.

There are several causes of hemolysis during the hemodialysis process. Table 6.4 shows related causes of hemolysis.

As previously stated, the roller pumps have a main role in some medical devices, especially in hemodialysis machines. Squeezing the blood tube and RBCs by roller pumps results in more hemolysis and this aspect shows that the quantity of rollers and the frequency of pumps can affect the hemolysis rate, also the needles and tubes can increase the hemolysis rate (see Section 6.4).

Table 6.4 Error Types and Causes of Hemolysis During Hemodialysis Procedure [48]

Error Type	Cause
Mechanical	High blood pump flow Single-needle dialysis Small-gauge cannula Kinked blood lines High negative arterial pressure Offset blood pump
Failure of rinsing, contamination with	Hydrogen peroxide Hypochlorite Formaldehyde
Priming error	Hypotonic saline
Dialysate error	Hypotonic dialysate Overheated dialysate
Dialysate contamination	Chloramines Nitrates Copper

Currently, many machines disinfect themselves by methods such as high temperature and chemical cycle. Generally, the temperature of the dialysate fluid is adjusted to provide a cooled dialysate to decrease intradialytic hypotension (most patients feel cold because of this condition). The temperature sensor can be reset, and subsequently, some patients have been dialyzed at 40°C in error, complaining of severe high temperature with some hemolysis. Additionally, due to an inadequate rinsing, the chemical materials, for instance hypochlorite, hydrogen peroxide, or formaldehyde, may remain in the system and can cause hemolysis problems with clinical symptoms during the following dialysis session [48]. Table 6.4 is a good reference for addressing hemolysis and its causes during the hemodialysis procedure.

6.6 HEMOLYSIS IN OXYGENATION AND LUNG MACHINES

Oxygenators and lung machines are made of several parts, including a blood pump, cannula (or graft), oxygenation membrane and heat exchanger. All these parts can affect the hemolysis rate during the oxygenation procedure. Sections 6.4 and 6.5 explain the role of some common parts in oxygenation and lung machines such as the needles, grafts and blood pumps in hemolysis.

The investigations revealed that the type of the blood pump (centrifugal vs roller pumps) can affect the hemolysis rate. The investigation results show

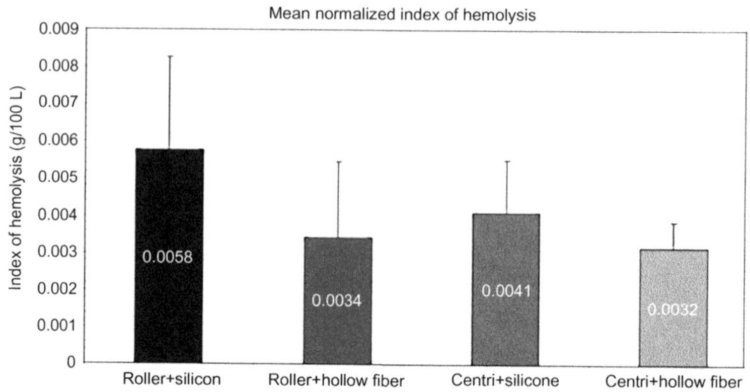

■ **FIGURE 6.9** The normalized index of hemolysis for roller and centrifugal pumps with hollow fiber membrane and silicon membrane in ECMO systems. *From A.D. Meyer, A.A. Wiles, O. Rivera, E.C. Wong, R.J. Freishtat, K. Rais-Bahrami, et al., Hemolytic and thrombocytopathic characteristics of extracorporeal membrane oxygenation systems at simulated flow rate for neonates, Pediatr. Crit. Care Med. 13 (4) (2012) e255—e261 with permission.*

a contrast for the blood pump type. Probably, this difference occurs due to the design and roughness of the internal surface of the pump. As explained before, the material and structure of the machine (here is an oxygenation membrane) can have an effect on hemolysis rate.

Fig. 6.9 represents the NIH for two kinds of membranes and blood pumps in extracorporeal membrane oxygenation (ECMO) systems. Although the results do not show any significant difference between the ECMO systems, the results help researchers to understand the hemolysis and its features. Considering Fig. 6.9, for both roller and centrifugal pumps, the hollow fiber membrane has a lower index of hemolysis and the roller pump can be more reliable than the centrifugal pump.

6.7 HEMOLYSIS IN TAH AND ARTIFICIAL HEART VALVES

Due to large flow rate of the blood in artificial cardiovascular system (heart chambers, heart valves and other component), hemolysis can be considered as one of most important issues in that system.

- *Total artificial heart*: TAH has been developed to support or replace a patient's failing heart. Generally, implantable artificial organs such as the TAH are planned to work for at least 5 years. Considering this period, blood issues (hemolysis etc.) can have an effect on the patient's body.

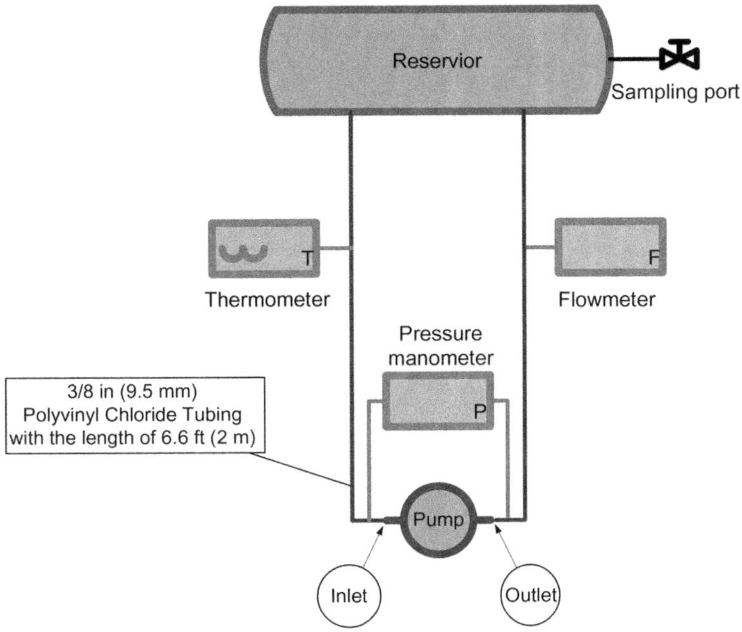

■ **FIGURE 6.10** Test setup for hemolysis measurement under the standard of ASTM "F 1841."

The American Society for Testing and Materials (ASTM) issued "F 1841" as a standard and guideline for the estimation of hemolysis induced by continuous flow blood pumps. Technically, there is not any general standard or equation to show hemolysis in the TAHs and the best method is an experiment test to analyze the hemolysis degree of the TAHs. Fig. 6.10 shows the schematic of the blood pump (TAH) and test loop under the standard of the ASTM.

The pump revolution per minute is adjusted to deliver 5 ± 0.25 L/minutes blood flow rate and a flowmeter controls the flow rate. Considering normal body temperature, all experiments are conducted at a $37 \pm 1°C$ by submerging portions of the test loop into a warm water bath. Blood samples of 1−2 mL (preferably 1 mL) are drawn from the blood reservoir before pumping and at every hour of pumping and test time. To reduce the errors, at least two blood samples at each sample time should be withdrawn and the second one has more reliability to show plasma free hemoglobin. The blood pump must work for at least 6 hours to take sufficient blood samples. This test should be repeated five times to increase the reliability of the results. Regarding the blood coagulation, some anticlotting materials may be applied for the experiment [50].

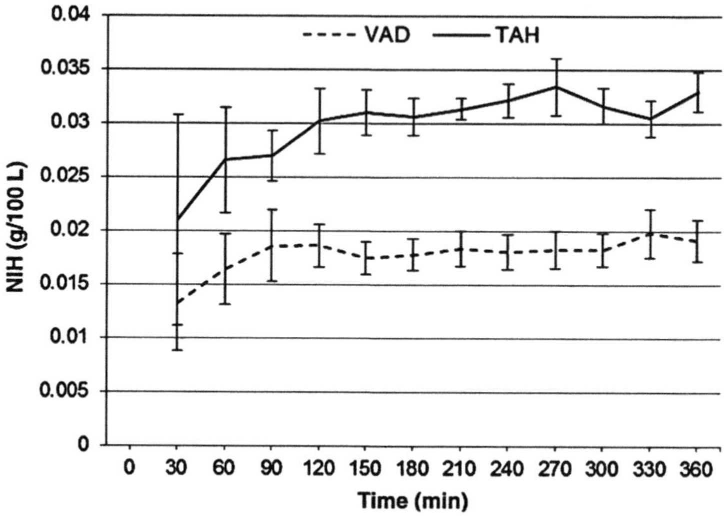

■ **FIGURE 6.11** A comparison between VAD and ReinHeart TAH for mean normalized index of hemolysis (NIH) values. *From F. Gräf, T. Finocchiaro, M. Laumen, I. Mager, U. Steinseifer, Mock circulation loop to investigate hemolysis in a pulsatile total artificial heart, Artif. Organs 39 (5) (2015) 416–422 with permission.*

In the following and as a case study, an experiment [51] will be studied. The experiment was conducted under ASTM F 1841 for testing the pulsatile ReinHeart TAH. As a control pump, the clinically approved Medos-HIA ventricular assist device (VAD) was used. The pumps were operated at 120 beats per minute with an aortic pressure of 120 to 80 mmHg and a mean atrial pressure of 10 mmHg, generating a blood flow rate of about 5 L/minutes. The average NIH values of the VAD and the TAH were measured and were 0.018 g/100 L and 0.03 g/100 L, respectively [51]. In this research, six simultaneous experiments with the TAH ventricle and the VAD were done and the results presented in Fig. 6.11.

- *Artificial heart valves*:

  The most common cause of major hemolytic anemia in patients who have an artificial heart valve is paraprosthetic valve regurgitation. The regurgitation is defined as an abnormal backward blood flow around the circumference of an artificial heart valve between the sewing ring and the annulus of the native valve. Considering several studies, the reported rate of paravalvular regurgitation varies between 0.3% and 13% [52].

  There are four goals for examining hemolysis in a patient with an artificial heart valve [52]:

  - confirmation of the existence of anemia;
  - determination of the existence and severity of hemolysis;

Table 6.5 Classification of Hemolysis Based on the Serum Levels of LDH and Haptoglobin

Degree	LDH (U/L)	Haptoglobin (mg %)
No hemolysis	<220	>37
Mild hemolysis	220–400	<37
Moderate hemolysis	400–800	<10
Severe compensate hemolysis	>800	0
Decompensated hemolysis	>1000	0

Adapted from M. Josa, M. Castella, C. Pare, J.L. Bedini, R. Cartana, C.A. Mestres, et al., Hemolysis in mechanical bileaflet prostheses: experience with the bicarbon valve, Ann. Thorac. Surg. 81 (4) (2006) 1291–1296.

- assessment of shortage states secondary to iron loss or enhanced hematopoiesis;
- consideration of factors contributing to increased hemolysis, mostly dysfunction of the artificial valve.

Serum levels of lactate dehydrogenase (LDH) is the best factor to show the presence of hemolysis and its degree [53]. Table 6.5 shows the severity (degree) of hemolysis based on the serum levels of LDH and haptoglobin.

The examinations demonstrate that mild hemolysis occurred frequently with artificial heart valves but severe hemolysis is uncommon. Furthermore, only valve position (not size) has an affect on hemolysis and hemolysis function is not time-dependent.

6.8 COLLECTION, STORAGE AND TRANSPORTATION OF WHOLE BLOOD

The main reasons for blood donation and transfusion are to help a patient's blood circulation maintain normal blood volume and blood oxygen level. Blood cold chain is a procedure system which starts with blood collection from the donor and continues with process, store, transport, completing with the transfusion of the blood to the recipient or patient. Several techniques in term of the blood cold chain may be defined to complete this system. Due to the fragility of the blood components, such as RBCs, quality and accuracy of these procedures are very vital.

These procedures and aspects can be presented as [54];

- Anticoagulant-preservatives
- Storage temperatures
- Storage equipment
- Blood transport containers

Anticoagulant-preservatives

Anticoagulant-preservatives are used to prevent the blood from clotting or coagulating after collecting the blood from the donor, and then these materials provide enough nutrients to maintain RBCs functionality and practicability during transportation and storage. The most vital elements in maintaining the functionality of RBCs are adenosine triphosphate (ATP) and glucose.

Acid citrate dextrose (ACD) was one of the earliest blood anticoagulants. The acidic pH does not maintain 2,3-bisphosphoglycerate (BPG) levels and it is no longer commonly used. Another solution is citrate phosphate dextrose (CPD), which is in common use in combination with RBC additive systems. CPD solution has a more alkaline pH and maintains BPG levels better than ACD [54]. Furthermore, ethylenediamine tetraacetic acid (EDTA) has been suggested as the anticoagulant for the hematological testing of blood cells [55]. The composition of these materials is given Table 6.6.

Table 6.6 Composition of Anticoagulant-Preservatives [54,55]

Material	Function
Sodium citrate	It is a calcium-chelating agent that delays the calcium-dependent steps in the clotting cascade and prevents coagulation
Dextrose	It supports adenosine triphosphate (ATP) generation and provides nutrients that are required by the red cells
Citric acid	It is used in conjunction with sodium citrate and dextrose to make the anticoagulant solution called acid citrate dextrose (ACD)
Sodium phosphate	When used in conjunction with citric acid, sodium citrate and dextrose, it comprises the anticoagulant citrate phosphate dextrose (CPD)
CPD to which adenine (A) is added, becomes CPDA-1 (the "1" signifies the formula used)	It improves the synthesis of ATP. Generally, CPDA-1 is used when the collected donation is to be stored as whole blood
EDTA is polyprotic acid containing four carboxylic acid groups and two amine groups with lone-pair electrons	It removes calcium from coagulation cascade to prevent blood clotting within the collection tube

Volume of anticoagulant (CPD and CPDA): Generally, the volume ratio of anticoagulant required to prevent clotting and preserve RBCs is 14:100 or 14 mL anticoagulant for each 100 mL of blood which is taken from the donor.

Additive solutions for RBCs: After removing plasma from whole blood by centrifugal filtration, a large portion of the anticoagulant and nutrients in CPD are filtered along with it. After this procedure, the RBCs need nutrients to survive and also a fluidic environment to keep normal flow features. Additive solutions can accomplish these needs. Additive solutions vary in composition depending on the supplier, such as saline, adenine, glucose and mannitol (SAGM). The volume ratio of additive solutions is 22.2:100 or 22.2 mL additive solutions for each 100 mL of blood [54].

Storage temperatures

Function and viability of collected blood will be normal, if the storage conditions remain within the correct range of temperature. Storage of the blood at low temperature reduces glycolytic activity and allows a long shelf life (the maximum allowable storage period) of the blood component, and also it delays the bacterial proliferation.

Donated blood and RBCs must be stored at $4 \pm 2°C$. Also transport temperature ranges of preprocessed and processed blood are 20–24°C, and 2–10°C, respectively. In addition, the storage time of collected blood is vital too. Table 6.7 summarizes the storage and transport conditions for preprocessed and processed blood.

Table 6.7 Storage and Transport Conditions for Whole Blood and Its Elements

Condition	Temperature Range	Storage Time
Transport of preprocessed blood	+20°C to +24°C	Less than 6 h
Storage of preprocessed or processed blood	+ 2°C to +6°C	Approx. 35 days
Transport of processed blood	+ 2°C to +10°C	Less than 24 h

The information is adapted from World Health Organization, Manual on management, maintenance and use of blood cold chain equipment. [Online]. Available: <http://www.who.int/bloodsafety/Manual_on_Management,Maintenance_and_Use_of_Blood_Cold_Chain_Equipment.pdf>, 2005.

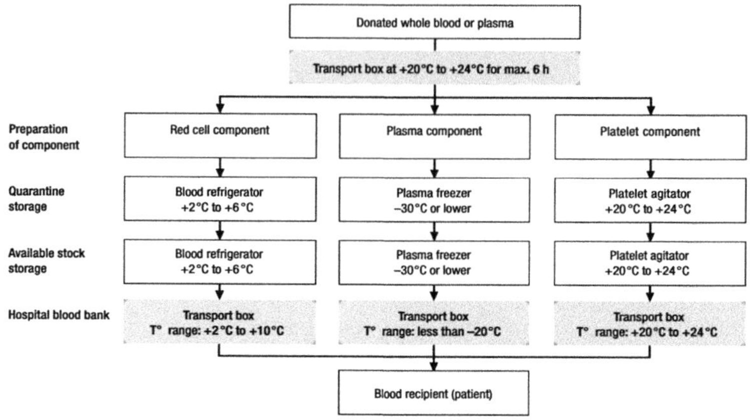

■ **FIGURE 6.12** The blood cold chain, procedures and the temperature conditions. *From World Health Organization, Manual on management, maintenance and use of blood cold chain equipment. [Online]. Available: <http://www.who.int/bloodsafety/Manual_on_Management,Maintenance_and_Use_of_Blood_Cold_Chain_Equipment.pdf>, 2005 with permission.*

Fig. 6.12 illustrates the temperature conditions of the blood cold chain and its procedures.

Storage equipment

Several devices are designed to control storage temperature of whole blood, plasma and bioparticles;

- *Blood storage refrigerators* are designed to maintain whole blood, RBC concentrates and other blood elements at $+4\pm2°C$.
- *Plasma storage freezers* are designed to produce a temperature of less than $-20°C$.
- *Room temperature storage devices* are manufactured to control the biofluids temperature at range of $+22\pm2°C$.

Transport containers of blood and its elements

The role of containers is to transport whole blood and its elements in transit from one location to another location under safe and reliable temperature conditions. The container is an insulated box with a sealed lid to provide an isolated volume from the ambient atmosphere. Generally, a frozen coolant (ice pack) will be placed inside the transport container to absorb the heat which is already in the container and increase the time of safe transportation. The new generation of containers are designed to get electrical energy from motor vehicles by a rechargeable battery to maintain the cold life for over 24 hours.

Ideal design features of a blood transport container are [56];

- Lightweight
- Robust
- Secure and lockable
- Cold life of at least 30 hours at +43°C.

CHAPTER SUMMARY

- The membrane of erythrocytes or RBCs is created by a phospholipid bilayer with protein inclusions, bicarbonate HCO^{-3} and ion channels for chloride Cl^-, sialoglycopeptides (blood groups), glucids, cytoskeleton meshing and mooring proteins.
- The oval of Cassini is the most popular and precise method to describe erythrocytes (RBC).
- Considering Fig. 6.2, *a* and *c; $d_1 d_2 = c^2$, $F_1F_2 = 2a$, $\varepsilon = \frac{a}{c}$*, Table 6.1 (from Ref. [4] with permission) shows the approximate formulas of RBC geometry parameters with respect to *c* and *ε*.
- Mechanical aspects of RBCs are governed by the elastic and viscous properties of the cell membrane. These aspects occur due to the viscose role of the lipid bilayer and the elastic role of the spectrin network, the bending resistance factor, and the viscosity of the fluids outside and inside of the RBC.
- To transport oxygen to the living tissues, the RBC needs to change its shape from its normal biconcave form to a parachute or bullet form through capillaries. This means a maximum allowable strain on the RBC membrane under the local biofluid pressure gradient.
- The simulation results for the mechanical parameters of RBC membrane (Spectrin network) are shown in Table 6.2
- Hemolysis is the damage of the RBCs' membrane, and it is assessed by the amount of hemoglobin released from inside the RBCs to the plasma. Several factors play key roles in hemolysis:
  - physiological diseases of the RBC membrane
  - shear stress
  - turbulence flow
  - pressure oscillations
  - flow features or geometry of the device
  - interaction with bio materials.
- The NIH is a standard clinical (experimental) indicator used to describe hemolysis in flow loop tests:

$$NIH\left(\frac{g}{100\,L}\right) = \Delta Hb \cdot V \cdot \frac{(100 - Hct)}{100} \cdot \frac{100}{Q \cdot t}$$

- Any misalignment or maladjustment in the blood pump can generate more pressure or high shear stress on informed elements of blood, such as RBCs, resulting in mechanical hemolysis.
- The investigations revealed that the type of the blood pumps (centrifugal vs roller pumps) can affect on the hemolysis rate.
- The most common cause of major hemolytic anemia in patients who have an artificial heart valve is paraprosthetic valve regurgitation.
- Classification of hemolysis based on the serum levels of LDH and haptoglobin. Table 6.5 is adapted from Josa et al. [53].
- Blood cold chain is a procedure system which starts with blood collection from the donor and continues with process, store, transport and completing with the transfusion of the blood to the recipient or patient.
- Anticoagulant-preservatives are used to prevent the blood from clotting or coagulating after collecting the blood from the donor, and then these materials provide enough nutrients to maintain RBCs functionality and practicability during transportation and storage. The famous anticoagulant-preservatives are acid citrate dextrose (ACD and ACD A), citrate phosphate dextrose (CPD and CPDA) and ethylenediamine tetraacetic acid (EDTA).
- Storage and transport conditions for whole blood and its elements. Table 6.7 is adapted from World Health Organization [56].
- The duty of containers is to transport whole blood and its elements in transit from one location to another location under safe and reliable temperature conditions.

PROBLEMS

1. Draw a schematic of an RBC membrane and explain the elements.
2. Which characteristics have a key role in the viscoelastic behavior of an RBC membrane?
3. Explain hemolysis and its causing factors.
4. Which formulas can be used to predict hemolysis? Explain their elements.
5. What is the relation between the RBCs concentrations and hemolysis?
6. Discuss hemolysis issues in blood pumps. Which pump is your recommendation, why?
7. Explain and classify the blood anticoagulant-preservatives and their compositions.
8. Classify the best temperature for storage and transport of whole blood and its elements.

REFERENCES

[1] J. Delaunay, P. Boivin, Le squelette du globule rouge [The skeleton of the red blood cell], La Recherche 223 (1990) 845–852.

[2] M. Thiriet, Biology and Mechanics of Blood Flows Part I: Biology, Springer Science + Business Media, LLC, New York, NY, 2008.

[3] W.T. Tse, S.E. Lux, Red blood cell membrane disorders, Br. J. Haematol. 104 (1999) 2–13.

[4] E.W. Weisstein, "Cassini Ovals." From MathWorld--A Wolfram Web Resource, Wolfram Research, Inc, 2015 [Online]. Available: <http://mathworld.wolfram.com/CassiniOvals.html>.

[5] B. Angelov, I.M. Mladenov, On the geometry of red blood cell, Geom. Integr. Quant. 300 (1999) 27–46. Available from: http://dx.doi.org/10.7546/giq-1-2000-27-46.

[6] Deleted in review.

[7] D.A. Fedosov, B. Caswell, G.E. Karniadakis, A multiscale red blood cell model with accurate mechanics, rheology, and dynamics, Biophys. J. 98 (2010) 2215–2225.

[8] M. Dao, J. Li, S. Suresh, Molecularly based analysis of deformation of spectrin network and human erythrocyte, Mater. Sci. Eng. C 26 (2006) 1232–1244.

[9] S. Suresh, J. Spatz, J. Mills, A. Micoulet, M. Dao, C. Lim, et al., Connections between single-cell biomechanics and human disease states: gastrointestinal cancer and malaria, Acta Biomater. 1 (2005) 15–30.

[10] D.H. Boal, Mechanics of the Cell, Cambridge University Press, Cambridge, 2002.

[11] J.P. Mills, L. Qie, M. Dao, C.T. Lim, S. Suresh, Nonlinear elastic and viscoelastic deformation of the human red blood cell with optical tweezers, Mech. Chem. Biosyst. 1 (3) (2004) 169–180.

[12] T. Yano, K. Sekine, A. Mitoh, Y. Mitamura, E. Okamoto, D.-W. Kim, et al., An estimation method of hemolysis within an axial flowblood pump by computational fluid dynamics analysis, Artif. Organs 27 (10) (2003) 920–925. *Blackwell Publishing, Inc.*

[13] A. Ostadfar, Design and experimental proof of selected functions in implantable artificial kidney. [Online]. Available: <summit.sfu.ca/system/files/iritems1/13866/etd8164_AOstadfar.pdf>, 2013.

[14] M. Giersiepen, L. Wurzinger, R. Opitz, H. Reul, Estimation of shear stress-related blood damage in heart valve prostheses in vitro comparison of 25 aortic valves, Int. J. Artif. Organs 13 (1990) 300–306.

[15] C. Bludszuweit, Three-dimensional numerical prediction of stress loading of blood particles in a centrifugal pump, Artif. Organs 19 (1995) 590–596.

[16] K.K. Yeleswarapu, J. Antaki, M. Kameneva, K. Rajagopal, Mathematical model for shear induced hemolysis, Artif. Organs 19 (1995) 576–582.

[17] M.V. Kameneva, G. Burgreen, K. Kono, B. Repko, J. Antaki, M. Umezu, Effects of turbulent stresses upon mechanical hemolysis: experimental and computational analysis, ASAIO J. 50 (2004) 418–423.

[18] L. Goubergrits, K. Affeld, Numerical estimation of blood damage in artificial organs, Artif. Organs 28 (2004) 499–507.

[19] R. Paul, J. Apel, S. Klaus, F. Schugner, P. Schwindke, H. Reul, Shear-stress-related blood damage in laminar couette flow, Artif. Organs 27 (2003) 517–529.

[20] Y. Orime, M. Shiono, S. Yagi, T. Yamamoto, H. Okumura, K. Nakata, et al., Jostra Rota Flow RF-30 pump system: a new centrifugal blood pump for cardiopulmonary bypass, Artif. Organs 24 (2000) 437–441.

[21] C. Gobel, A. Arvand, R. Eilers, O. Marseille, C. Bals, B. Meyns, et al., Development of the MEDOS/HIADeltaStream extracorporeal rotary blood pump, Artif. Organs 25 (2001) 358–365.

[22] K. Araki, Y. Taenaka, T. Masuzawa, E. Tatsumi, Y. Wakisaka, M. Watari, et al., Hemolysis and heat generation in six different types of centrifugal blood pumps, Artif. Organs 19 (1995) 928–932.

[23] T. Yasuda, K. Shimokasa, A. Funakubo, T. Higami, T. Kawamura, Y. Fukui, An investigation of blood flow behavior and hemolysis in artificial organs, ASAIO J. 46 (2000) 527–531.

[24] Y. Takami, T. Nakazawa, K. Makinouchi, E. Tayama, J. Glueck, R. Benkowski, et al., Hemolytic effect of surface roughness of animpeller in a centrifugal blood pump, Artif. Organs 21 (1997) 686–690.

[25] K. Yamazaki, R. Kormos, T. Mori, M. Umezu, M. Kameneva, J. Antaki, et al., Anintraventricular axial flow blood pump integrated with a bearing purge system, ASAIO J. 41 (1995) M327–M332.

[26] E. Brewster, S. Collins, G. Funnell, E. Smith, The effect of high pressure on the hemolysis of red blood cells, Under Sea Biomed. Res. 3 (1976) 151–155.

[27] T. Yamaguchi, S. Terada, H. Ideguchi, High-pressure-induced hemolysis of hereditary spherocytic erythrocytes is not suppressed by DIDS labeling, Jpn. J. Physiol. 47 (1997) 571–574.

[28] H. Kitajima, T. Yamaguchi, E. Kimoto, Hemolysis of human erythrocytes under hydrostatic pressure is suppressed by cross-linking of membrane proteins, J. Biochem. 108 (1990) 1057–1062.

[29] S.D. Chambers, K. Laberteaux, S. Merz, J. Montoya, R. Bartlett, Effects of static pressure on red blood cells on removal of the air inter-face, ASAIO J. 42 (1996) 947–950.

[30] R. Beissinger, J.F. Laugel, Low stress shear-induced hemolysis in capillary flow, Trans. Am. Soc. Artif. Intern. Organs 29 (1983) 158–162.

[31] A. Mitoh, T. Yano, K. Sekine, Computational fluid dynamics analysis of an intracardiac axial flow pump, Artif. Organs 23 (2003) 34–40.

[32] E. Okamoto, T. Hashimoto, T. Inoue, Y. Mitamura, Blood compatible design of a pulsatile blood pump using computational fluid dynamics and computer-aided design and manufacturing technology, Artif. Organs 27 (2003) 61–67.

[33] W.K. Chan, Y. Wong, Y. Ding, L. Chua, S.C. Yu, Numerical investigation of the effect of blade geometry on blood trauma in a centrifugal blood pump, Artif. Organs 26 (2002) 785–793.

[34] X. Song, A. Throckmorton, H. Wood, J. Antaki, D. Olsen, Computational fluid dynamics prediction of blood damage in a centrifugal pump, Artif. Organs 27 (2003) 938–941.

[35] D. Arora, M. Behr, M. Pasquali, A tensor-based measure for estimating blood damage, Artif. Organs 28 (11) (2004) 1002–1015.

[36] L. Gu, W.A. Smith, Evaluation of computational models for hemolysis estimation, ASAIO J. 51 (2005) 202–207.

[37] L.B. Leverett, J. Hellums, C. Alfrey, E. Lynch, Red blood cell damage by shear stress, Biophys. J. 12 (1972) 257–273.

[38] H. Mizunuma, S. Sakai, Intermittent stops of shear reduce the mechanical damage of red blood cells, Artif. Organs 31 (2007) 472–475.

[39] O.K. Baskurt, Handbook of Hemorheology and Hemodynamics, IOS Press, Amsterdam, Netherlands, 2007.

[40] H. Polaschegg, Red blood cell damage from extracorporeal circulation in hemodialysis, Semin. Dial. 22 (5) (2009) 524–531.

[41] Y. Takami, K. Makinouchi, T. Nakazawa, J. Glueck, R. Benkowski, Y. Nose, Effect of surface roughness on hemolysis in a pivot bearing sup-ported Gyro centrifugal pump (C1E3), Artif. Organs 20 (1996) 1155–1161.

[42] P.L. Blackshear, F. Dorman, J. Steinbach, E. Maybach, A. Singh, R. Collingham, Shear, wall interactions and hemolysis, Trans. Am. Soc. Artif. Intern. Organs 12 (1966) 113–120.

[43] M. Grigioni, C. Daniele, U. Morbiducci, G. D'Avenio, G. Di Benedetto, V. Barbaro, The power law mathematical model for blood damage prediction: analytical developments and PhysicalInconsistencies, Artif. Organs 28 (5) (2004) 467–475.

[44] P. Blackshear, R. Andreson, Hemolysis thresholds in microporous structure, Red Cell Rheol. (1978) 373–391.

[45] A.L. Zydney, A Cross Flow Membrane Plasmapheresis—An Analysis of Flux and Hemolysis, Thesis, Massachusetts Institute Technology, 1985.

[46] R.A. Bowen, G.L. Hortin, G. Csako, O.H. Otanez, A.T. Remaley, Impact of blood collection devices on clinical chemistry assays, Clin. Biochem. 43 (2010) 4–25.

[47] M.K. Sharp, S.F. Mohammad, Scaling of hemolysis in needles and catheters, Ann. Biomed. Eng. 26 (1998) 788–797.

[48] A. Davenport, Intradialytic complications during hemodialysis, Hemodial. Int. 10 (2006) 162–167.

[49] Deleted in review.

[50] American Society for Testing and Materials, Standard Practice for Assessment of Hemolysis in Continuous Flow Blood Pumps, American Society for Testing and Materials, West Conshohocken, 2013.

[51] F. Gräf, T. Finocchiaro, M. Laumen, I. Mager, U. Steinseifer, Mock circulation loop to investigate hemolysis in a pulsatile total artificial heart, Artif. Organs 39 (5) (2015) 416–422.

[52] R. Maraj, L.E. Jacobs, A. Ioli, M.N. Kotler, Evaluation of hemolysis in patients with prosthetic heart valves, Clin. Cardiol. 21 (1998) 387–392.

[53] M. Josa, M. Castella, C. Pare, J.L. Bedini, R. Cartana, C.A. Mestres, et al., Hemolysis in mechanical bileaflet prostheses: experience with the bicarbon valve, Ann. Thorac. Surg. 81 (4) (2006) 1291–1296.

[54] J. Hardwick, Blood storage and transportation, ISBT Sci. Ser. 3 (2008) 177–196.

[55] G. Banfi, G.L. Salvagno, G. Lippi, The role of ethylenediamine tetraacetic acid (EDTA) as in vitro anticoagulant for diagnostic purposes, Clin. Chem. Lab. Med. 45 (5) (2007) 565–576.

[56] World Health Organization, Manual on management, maintenance and use of blood cold chain equipment. [Online]. Available: <http://www.who.int/bloodsafety/Manual_on_Management,Maintenance_and_Use_of_Blood_Cold_Chain_Equipment.pdf>, 2005.

Chapter 7

Reaction of Blood and Biomaterials

LEARNING OBJECTIVES

- Description of the immune system within humans.
- Characterization of antibodies and their activity.
- Explanation of the hemostatic coagulation factors and the activation mechanism.
- Biomechanics of clotting.
- Definition of thrombosis and its classification.
- Characterization of rolling adhesion.
- Explanation and calcification of biocompatibility and its aspects.
- Biomaterial aspects and its classification.

7.1 IMMUNE SYSTEM

Generally, the human body can resist almost all kinds of organisms or toxins that have a tendency to damage the living tissues and human organs. This ability is known as immunity. If the body is attacked by a bacterium, virus, toxin or foreign tissues, much of the immunity develops as acquired immunity and requires weeks or months to develop itself. Acquired immunity is created by a particular immune system that forms antibodies and/or activated lymphocytes that attack and eliminate the invading organism or toxin [1].

The human body can produce two types of acquired immunity [1];

- B-cell immunity (or humoral immunity) which produces the antibodies by B lymphocytes, and
- T-cell immunity (or cell-mediated immunity), which produces the activated lymphocytes by T lymphocytes.

Originally both B and T lymphocytes are derived in the embryo from stem cells of pluripotent hematopoietic. To form activated T lymphocytes, the lymphocytes migrate to and are preprocessed in the thymus

Biofluid Mechanics. DOI: http://dx.doi.org/10.1016/B978-0-12-802408-9.00007-7

gland, and therefore they are called T lymphocytes. The other lymphocytes, are preprocessed in the liver and in the bone marrow and they are called B lymphocytes [1]. Fig. 7.1 illustrates the two lymphocyte systems for the formation of the activated T lymphocytes and the antibodies. The antibodies act as a defense mechanism to eliminate antigens (invaders agents) from the body. Fig. 7.2 shows the schematic of antibody–antigen binding and the structure of the typical IgG antibody with polypeptide chains.

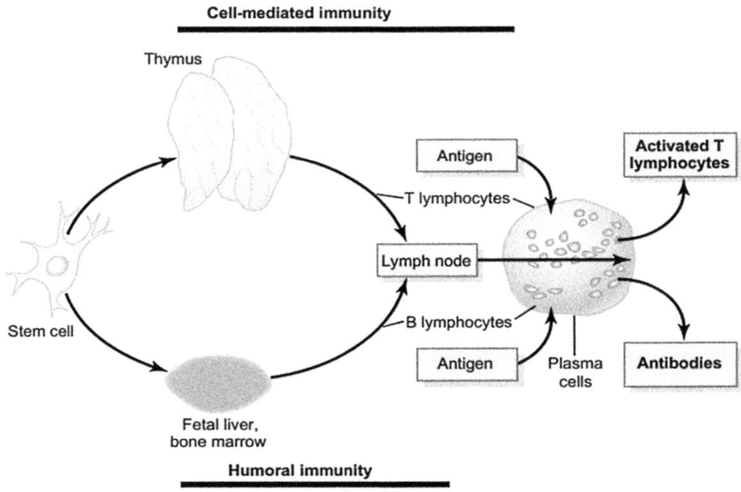

■ **FIGURE 7.1** Formation schematic of antibodies and sensitized lymphocytes; also it illustrates the origin of T and B lymphocytes that are responsible for the cell-mediated and humoral immune processes, respectively. *Adapted from A.C. Guyton, J.E. Hall, Textbook of Medical Physiology, Elsevier Saunders, Philadelphia, PA, 2006 with permission.*

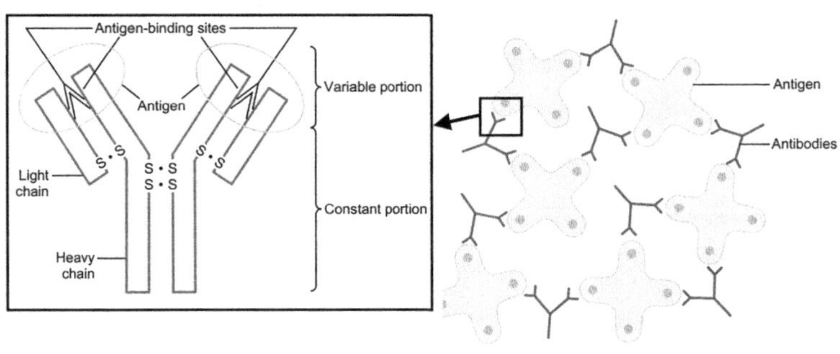

■ **FIGURE 7.2** Right: binding of antigen molecules and antibodies. Left: structure of the typical immunoglobin G (IgG) antibody with two heavy and two light polypeptide chains. *Modified from A.C. Guyton, J.E. Hall, Textbook of Medical Physiology, Elsevier Saunders, Philadelphia, PA, 2006 with permission.*

Generally, antibodies act by two methods to defend the body against invaders [1]:

1. By direct attack on the invading agents and
2. By activation of the "complement system," a system of about 20 proteins, to destroy the invading agents.

The direct attacks often are not strong enough to have a key role in defending the human body against the invader. The majority of the defense system derives through the complement system.

Another mechanism to destroy invading bacteria or viruses is phagocytosis. In the phagocytosis method, a cell absorbs the invader agents, including particles, molecules and bacteria, by engulfing it. The best example for this method is engulfing and destroying a bacterium by a white blood cell (WBC).

7.2 COAGULATION AND THROMBOSIS

Hemostasis is a mechanism to control and stop bleeding from damaged blood vessels and immediately after a blood vessel has been injured or cut, the mechanism will be activated.

The hemostasis is completed by several procedures [1]:

1. Constriction of blood vessel: this feature immediately reduces the blood flow from the injured vessel.
2. Platelet plug formation: platelets stick together to shape a sealing plug to cover the hole in the blood vessel wall.
3. Blood clot formation as a result of blood coagulation.
4. Growing fibrous tissue into the blood clot to block the hole in the vessel.

The clot begins to grow in 15–20 seconds (for a severe injury) or 1–2 minutes (for a minor injury). Several activation factors from the injured vessel wall, platelets and blood proteins work to initiate the clotting procedure.

If the vessel opening is small, the injured area of the blood vessel is filled with a clot in 3–6 minutes after rupture. Between 20 minutes to an hour, the clot retracts.

Platelets have two main functions in the blood circulation system [2]:

1. Decrease and control of bleeding by the formation of platelet plugs;
2. Control of platelet plugs by catalyzing coagulation reactions, leading to fibrin formation.

Normally, platelets are very sensitive cells among blood cells and respond to minimal stimulation. The activation procedure causes platelet cells to become very adhesive bio cells with irregular sphere shapes and spiny pseudopods. Platelet cells adhere to injured blood vessels and artificial surfaces. In the case of a vessel injury, the bond procedure contains the interaction of platelet glycoprotein Ib (GP Ib) and components of connective tissue which become exposed and need plasma von Willebrand factor (vWF) as an important cofactor. The genetic absence of GP Ib or vWF results in a malfunction in platelet adhesion and serious unusual bleeding [2].

In the case of an artificial surface, the procedure of platelet adhesion to the surface may also be mediated through platelet glycoprotein IIb/IIIa, likewise the GP Ib-vWF interaction. GPIIb/IIIa is known as a platelet receptor for adhesive plasma proteins, including fibrinogen, vWF, fibronectin and vitronectin, to support cell attachment. It is important to know that the inactive platelets do not bind these proteins, which generally occur only after platelet activation and a structural change in GP IIb/IIIa [2].

Platelet aggregation involves platelet to platelet adhesion and is important for an effective hemostasis. Platelet adhesion and platelet aggregation are two sides of a coin in the hemostasis mechanism.

Several reactions are initiated to create platelet aggregation [2]:

1. the release of dense granular adenosine diphosphate,
2. the creation of small amounts of thrombin and
3. the activation of platelet biochemical procedures to produce thromboxane A_2.

These reactions (1–3) work to create and increase platelet aggregation in hemostasis function.

7.2.1 Coagulation

Around 50 elements in the blood and in human tissues affect the blood coagulation mechanism. These elements can be classified into two groups:

1. Procoagulant elements which promote coagulation and
2. Anticoagulants that prevent coagulation.

The procoagulants will be activated by a ruptured region of a vessel and then a clot starts to grow. Table 7.1 classifies the coagulation factors and their function in hemostasis.

Table 7.1 Coagulation Factors in Blood and Their Action and Plasma Concentrations [2,3]

Coagulation Factors	Actions [3]	Normal Plasma Concentration (μg/mL) [2]
Factor I, Fibrinogen	Substrate for thrombin and precursor of fibrin. Its role is to be converted into an insoluble protein and then back to dissolved elements.	2500
Factor II, Prothrombin	Precursor to thrombin. In the presence of calcium, it is changed to thrombin (IIa), which in turn stimulates platelet aggregation and activates cofactors protein C and factor XIII.	100
Factor III, Thromboplastin	Tissue factor activates factor VII when blood is exposed to tissue fluids.	0
Factor IV, Ionized calcium	This active form of calcium is needed for the thromboplastin activation and for conversion of prothrombin to thrombin.	
Factor V, Proaccelerin or labile factor	This is consumed during clotting and accelerates the transformation of prothrombin to thrombin.	7
Factor VI, Nonexistent	This is activated by tissue thromboplastin, which in turn activates factor X.	0.5
Factor VII, Proconvertin or stable factor		
Factor VIII, Antihemophilic	This cofactor is used for the cleavage of factor X-Xa by IXa. Factor VIII is described as VIII/vWF:VIII:C active portion, measured by clotting, VIII:Ag is the antigenic portion, vWF:Ag measures antigen that binds to endothelium for platelet function; it is deficient in hemophilia A.	0.1
Factor IX, Plasma thromboplastin component	A component of the thromboplastin generating system, it influences amount as opposed to rate. It is deficient in hemophilia B, also known as Christmas disease. It is sex linked and vitamin K–dependent.	6
Factor X, Stuart-Prower	Final common pathway merges to form conversion of prothrombin to thrombin, activity also related to factors VII and IX.	10
Factor XI, Plasma thromboplastin antecedent	Essential to intrinsic thromboplastin generating of the cascade. Bleeding tendencies vary, but there is the risk of postoperative hemorrhage.	4
Factor XII, Hageman factor	This surface contact factor is activated by collagen. Patients do not bleed but have a tendency to thrombosis.	30
Factor XIII, Fibrin stabilizing factor	In the presence of calcium, this transaminase stabilizes polymerized fibrin monomers in the initial clot. This is the only factor that is not found in circulating plasma.	15
High-Molecular-Weight Kininogen (HMWK)	This surface contact factor is activated by kallikrein.	70
Prekallikrein, Fletcher factor	This is a surface contact activator, of which 75% is bound to HMWK.	50

One of the best methods to describe the coagulation mechanism is the cascade theory. This theory can be classified under two conditions: in vivo and in vitro conditions.

The coagulation cascade theory determined two starting points for the creation of thrombin: the intrinsic system with factor XII and surface contact; and the extrinsic system with factor VIIa and tissue factor. Both systems have a common pathway, where they produce factor Xa from X, thrombin from prothrombin and fibrin from fibrinogen.

Figs. 7.3 and 7.4 illustrate the cascade theory in vivo and in vitro (by the intrinsic and the extrinsic systems).

The extrinsic system is started by the activation of factor VII after damage to a vessel and due to the release of tissue thromboplastin. The result is a complex with calcium and tissue thromboplastin. Factors X and Xa are converted by this complex and then sequentially prothrombin is converted to thrombin by these factors (X and Xa). This procedure takes between 10 and 15 seconds from the initial point to the end point and shows the sensitivity and reaction time of the human body to damage.

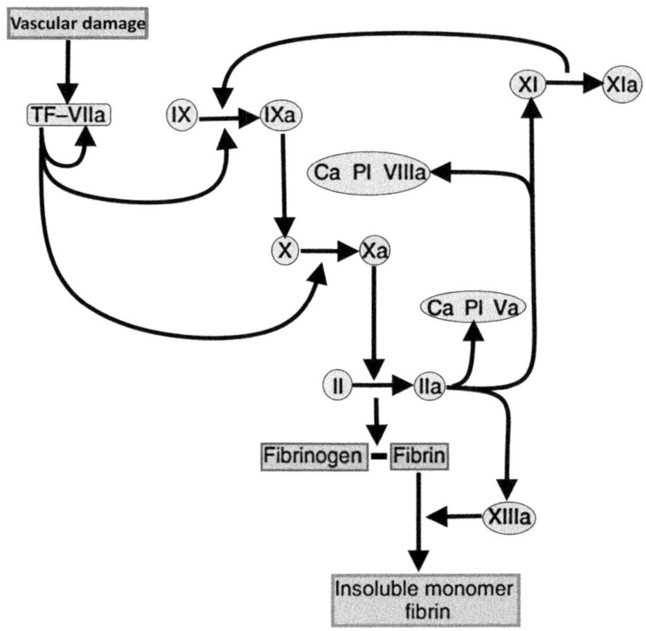

■ **FIGURE 7.3** In vivo coagulation cascade. *Modified from B. Ciesla, Hematology in Practice, F. A. Davis Company, Philadelphia, PA, 2007.*

In the intrinsic system, contact activation is begun by changes induced by vascular damage and the negatively charged surface of the tissue. A cofactor such as Prekallikrein is involved for the activation of factor XII by factor XIIa. High-Molecular-Weight Kininogen (HMWK) supports the activation of XI. Then XIa activates IX to IXa and also X converts to Xa (in the presence of VIIIa). At this point the reaction enters the common pathway and both intrinsic and extrinsic systems contain I, II, V and X factors. The procedures will end with in a fibrin monomer polymerizing into a fibrin clot and Factor XIII follows activation by thrombin [3], see Figs. 7.3 and 7.4. The figures show that calcium plays key role for the activation process.

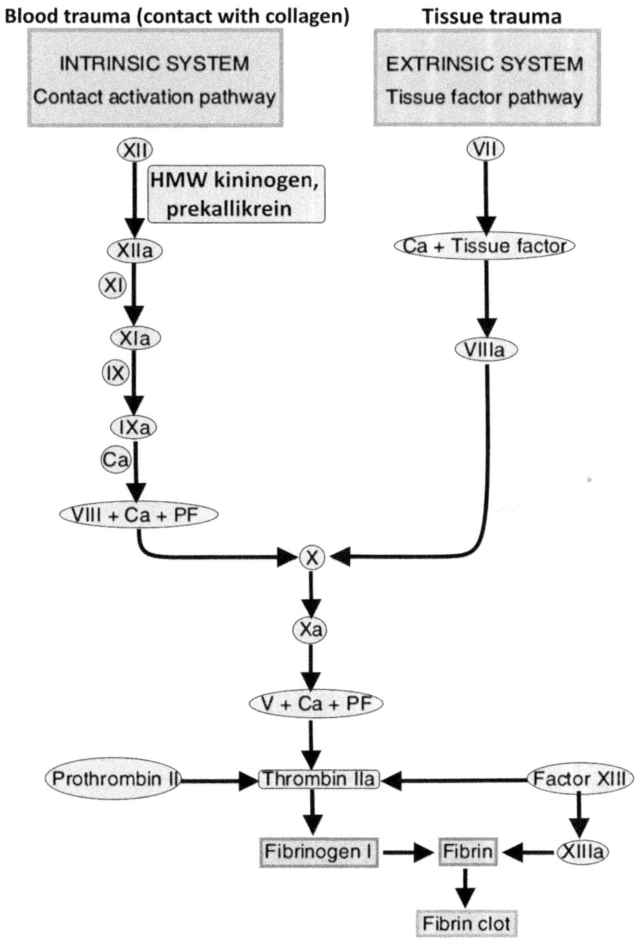

■ **FIGURE 7.4** In vitro coagulation cascade. *Modified from B. Ciesla, Hematology in Practice, F. A. Davis Company, Philadelphia, PA, 2007.*

7.2.2 Coagulation Inhibitors

Generally, inhibitors are soluble plasma proteins and perform as anticoagulants. The duty of these anticoagulants is to make a malfunction or error in the clotting cascade to prevent clotting or to keep it under control. Two major inhibitors are classified in plasma [3]:

1. Protease inhibitors (inhibitors of coagulation factors):
   - Antithrombin
   - Heparin cofactor II
   - Tissue factor pathway inhibitor
   - Alpha-2-antiplasmin
   - C1
2. The protein C pathway (inactivation of activated cofactors):
   - Protein C and protein S

7.2.3 Biomechanics of Clot

The mechanical properties of clots and fibrin are important due to their role in hemostasis. In the case of bleeding, the clot must form a plug to stop and control bleeding and also the material strength of the plug must be strong to resist the blood flow pressure. Also in the situation of thrombus development (see next subsection), the mechanical properties are essential since it responds to treatments, such as artery angioplasty.

Epidemiological investigations have demonstrated a connection between myocardial infarction and clot mechanical properties. If the clot is very stiff, it could be more friable or if the clot is more viscous or plastic, it could have a greater tendency to embolize. Fibrinogen—one of key factors in clot formation—is known as an elongated protein with an average concentration of about 2.5 g/L in the blood, with average length of 45 nm, and is made up of spherical domains at both ends connected by helical coils to a smaller spherical region in the middle. Considering Hooke's and Newton's laws, fibrin mechanical behavior can be classified as a viscoelastic polymer [4].

The elastic modulus of clots is varied and depends on the clot structure, and it can be changed by physiological conditions. For instance the elastic modulus is about two to four times greater for clots formed from thrombin than for those formed from the cleavage of only the A fibrinopeptides. The reported range of modulus varies, from less than 1 dyne/cm^2 to about $\sim$23,000 dyne/cm^2 [4–6].

The branching of fibers creates a network in the clot. Fig. 7.5 shows a scanning electron micrograph of a fibrin clot.

■ **FIGURE 7.5** Scanning electron micrograph (SEM) of a clot with thin fibers and numerous branch points at high thrombin concentration. Magnification bar = 5 μm. *From J.W. Weisel, The mechanical properties of fibrin for basic scientists and clinicians, Biophys. Chem. 112 (2004) 267–276 with permission.*

Table 7.2 Mechanical Properties of Fibrin Fibers (Cross-Linked, Uncross-Linked)

Fiber Type	Cross-Linked	Uncross-Linked
ε_{max} (extensibility)	147% ± 5%	226% ± 8.7%[a]
$\varepsilon_{elastic}$ (elastic limit)	≤ 50%	120%[a]
E_0 (MPa) (total elastic modulus)	8.0 ± 1.0	3.9 ± 0.6
E_∞ (MPa) (relaxed elastic modulus)	4.0 ± 0.6	1.9 ± 0.3
τ_1 (s) fast relaxation time	2.1 ± 0.2	3.6 ± 1.3
τ_2 (s) slow relaxation time	49 ± 4	57 ± 8
h (strain hardening factor)	1.9 ± 0.3	3.0 ± 0.3
E_{loss} (percentage energy loss per stretch cycle)	≤ 40–70%	0–70%

Adapted from W. Liu, C.R. Carlisle, E.A. Sparks, M. Guthold, The mechanical properties of single fibrin fibers, J. Thromb. Haemost. 8 (2010) 1030–1036.
[a]Values from [8]

Some studies have been carried out to measure the elastic modulus of an individual fibrin fiber. The results revealed that the maximum elastic modulus for individual cross-linked and uncross-linked fibrin fibers are around 9 and 4.5 Mpa, respectively. Also, the experiments reveal that the elastic limit ($\varepsilon_{elastic}$) of fully cross-linked fibers should be less than 50% strain. Cross-linked fibers are stiffer, with less extensibility and less pronounced strain hardening [7].

Table 7.2 summarizes the mechanical properties of cross-linked and uncross-linked fibrin fibers. As the table shows, crosslinking increases the fiber stiffness of fibrins by an approximate factor of two, for instance the total elastic modulus, E_0, increases from 3.9 to 8.0 MPa and the relaxed elastic modulus, E_∞, increases from 1.9 to 4.0 MPa upon cross-linking. Furthermore, fibers will be hardened with increasing strain (*h*, strain hardening), as E_0 increases by a factor of 1.9 (cross-linked) and 3.0 (uncross-linked) at strains greater than 110% [7].

7.2.4 Thrombosis

Thrombosis is the development of a clot inside blood vessels and is thought to begin with a malfunctioning endothelium or vessel damage and the outcome is the clot activation procedure on the blood vessel wall. Then, this clot can block the blood vessel or can move and block the blood vessels inside the blood circulation system (see Fig. 7.6). Whether thrombosis develops depends on several factors including biochemical factors, blood flow and the geometry of vessel and damaged area.

Thrombosis is classified into two major types:

- Arterial thrombosis: Generally, arterial thrombosis occurs after the destruction or rupture of an atherosclerotic plaque, and can cause ischemic injuries, especially in organs or tissues with a terminal vascular bed. Stroke and cardiac ischemia are the most acute clinical manifestations of atherothrombosis [9].
- Venous thrombosis: It is a blood clot that forms inside a vein. Venous thromboembolism (VTE) is known as the most common vascular disease after acute myocardial infarction and stroke. Usually, VTE is characterized by two main clinical events: deep venous thrombosis (DVT); and pulmonary embolism (PE). Clinical studies have showed that PE generally follows DVT [10].

Biomechanical measurements of thrombosis in in vivo conditions are not easy. A method was introduced to estimate the viscoelastic properties of a thrombus by using an ultrasonic shear-wave [11].

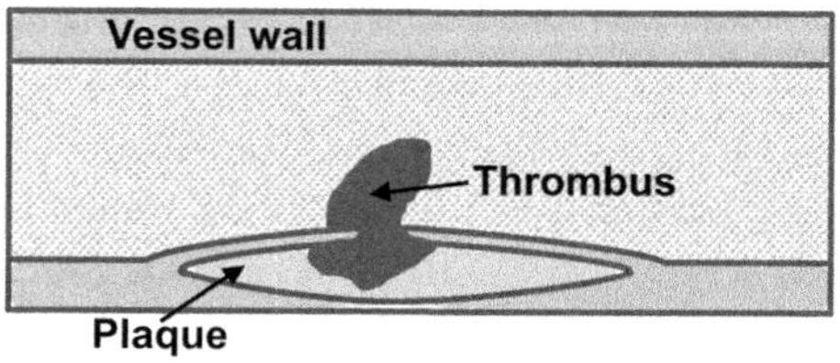

■ **FIGURE 7.6** Schematic of a thrombus inside blood vessel.

The hemodynamics aspects, such as shear rate, have a main role in thrombosis formation and in its growth. The thrombosis formation is a time-dependent function. The following equations describe thrombus growth rate, J [12]:

$$J_i = \frac{1}{2\pi(r_{i,0} - h_i(t))w_i}\frac{d}{dt}\pi(r_{i,0}^2 - (r_{i,0} - h_i(t))^2)w_i \tag{7.1}$$

where r_0 is the initial vessel radius, h is the time-dependent thrombus thickness, t is time and w is the axial width being sampled at a given location, based on rectangular sectioning. Also, the upwind differentiating scheme can be used every second to define the local thrombus growth rate, J, derived from reduction of Eq. (7.1) [12]:

$$J_i = \frac{(h_i^n - h_i^{n-1})}{\Delta t} \tag{7.2}$$

The superscript n and i denote the time step and the axial point, respectively.

Research was carried out in vitro (by porcine whole blood and a Pyrex stenosis setup) and a positive correlation was found between thrombus accumulation rates and wall shear rate up to 6000 seconds^{-1}, with a decrease in growth rates at wall shear rate >6000 seconds^{-1}. In addition, two series of equations were derived to show the correlation between shear rate and thrombus growth in an in vitro condition [12]. Fig. 7.7 illustrates the results of the research.

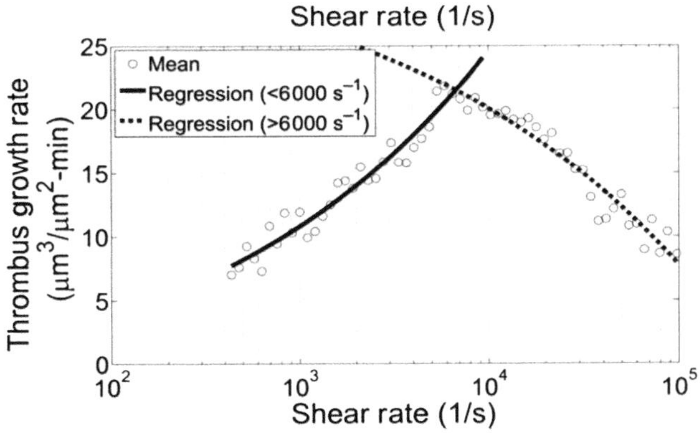

FIGURE 7.7 Relation between thrombus growth rates and wall shear rates. Plot of power law regressions for low (<6000 s^{-1}, solid line $r^2 = 0.92$) and high shear rate (<6000 s^{-1}, dotted squares $r^2 = 0.94$). *Reprinted from D.L. Bark Jr., A.N. Para, D.N. Ku, Correlation of thrombosis growth rate to pathological wall shear rate during platelet accumulation, Biotechnol. Bioeng. 109 (10) (2012) 2642–2650 with permission.*

7.3 ROLLING ADHESION AND EFFECT OF FORCES

Blood cells adhesion and rolling on the vessel walls are particularly important for circulation diseases, such as stenosis and thrombosis.

Generally, the adhesion of blood cells to other cells or to the extracellular matrix is a dynamic action. Cells' adhesion will be changed by environmental cues during morphogenesis, tissue remodeling or other responses. Blood flow is the best medium to understand the dynamic behavior of blood cells. Rolling adhesion on a vessel wall is the first phase for some biofluids aspects such as circulating leukocytes, hematopoietic progenitors, or platelets activation and attachment to specific organs or to sites of infection or injury. Cell rolling requires the rapid formation, attachment and detachment of adhesive bonds in the blood flow between two surfaces or molecules. The adhesive bond is known as the sum of noncovalent interactions, for instance, van der Waals forces, hydrogen bonds, electrostatic interactions, dipole–dipole interactions. The role of adhesion receptors is vital in this procedure. They diffuse laterally in the cell membrane and bind ligands on another membrane or in the extracellular matrix. The adhesion receptors are carried by moving cell to ligands on the surface of another cell or in the matrix in normal and tangential directions. Blood flow imposes extra transport and mechanical limitations on interacting molecules as one of them resides on a moving cell [13].

Generally, after an infection of the body by a virus or bacterium, the WBCs try to remove the foreign object. In the first step, the foreign object induces the capillary endothelium to express molecules of adhesion on its cell surfaces. The WBCs bind to receptors on the endothelium of the vessel wall and then start the rolling procedure which involves translation and irregular rotation of the WBCs, then they adhere to the infected area to remove it [14].

Fig. 7.8 shows the rolling adhesion of a WBC on a surface and its mechanical variables. F_r and T are the resultant force (along the x axis) and torque

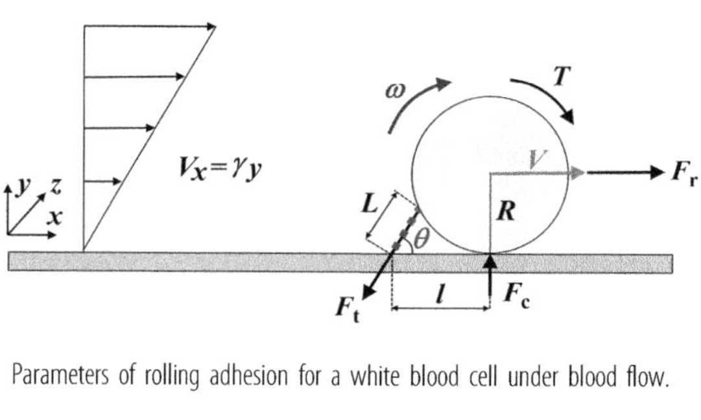

■ FIGURE 7.8 Parameters of rolling adhesion for a white blood cell under blood flow.

(along the z axis about the cell center), respectively. The rolling motions of a WBC of radius R are governed by the F_r, the T applied by the blood flow and also the tether force F_t applied through the receptor–microvillus bond, and the contact force F_c. The length of the microvillus is L and l is the lever arm between the center of the blood cell and the microvillus. The blood flow velocity, V, bordered with a vessel wall (or solid surface) is parallel to the surface with no slip condition (it increases from zero on the surface to the maximum velocity in the center line of the vessel or tube). In this rolling motion, the translational velocity and angular velocity are V and ω, respectively, and the shear rate of blood flow, γ, is the ratio of local flow velocity to the distance from the surface (see Fig. 7.8).

Tether formation affects the force acting on the blood cells. Mathematically, the governing equations of rolling adhesion can be written as [14]:

$$F_t \cos\theta = 32.054\, R^2 \tau_w \tag{7.3}$$

$$F_t l \sin\theta = 43.916\, R^3 \tau_w \tag{7.4}$$

$$\tau_w = \mu\,\gamma \tag{7.5}$$

where γ is the shear rate, μ is the blood viscosity, τ_w is the wall shear stress and l is lever length (see Fig. 7.8). Also the theta is [14]:

$$\theta = \tan^{-1}\left(\frac{R}{l}\right) + \cos^{-1}\left(\frac{l^2 + L^2}{2L\sqrt{l^2 + R^2}}\right) \tag{7.6}$$

where R is cell radius and L is microvillus length.

A study presented a technique to measure the adhesion bond force of the tether formation from WBC membranes. The result revealed that a minimum force of 45 pN (Pico Newton) is necessary for tether formation [14–16].

Considering the microvillus as a spring with a spring constant of k_1, for forces below 45 pN, the force on the bond is defined by [14,15]:

$$F_t = k_1(L - L_0) \tag{7.7}$$

where k_1, microvillus spring constant, is 43 pN/μm and L_0 is the length of unstressed microvillus and is equal to 0.35 μm. With regard to the initial bond force, F_0, in the process of microvillus tether formation, F_t can be written as [14–16]:

$$F_t = F_0 + k_2 \frac{dL}{dt} \tag{7.8}$$

where k_2 is related to variations of the spring constant in time and is equal to 11 pN s/μm, $\frac{dL}{dt}$ is the pulling velocity of the tether off the surface and F_0 is 45 pN (in the case of a Neutrophil).

The F_t can be calculated by solving the coupled system of governing Eqs. (7.7) and (7.8). The anticipated tether length, L, and moment arm, l, for shear stress of 0.08 pN/m against time are presented in Fig. 7.9. The figure shows that the F_t decreases in time, (by 50% in 0.2 seconds) and then it decreases with a low slope (close to a plateau). The moment arm and tether length increase rapidly, especially before 0.2 seconds.

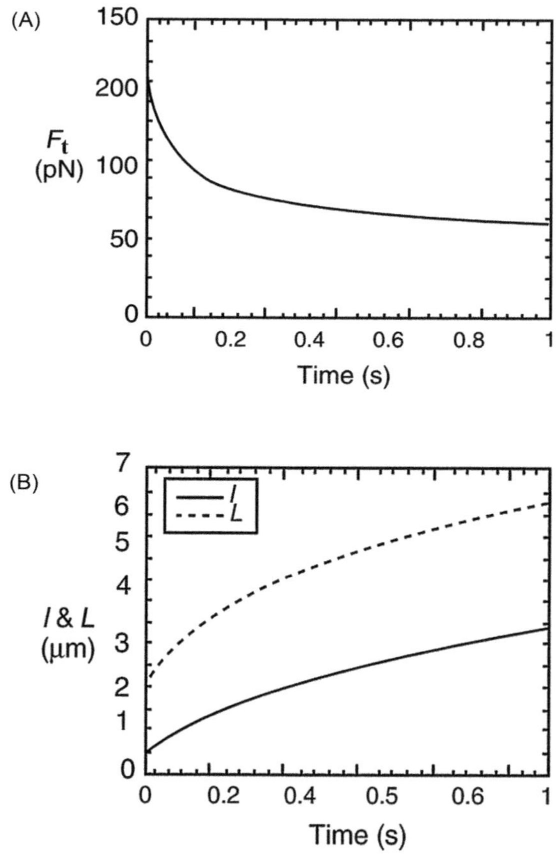

■ **FIGURE 7.9** Time evolution of (A) the tether bond force, F_t, and (B) moment (lever) arm, l, and tether length, L in time. *From J.Y. Shao, H.P. Ting-Beall, R. Hochmuth, Static and dynamic lengths of neutrophil microvilli, Proc. Natl. Acad. Sci. USA, 95 (12) (1998) 6797–6802 with permission.*

7.4 BIOCOMPATIBILITY AND BIOMATERIALS

For thousands of years, humans have used materials such as gold, silver, glass, iron, etc. for medical purposes. Archeologists discovered the world's oldest artificial eye that belonged to a woman of 28–32-years-old in Shahr-e Sūkhté (the burnt city) in Iran. They estimated the age of the eye to be around 3000 BC and was made of bitumen paste and gold. This discovery is the first known ocular prosthesis in medical history [18]. Probably, similar ancient using artificial materials as an assistive or prosthesis parts may have been used throughout the world.

For more than five decades medicine and biotechnology have used a wide range of biomaterials including metals, polymers and ceramics in medical devices. For decades, there were a few developments in the biomaterials industry but during recent decades, scientists and researchers have made and introduced a wide range of biomaterials.

7.4.1 Biocompatibility

Biocompatibility is the main precondition for any material intended to be used as components for medical devices. Using incompatible materials generates biological reactions from the patient's body and even creates fatal damage to their user. Considering these concerns, international and domestic regulations have been applied to ensure that manufacturers of medical devices follow the standards and biocompatibility requirements for their products.

One of the famous standards for assessment of biocompatibility for medical devices is International Standard ISO-10993-1, Biological Evaluation of Medical Devices—Part 1: Evaluation and Testing within a Risk Management Process. This standard is the most used standard for evaluating biocompatibility and provides a road map with several steps to plan an assessment for biological evaluation of medical devices.

The standard describes [19]:

- test selection;
- sample preparation;
- test considerations for cytotoxicity, sensitization, hemocompatibility, pyrogenicity, implantation, genotoxicity, carcinogenicity, reproductive and developmental toxicity and biodegradation;
- animal safety rules to validate omission of specific biocompatibility tests;
- evaluation of known or potentially toxic chemical entities and
- filing of a biocompatibility test report.

The ISO 10993 standard and US Food and Drug Administration (FDA) have introduced an evaluation system for test requirements which are

Table 7.3 Classification of Contact Durations and Tissue Contact Types for Medical Devices

Contact Duration	
Limited ≤24 h	
Prolonged >24 h and <30 days	
Permanent >30 days	
Tissue Contact	
Surface devices	Intact skin
	Mucosal membrane
	Breached or compromised surface
External communicating devices	Blood path, indirect
	Tissue/bone/dentin
	Circulating blood
Implant devices	Tissue/bone
	Blood

based on contact duration and types of tissue contact. Table 7.3 represents the classification of medical devices according to tissue contact and contact duration.

Generally, biocompatibility of medical devices depends on their application inside the human body. The best example is the differences between internal use (in vivo) and external use (ex vivo) biomaterials. ISO 10993 and FDA recommended several biological control tests for biocompatibility and one of them is hemocompatibility.

Hemocompatibility: The purpose of hemocompatibility tests is to evaluate the effects and changes on blood and its components by medical devices or chemical leaching from biomaterials. This test covers medical devices which have a very long-term and active contact with blood in the circulation system.

The ISO considers two levels for each tissue contact category (see Table 7.3), main (level 1) and optional (level 2). Furthermore, ISO 10993-4 defines five categories for hemocompatibility tests for each level:

- Thrombosis, by light microscopy or scanning electron microscopy
- Coagulation, by partial thromboplastin time
- Platelets, by platelet count
- Hematology, by leukocyte count and differential; hemolysis
- Immunology, by C3a, C5a, TCC, Bb, iC3b, C4d, SC5b-9.

In addition to these tests, hemolysis testing is recommended by using the ASTM F756 standard (both direct and indirect methods) or an equivalent methods.

One of the challenging issues for hemocompatibility tests is the blood source. Some standards accept both human and nonhuman bloods but some standard systems prohibit nonhuman blood for testing (eg, European systems).

Tables 7.4 and 7.5 classify the biological evaluation tests of medical devices regarding the ISO and FDA standards for initial and supplementary tests. As Table 7.4 shows, for external communicating and implant devices which have direct contact with human blood, the following tests are required for most cases (except for a few "A-limited" (or $\leq$24 hours) contact times):

- Cytotoxicity test, is the first test to evaluate the toxic effects of the material to mammalian cells
- Sensitization test, is the evaluation of allergic response to materials by cells and the immune system mechanism
- Irritation or Intracutaneous reactivity test, is the assessment of local inflammatory response to materials
- Systemic toxicity (acute) test, is the evaluation of the potential adverse of materials on organs or tissues which are far from the contact area for short time ($t \leq 28$ days)
- Subchronic toxicity (subacute toxicity), is the evaluation of the potential adverse of materials on other organs or tissues which are far from the contact area for long time ($1 \leq t \leq 3$ months)
- Genotoxicity test, is the evaluation of DNA damage (for over 30 days)
- Implantation test, is implanting the material to ascertain the local tissue reaction (on animal, for less than 13 weeks)
- Hemocompatibility (explained before).

7.4.2 Biomaterials

Biomaterials are defined as the materials which have a positive or neutral interaction with living organs or tissues. Biomaterials can be classified into three main types: metals; polymers and ceramics. Metals are known as the most economic material with a wide range of applications in biomedical engineering fields, including surgical tools, implants (orthopedic, dentistry and etc.), medical devices, and laboratory equipment, etc. The typical implant metals are 316 L stainless steel (L denotes its low carbon content), Cobalt–Chrome (Co-Cr)

Table 7.4 Initial Evaluation Tests for Consideration

Device Categorization by			Biologic Effect								
Nature of Body Contact		Contact Duration (See 5.3) A—Limited (≤24 h) B—Prolonged (>24 h to 30 d) C—Permanent (>30 d)									
Category	Contact		Cytotoxicity	Sensitization	Irritation or Intracutaneous Reactivity	Systemic Toxicity (Acute)	Subchronic Toxicity (Subacute Toxicity)	Genotoxicity	Implantation	Hemo-compatibility
Surface device	Intact skin	A	X	X	X					
		B	X	X	X					
		C	X	X	X					
	Mucosal membrane	A	X	X	X					
		B	X	X	X	O	O		O	
		C	X	X	X	O	X	X	O	
	Breached or compromised surface	A	X	X	X	O				
		B	X	X	X	O	O		O	
		C	X	X	X	O	X	X	O	
External communicating device	Blood path, indirect	A	X	X	X	X				X
		B	X	X	X	X	O			X
		C	X	X	O	X	X	X	O	X
	Tissue/bone/dentin+	A	X	X	X	O				
		B	X	X	O	O	O	X	X	
		C	X	X	O	O	O	X	X	
	Circulating blood	A	X	X	X	X		Ô		X
		B	X	X	X	X	O	X	O	X
		C	X	X	X	X	X	X	O	X
Implant device	Tissue/bone	A	X	X	X	O				
		B	X	X	O	O	O	X	X	
		C	X	X	O	O	O	X	X	
	Blood	A	X	X	X	X	X		X	X
		B	X	X	X	X	O	X	X	X
		C	X	X	X	X	X	X	X	X

Adapted from U. F. a. D. Administration, "Initial Evaluation Tests for Consideration," 28 5 2015 [Online]. Available: http://www.fda.gov/MedicalDevices/DeviceRegulationandGuidance/GuidanceDocuments/ucm080742.htm with permission.
X = ISO Evaluation Tests for Consideration.
O = Additional Tests which may be applicable.
Note+: Tissue includes tissue fluids and subcutaneous spaces.
Note^: For all devices used in extracorporeal circuits.

Table 7.5 Supplementary Evaluation Tests for Consideration

Device Categorization by			Biologic Effect			
Nature of Body Contact		Contact Duration (See 5.3) A—Limited (≤24 h) B—Prolonged (>24 h to 30 d) C—Permanent (>30 d)				
Category	**Contact**		**Chronic Toxicity**	**Carcinogenicity**	**Reproductive/ Developmental**	**Biodegradable**
Surface device	Intact skin	A				
		B				
		C				
	Mucosal membrane	A				
		B				
		C	O			
	Breached or compromised surface	A				
		B				
		C	O			
External communicating device	Blood path, indirect	A				
		B				
		C	X	X		
	Tissue/bone/ dentin	A				
		B				
		C	O	X		
	Circulating blood	A				
		B				
		C	X	X		
Implant device	Tissue/bone	A				
		B				
		C	X	X		
	Blood	A				
		B				
		C	X	X		

Adapted from FDA, "510(k) Memorandum - #G95-1 Attachment B - Table 2 Supplementary Evaluation Tests for Consideration," FDA, 28 5 2015 [Online]. Available: http://www.fda.gov/MedicalDevices/DeviceRegulationandGuidance/GuidanceDocuments/ucm080752.htm with permission.
X = ISO Evaluation Tests for Consideration.
O = Additional Tests which may be applicable.

alloys, Titanium (Ti) alloys, Cobalt–Chromium–Molybdenum (Co-Cr-Mo) alloy, Cobalt–Chromium–Tungsten–Nickel (Co-Cr-W-Ni) and Co-Ni-Cr-Mo-Ti.

Polymers are useful in medicine fields due to their flexibility in chemical structure and physical properties. The application range of polymeric biomaterials are very wide like metals and they have uses from a pump chamber for drug delivery to scaffolds for tissue engineering field. The standard polymeric biomaterials are polyethylene, polypropylene, polyurethane, polymethyl-methacrylate polyvinyl chloride, polycarbonate, polyethersulfone (PES), etc. PES polymer is one of the most important materials (as a filter membrane) in the field of blood filtration and hemodialysis, in particular as a PES membrane in hemodialysis machines.

In the biomaterial industry, ceramics and glasses are recognized as highly biocompatible materials. These materials have uses from an eyeglass to a femoral head of a total hip joint replacement. Clinical applications require many forms of bioceramics or glasses, from powders to solid bodies. Calcium phosphate group, alumina, titania, silicon nitride, silicon carbide, bioactive glasses are well-known materials in this category of biomaterials.

CHAPTER SUMMARY

- The human immune system can resist almost all kinds of organisms or toxins that have a tendency to damage living tissues and human organs.
- Antibodies act in two ways to defend the body against invaders: (1) by direct attack on the invading agents and (2) by activation of the "complement system," a system of about 20 proteins, to destroy the invading agents.
- The hemostatic is a mechanism to control and stop bleeding from damaged blood vessels.
- The hemostasis is completed by several procedures: constriction of blood vessels, platelet plug formation, blood clot formation and growing fibrous tissue into the blood clot to block the hole in the vessel.
- Platelets have two main functions in the blood circulation system: (1) to decrease and control bleeding by formation of platelet plugs; (2) control of platelet plugs by catalyzing coagulation reactions, leading to the fibrin formation.
- There are 50 elements involved in the coagulation procedure. They can be classified into two groups: (1) procoagulant elements which promote coagulation and (2) anticoagulants that prevent coagulation.
- Epidemiological investigations have demonstrated a connection between myocardial infarction and clot mechanical properties.

- Thrombosis is the development of a clot inside blood vessels and is thought to begin with a malfunctioning endothelium or vessel damage leading to the outcome of a clot activation procedure on the blood vessel wall.
- Thrombosis is classified into two major types: arterial and venous thrombosis.
- Blood cells adhesion and rolling on the vessel walls are particularly significant for body infections, such as viral and bacterial diseases, or circulation diseases, such as stenosis and thrombosis.
- Biocompatibility is the main precondition for any material intended to be used as components for medical devices.
- The ISO 10993 standard and US FDA have introduced an evaluation system for test requirements which are based on contact duration and types of tissue contact.
- The purpose of hemocompatibility tests are to evaluate the effects and changes on blood and its components by medical devices or chemical leaching from biomaterials.
- Biomaterials can be classified into three main types: metals; polymers and ceramics.
- PES polymer is one of the most important materials (as a filter membrane) in the field of blood filtration and hemodialysis.

PROBLEMS

1. What is the difference between an antigen and antibody?
2. How many factors should work in a coagulation process?
3. Discuss Fig. 7.3.
4. What is the difference between Figs. 7.3 and 7.4?
5. Explain thrombus formation inside blood vessels.
6. What are the most acute clinical manifestations of atherothrombosis?
7. What is your suggestion to measure thrombosis degree in vivo?
8. Discuss rolling and rolling adhesion in biofluid flow?
9. Discuss the role of shear rate in rolling adhesion of WBCs.
10. Imagine you are designer of an implantable artificial kidney, which standards and tests must be considered by you to pass ISO and FDA requirements?

REFERENCES

[1] A.C. Guyton, J.E. Hall, Textbook of Medical Physiology, Elsevier Saunders, Philadelphia, PA, 2006.

[2] B.D. Ratner, A.S. Hoffman, F.J. Schoen, J.E. Lemons, Biomaterials Science, an Introduction to Materials in Medicine, Academic Press, London, 1996.

[3] B. Ciesla, Hematology in Practice, F. A. Davis Company, Philadelphia, PA, 2007.

[4] J.W. Weisel, The mechanical properties of fibrin for basic scientists and clinicians, Biophys. Chem. 112 (2004) 267–276.

[5] M.D. Bale, M.F. Muller, J.D. Ferry, Rheological studies of creep and creep recovery of unligated fibrin clots: comparison of clots prepared with thrombin and ancrod, Biopolymers 24 (1985) 461–482.

[6] M.E. Carr, A. Krishnaswami, E.J. Martin, Platelet contractile force (PCF) and clot elastic modulus (CEM) are elevated in diabetic patients with chest pain, Diabet. Med. 19 (2002) 862–866.

[7] W. Liu, C.R. Carlisle, E.A. Sparks, M. Guthold, The mechanical properties of single fibrin fibers, J. Thromb. Haemost. 8 (2010) 1030–1036.

[8] W. Liu, L. Jawerth, E. Sparks, M. Falvo, R. Hantgan, R. Superfine, et al., Fibrin fibers have extraordinary extensibility and elasticity, Science 634 (2006) 313.

[9] R.E. Rumbaut, P. Thiagarajan, Platelet-Vessel Wall Interactions in Hemostasis and Thrombosis, Morgan & Claypool Life Sciences, San Rafael, CA, 2010.

[10] E. Previtali, P. Bucciarelli, S.M. Passamonti, I. Martinelli, Risk factors for venous and arterial thrombosis, Blood Transfus. 9 (2011) 128–138.

[11] C.-C. Huang, P.-Y. Chen, C.-C. Shih, Estimating the viscoelastic modulus of a thrombus using an ultrasonic shear-wave approach, Am. Assoc. Phys. Med. 40 (4) (2013) 042901–042907.

[12] D.L. Bark Jr., A.N. Para, D.N. Ku, Correlation of thrombosis growth rate to pathological wall shear rate during platelet accumulation, Biotechnol. Bioeng. 109 (10) (2012) 2642–2650.

[13] R.P. McEver, C. Zhu, Rolling cell adhesion, Annu. Rev. Cell Dev. Biol. 26 (2010) 363–396.

[14] G.A. Truskey, F. Yuan, D.F. Katz, Transport Phenomena in Biological Systems, Pearson, Upper Saddle River, NJ, 2010.

[15] C. Pozrikidis, Modeling and Simulation of Capsules and Biological Cells, CRC Press, Boca Raton, FL, 2005.

[16] R.M. Hochmuth, J.Y. Shao, Micropipette suction for measuring piconewton forces of adhesion and tether formation from neutrophil membranes, Biophys. J. 71 (1996) 2892–2901.

[17] Deleted in review.

[18] A.N. Moghadasi, "Artificial Eye in Burnt City and Theoretical Understanding of How Vision Works," Iran J Public Health vol. 43 (no. 11) (2014) 1595–1596.

[19] U. D. o. H. a. H. S. Food and Drug Administration, "FDA.gov," 2013 [Online]. Available: <http://www.fda.gov/downloads/medicaldevices/deviceregulationandguidance/guidancedocuments/ucm348890.pdf>.

[20] Deleted in review.

[21] Deleted in review.

Chapter 8

Real Time Measurement Techniques of Biofluids

LEARNING OBJECTIVES

- Description of the sensors and transducers.
- Characterization of Biosensors.
- Characterization and classification of sensors and transducers in medical systems.
- Characterization and classification of the measurement techniques for body temperature.
- Characterization and classification of the measurement techniques for the flow rate and volume of blood.
- Characterization and classification of the measurement techniques for blood pressure and sound.
- Explanation of cardiac catheterization (Heart-Cath).
- Description and classification of measurement techniques for the respiratory system.
- Description and classification of the medical imaging systems.

8.1 INTRODUCTION

There is no boundary for technology. During last four decades, a huge revolution has occurred for the electronics industries including biomedical engineering, although the technology revolution alone cannot make medicine humane. The most important aspect in biomedical devices is safety and reliability.

Around the world there are inclusive regulations, rules and standards which apply to the planning, designing, manufacturing, testing and installation of

Biofluid Mechanics. DOI: http://dx.doi.org/10.1016/B978-0-12-802408-9.00008-9

medical devices. The ISO, FDA, DIN, ASTM, ANSI and AAMI represent the comprehensive standards, rules and requirements for the medical devices industry, from the initial materials to quality control in production line.

Medical devices are made of several parts including electrical, mechanical, or electromechanical parts, such as sensors. There are two sensing systems for biomedical purposes: regular transducers and biosensors. Sensors, transducers and biosensors have a major role in the diagnosis and treatment of diseases in the health care system, from the simple stethoscope or otoscope to advanced image processing or monitoring systems.

8.1.1 Sensors and Transducers

What is a sensor? A sensor is a device to sense a physical quantity such as: voltage, current, temperature, pressure, mass, strain, humidity, light, flow, etc. A signal sensor is needed to sense and convert these variables to indicate the value or quantity of the variable. Typically, a time-dependent analog voltage is the best signal for this purpose and is related to the physical variable. The procedure of signal transfer is made by a transducer. The transducer transfers this signal to an analog/digital converter before it is transferred to a microprocessor which only recognizes digital signals, then it converts the signal to analog again using a digital/analog converter to show the output on screen. Fig. 8.1 shows the schematic of signal processing in a sensor system.

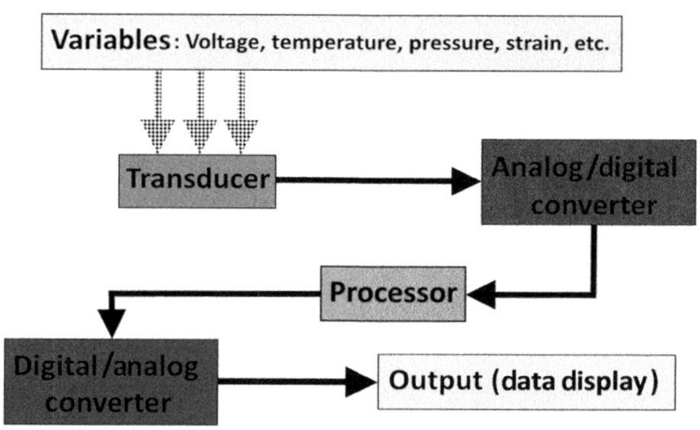

■ **FIGURE 8.1** Schematic of a simple measurement system to collect and analyze the physical variables.

8.1.2 Biosensor Classification

A biosensor is a sensor to detect an analyte and is made of a biological element which acts as a sensor or a transducer for detecting and transmitting the signal. Biosensors can be classified by the origin of their main components: biochemical components, transducer and analytes. Table 8.1 explains the main components of the biosensors. From the table, it can be seen that the biosensor construction has two main parts (or layers): (1) transducer or electrode and (2) bioreaction material or biointerface. Fig. 8.2 shows a simple view of biosensors. A bioreceptor layer detects the biochemical components, protein, tissue, virus and bacterium. After detection, the transducer transfers a signal to a microprocessor to analyze the data. The transducer can work based on electrochemical, optical,

Table 8.1 Classification of Sensors and Transducers in Medical Systems

Biosensors		
Bioreceptors	**Transducers**	**Analytes**
■ Antibodies ■ Nucleic acids ■ Enzymes ■ Tissues ■ Cells ■ Microorganisms	■ Electrochemical ■ Microelectromechanical (MEMS) ■ Optical ■ Piezoelectric ■ Mass based ■ Thermometric	■ Disease biomarkers ■ Drugs ■ Metabolites ■ Pollutants ■ Vitamins and antioxidants ■ Blood components ■ Others

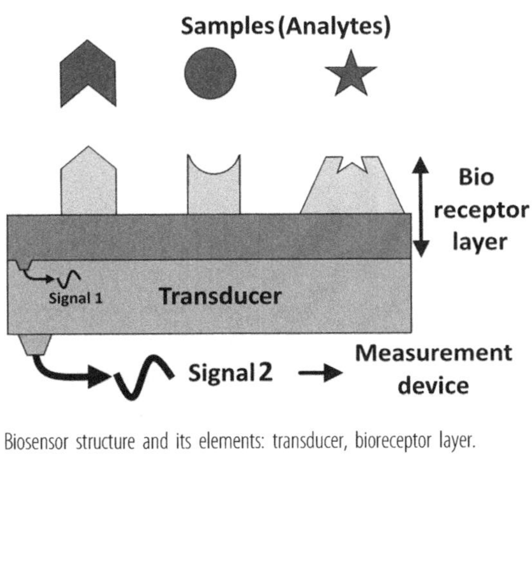

■ **FIGURE 8.2** Biosensor structure and its elements: transducer, bioreceptor layer.

thermal or piezoelectric approaches. The bioreceptor will bind to the analyte and cause the transducer to produce and transfer a signal. There are two signals, the primary signal from the bioreceptor to the transducer, and the second signal from the transducer to the processor system. Table 8.1 and Figs. 8.1 and 8.2 further clarify the biosensor, transducer structure and its components.

8.2 MEASUREMENT OF TEMPERATURE

A patient's body temperature is one of the most important aspects among their disease symptoms. Some medical problems, such as infections, shocks, etc., are revealed by the abnormal temperature of a patient's body (or skin). Image processing technology can reveal some internal problems, such as inflammation, arthritis or even the emotional behavior of patients, by measuring and detecting temperature differences inside the human body. Several units exist for measuring the temperature: Celsius or centigrade (°C); Fahrenheit (°F) and Kelvin (K). These units can be converted to each other;

$$°C = (°F - 32) \times 1.8 \tag{8.1}$$

$$°C = K - 273.15 \tag{8.2}$$

$$K = (°F - 32) \times 1.8 + 273.15 \tag{8.3}$$

The average normal body temperature in humans is about 37°C or 98.6°F.

Several types of measurement are used to collect thermal information from a patient‘s body [1]:

- Fluid/glass thermometer
- Thermocouple
- Thermistor
- Radiation and optical system

8.2.1 Fluid/Glass Thermometer

The fluid/glass thermometer is the most popular thermometer used around the world to measure body temperature. It consists of a sealed glass tube with a small internal diameter (like a lumen) connected to a reservoir. Increasing the temperature expands and pushes up the fluid in the reservoir and tube. A scale on the tube and the location of the expanded fluid indicate the temperature, see Fig. 8.3A.

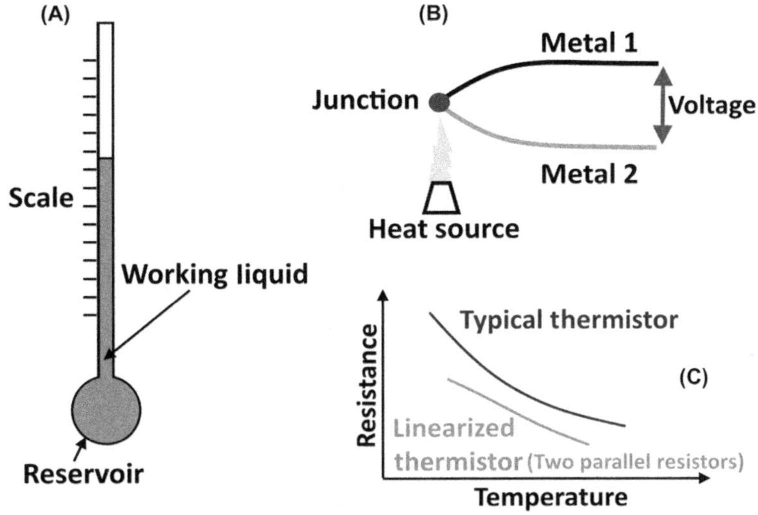

■ **FIGURE 8.3** (A) Fluid/glass thermometer. (B) Thermocouple and (C) Resistance/Temperature graph for typical thermistor (one resistor) and linearized (two parallel resistors) thermistor.

Usually, mercury and petrolate are the most common fluids in this kind of thermometer. Mercury and petrolate usually can measure temperature in a range of −40 to +900°C and −200 to +260°C, respectively [2].

8.2.2 Thermocouple

A thermocouple is an electrical circuit with two dissimilar metal wires which are joined together. If one end of this circuit is at a higher temperature than the other side (junction of two wires), there is an electron flow (or current) from one side to the other side. A voltmeter can measure the voltage across the two metal wires (see Fig. 8.3B). This voltage effect is known as the Seebeck voltage after the man who discovered it in 1821.

The voltage is at very low scales and the sensitivities of common thermocouples range from 6.5 to 80 μV/°C at 20°C with accuracy from 0.25% to 1% [1].

8.2.3 Thermistor

A thermistor is a semiconductor made of ceramic materials and it reacts like a resistor which is sensitive to temperature. Usually, bioapplication thermistors have a high negative temperature coefficient, which means that their resistance increases as temperature decreases, and vice versa. In biomedical applications, the resistivity of a thermistor is between 0.1 and 1000 $\Omega \cdot m$ [1].

One the most common thermistors in medical applications is the YSI 400 series thermistor. One of the issues in thermistors is their high nonlinear behavior and this is solved connecting a parallel resistor to the thermistor resulting in a linear behavior (close to linear) for the thermistor [2], see Fig. 8.3C.

8.2.4 Radiation and Optical Systems

This method is associated with the heat radiation from a heat source based on heat transfer principles. This technique makes it possible to measure the body temperature without physical contact. The name of this method is medical thermography.

The heat transfer inside the human body can vary due to physiological conditions, for example, cellular or blood circulation characters. Medical infrared imaging is one of the most advanced medical thermal imaging methods in the world. An infrared camera is equipped to the microbolometer to detect the heat sources. The high resolution of this system helps to show the details of human tissues such as blood vessels. Fig. 8.4 shows a thermogram from a patient's chest by using an infrared camera from Spectron IR.

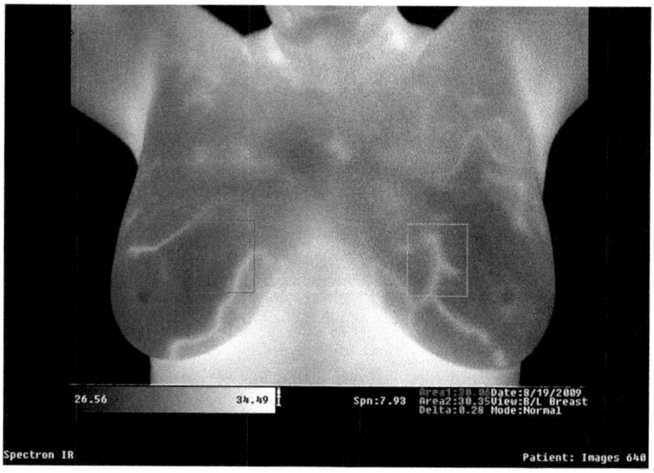

■ **FIGURE 8.4** A breast thermogram by a spectron IR infrared camera. *From [3] with permission.*

8.3 MEASUREMENT OF FLOW RATE AND VOLUME OF BLOOD

The measurement of flow rate and blood volume is one of the most important factors in biological measurements and is used, for example, for cardiac output, oxygen concentration or drug delivery.

Several techniques are employed to measure the flow rate and blood volume [1]:

- *Indicator-Dilution method by using continuous infusion*:

  This method can measure flow average over number of heart beats. This technique uses concentration principles and Fick's method to analyze and measure the blood flow rate.

  Concentration and Fick technique: An indicator with the quantity of m_0 is added to a fluid volume, V. The concentration of the indicator is [1]:

$$C = \frac{m_0}{V} \tag{8.4}$$

  Then if an additional quantity m of indicator is added, the increase in concentration is:

$$\Delta C = \frac{m}{V} \tag{8.5}$$

  If the volume of biofluid in the measurement control volume (measurement space) is continuously replaced or removed, the clinicians have to add a fixed quantity of the indicator per unit of time. So Eq. (8.5) can be written as:

$$\Delta C = \frac{\left(\frac{dm}{dt}\right)}{\left(\frac{dV}{dt}\right)} \tag{8.6}$$

  Considering Eq. (8.6), the flow rate, F, can be written as:

$$F = \frac{dV}{dt} = \frac{\left(\frac{dm}{dt}\right)}{\Delta C} \tag{8.7}$$

  Considering Eq. (8.7), the following equation can be written to measure cardiac output [1,4]:

$$F = \frac{\left(\frac{dm}{dt}\right)}{C_a - C_v} \tag{8.8}$$

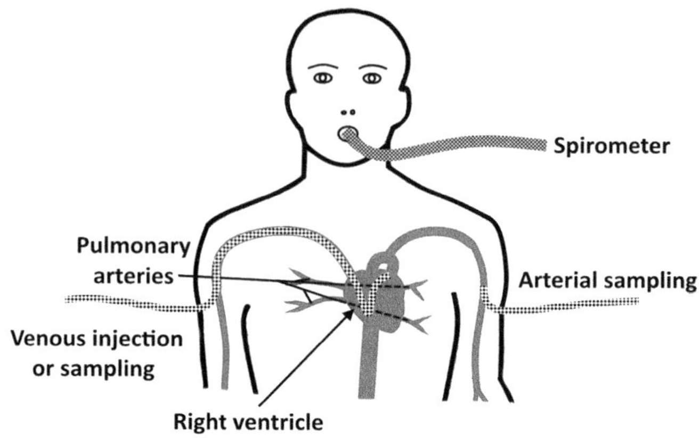

■ **FIGURE 8.5** Schematic of injection or sampling and measurement requirements for Indicator-Dilution methods.

where C_a and C_v are O_2 concentration for arterial and venous vessels (liters/liter), respectively, dm/dt is consumption of O_2 (L/minute), and F is the blood flow rate (L/minute). In Fick's method, the consumption of O_2 is measured by a spirometer and the concentration difference is measured and recorded by drawing blood samples from the pulmonary artery because this artery has a more accurate oxygen concentration in comparison to the other veins. For arterial sampling there is no limitation and all arteries have enough well mixed blood without any consumption of oxygen. Fig. 8.5 illustrates the location of injection or sampling and measurement requirements for Indicator-Dilution methods.

- *Indicator-Dilution method by using rapid injection*:

  This technique is more common than the continuous infusion method. There are two procedures under this category [1]: the dye-dilution and the thermodilution methods.

  In the *dye-dilution method*, which is a clinically common method to measure cardiac output, the dye is injected into the pulmonary artery and blood samples are drawn from an artery through a colorimeter cuvette. The blood sample is continuously measured by using the principle of absorption photometry to analyze the dye concentration.

  The *thermodilution method* is the most common method in this field. Cold saline, as an indicator, is injected into the right atrium and blood temperature is measured in the pulmonary artery. A thermistor (or sensor) located near the pulmonary artery detects the temperature differences (drop) in the blood. In this method bioheat and biomass transfer play a major role in measuring blood flow rate in the cardiovascular system. The blood flow rate can be calculated by the following equation [1]:

$$F = \frac{Q}{\rho_b c_b \int_0^{t_1} \Delta T_b(t) dt} \tag{8.9}$$

where Q is heat content of injectate (joule, $= V_i \Delta T_i \rho_i c_i$), c_b is specific heat of blood J/(kg K) and ρ_b is density of blood (kg/m^3). The following issues produce the measurement errors for the calculation of the blood flow rate in this method:

- Heat transfer through the catheter walls at the injection site;
- Adequate mixing between the injection site and sampling site;
- Heat transfer between the heart wall and the blood.

8.3.1 Electromagnetic Flowmeter

This type of flowmeter is based on the electromagnetic principle, when a fluid containing particles with electric charges (such as ions) flows in a magnetic field, an electromagnetic force is generated. If a charge particle, q, moves with a velocity, U, in a magnetic field with magnetic flux density, B (Tesla), then a force, F, will be applied on the charge particle. The F is expressed in form of vector:

$$F = q(U \times B) \tag{8.10}$$

If blood with positive and negative particles flows through the magnetic field, they will move in opposite directions. Due to this arrangement, an electrical field, E, will be generated. If two electrodes are placed along this electrical field, an electromagnetic field with a potential difference, V (volt), will appear between two electrodes. The governing equation for this effect is [5]:

$$V = UBd \tag{8.11}$$

where d is the distance between two electrodes or vessel diameter. Fig. 8.6 illustrates the function and geometry of electromagnetic principle to measure the blood flow rate in a vessel.

Eq. (8.11) reveals that the voltage between electrodes has a direct relation to the blood velocity. Considering the flow rate formula in fluid mechanics (see chapter: Fluid Mechanics and Biofluids Principles) and using the mean velocity, the blood flow rate Q is formulated as:

$$Q = UA = U\frac{\pi d^2}{4} = \frac{\pi V d}{4B} \tag{8.12}$$

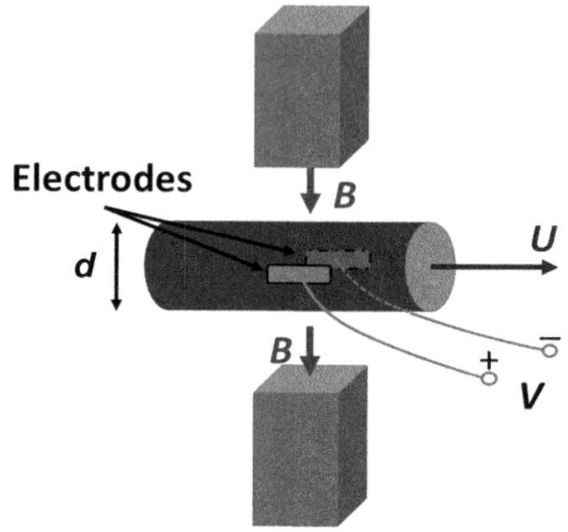

■ **FIGURE 8.6** Schematic of an electromagnetic flowmeter: U is blood flow velocity, V is electrical potential difference, B is magnetic flux density and d is diameter of vessel.

There are two options for power sources in an electromagnetic flowmeter: direct current (DC) design and alternative current (AC) design. If the flowmeter (Fig. 8.6) uses a DC magnetic field, the output voltage shows continuously the blood flow. An AC design usually is more common than the DC design due to fewer problems in operation.

8.3.2 **Ultrasonic Flowmeter**

An ultrasonic flowmeter can measure instantaneous blood flow. An ultrasonic probe can create sound waves through the living tissues to analyze tissue biocharacters, such as blood flow rate. Advanced models of ultrasonic flowmeters can even measure and analyze the blood profile. Transducers have the key role of converting electrical signals to acoustic waves in ultrasonic devices. Since the transducers have a fixed diameter, they can produce diffraction shapes which are similar to a hole in optics. The differences in diameters and shapes help to provide near and far detection fields [1].

Fig. 8.7 shows the beam shapes for two transducer diameters, D and $D/2$, at the same level and frequency on a blood vessel. The figure represents different penetration fields (near and far field) in respect to the diameter of the transducers which have positive roles in the range of penetration field. Technically, the best results occur in range of the near field or the initial distances. Eq. (8.13) explains this aspect of the ultrasonic principle

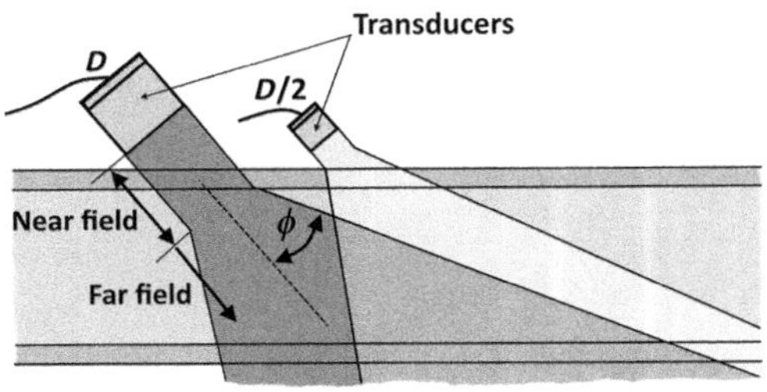

■ FIGURE 8.7 Schematic of two ultrasonic transducers with two different diameters and same operational frequency which cause different penetration fields.

for the near field and Eq. (8.14) represents the angle of beam divergence, ϕ, for the far field [1];

$$d_{nt} = \frac{D^2}{4\lambda} \tag{8.13}$$

$$\sin\phi = \frac{1.2\lambda}{D} \tag{8.14}$$

where d_{nt} is distance of near field, D is transducer diameter and λ is wavelength. These characters and geometries are shown in Fig. 8.7.

Transit-time flowmeters and continuous-wave Doppler flowmeters are other types of sonic instrumentation for measuring blood flow rate. There are two arrangements for transit-time flowmeter technology: (1) two transducers are located diagonally on two sides of tube (blood vessel) or (2) two transducers are located on one side of a vessel and a reflector is located on the other side in the middle of them and role of the reflector is to reflect the waves between the transducers.

The transit-time flowmeter develops a precise measure of the transit time for a sonic wave to travel from one transducer to another one. The difference between the integrated transit times of upstream and downstream derives a measurement for the blood flow rate and the velocity inside the blood vessels. Fig. 8.8 shows the schematic of the transit-time flowmeter for both systems (Fig. 8.8A and B).

The continuous-wave Doppler flowmeter transmits an ultrasonic signal from within the flow stream. The signal is reflected off blood particles such as blood cells and returns to the receiver. There is a difference between the sending and receiving frequency due to the effect of the Doppler frequency shift. The instrument and analysis system

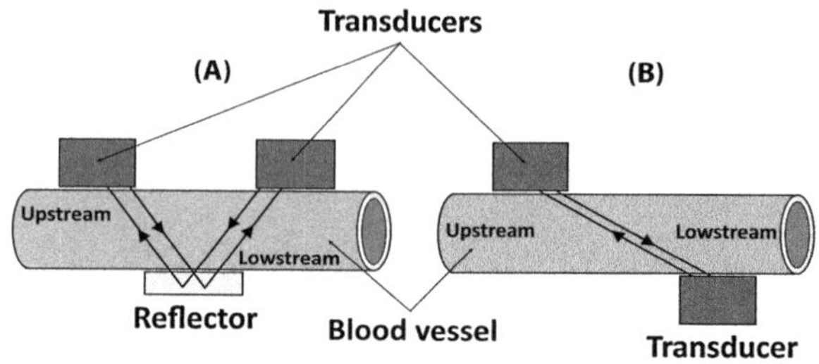

■ **FIGURE 8.8** Arrangement schematic of transit-time flowmeters for blood flow detection. (A) One side transduces, the other side is the reflector. (B) Diagonal transducers.

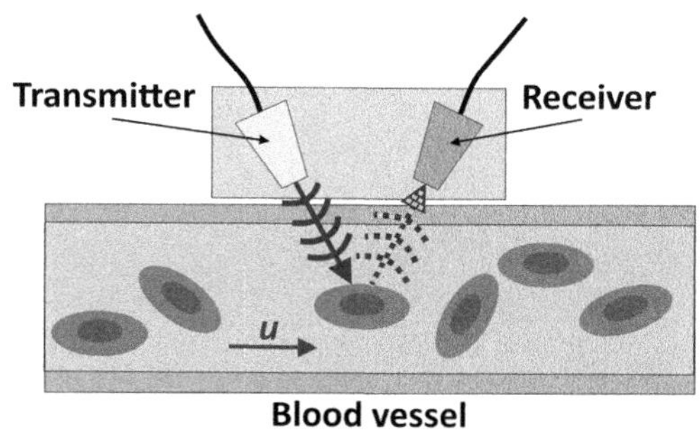

■ **FIGURE 8.9** Schematic of continuous-wave Doppler flowmeter. The sound wave is beamed through the blood vessel wall, reflected by the red blood cell and received by a receiver. The output signal transfers to a microprocessor to calculate the blood velocity and the blood flow rate.

measure and calculate the average velocity of flow. Furthermore, the flow area can be measured by liquid depth to determine the flow area. Flow rate can be calculated by multiplying the flow area by the average velocity of the blood. The following equation shows the relation between frequency and velocity in a continuous-wave Doppler instrument [1]:

$$\frac{f_{\mathrm{d}}}{f_0} = \frac{u}{c} \tag{8.15}$$

where f_{d} and f_0 are Doppler frequency shift and source frequency respectively, u is target velocity and c is velocity of sound.

Fig. 8.9 shows the simplest instrumentation of a continuous-wave Doppler flowmeter that consists of a transmitter and receiver to measure the blood velocity and the blood flow rate.

8.3.3 Blood Flow Analyzing and Recording by a Laser Doppler Flowmeter

In this method three approaches have been used:

1. Airborne beams, to study the velocity of red blood cells in the small vessels, such as the retina.
2. Catheters of fiber-optic have been inserted into blood vessels to analyze and measure the optical data from blood vessels.
3. Applying the laser Doppler principle in microscopic level. In this method, the objects pass through the microscope and the light and laser beam reflect from the objects. This system can measure blood flow velocities in the range from 50 μm/s to 50 cm/s.

The blood flow rate can be calculated by the instrument microprocessor and by multiplying the measured blood velocity by the internal area of the blood vessels.

8.3.4 Implantable Blood Sensors

Many applications, such as the cardiac pacemaker, implantable cardioverter defibrillator (ICD) or similar implantable devices, need to be able to control hemodynamics aspects of blood, such as blood velocity, with a precise measurement. These sensors have to be sensitive to the heart rhythm disorders and have long-term stability.

An implantable multisensor system for in vivo monitoring of cardiovascular parameters was presented. In this method various hemodynamic aspects, such as oxygen saturation level, pressure, etc., can be measured in vivo by electromechanical system (MEMS) technology [6].

8.3.5 Plethysmography

The plethysmography technique measures and records the blood volume change in tissues. This method is known as one of the most accurate techniques to measure changes in blood volume in limbs, organs and tissues. The rate of these changes to time results in the blood flow rate in the designated area (see Eq. 8.7).

The venous occlusion (VO) technique is a major method to measure blood flow in the limb by the plethysmographic measurement. In this technique, two cuffs are attached to the proximal and distal locations of the segment. The distal cuff pressure is more than arterial pressure to occlude the blood flow in all arteries and veins under the distal cuff. Then a pressure lower than the minimal arterial pressure is applied to the

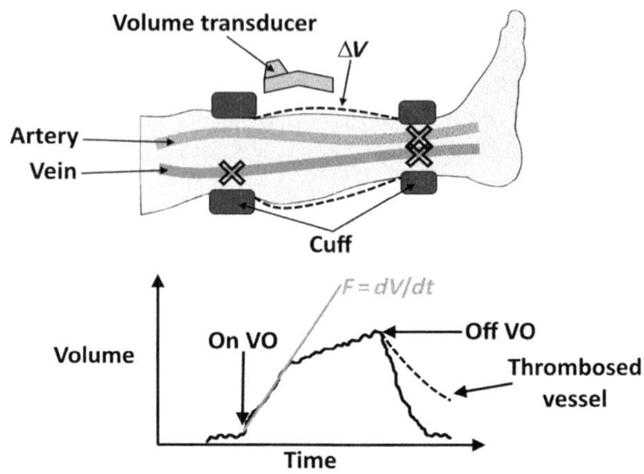

■ **FIGURE 8.10** Schematic of the Plethysmography technique. Top: setup test; Bottom: volume-time graph for the Plethysmography test, during on and off of VO by cuffs.

proximal cuff which results the occluded blood flow in veins and nonblocked flow in arterial vessels. The next stage is measurement of the volume change in the segment. As a result, the blood flow rate is determined by the rate of volume increase or dv/dt. For regular subjects, the proximal cuff pressure for VO is selected at around 6.7 kPa (50 mmHg), and the distal cuff pressure is around 20 kPa (150 mmHg). Fig. 8.10 shows the principle of the VO method applied to a part (tibia bone) of the leg. After VO cuff pressure is turned on, the volume increases because of arterial flow. When the clinician releases the cuff pressure the volume returns to the normal pretest condition. If the return time is slower than the normal value, probably there is a thrombosis inside the vessels in that area—this is an alternative technique for the diagnosis of venous thrombosis (see Fig. 8.10).

8.3.6 **Heat Transport Blood Flowmeter**

This technique is based on the heat and mass transport principle. The blood flow can transfer heat to the tissues and it can be considered as a mechanism for blood flow estimation. Therefore, the blood flow rate of tissue can be measured from the heat transfer rate when a constant temperature difference is maintained between the arterial blood and the tissue region. The heat transfer rate between the arterial blood of temperature T_a and the tissue of temperature T_t can be calculated by [5]:

$$H = \rho c q(T_t - T_a) \tag{8.16}$$

where q is tissue blood flow rate, ρ and c are the density and specific heat of the blood, respectively.

8.3.7 Nuclear Magnetic Resonance Tissue Blood Flowmeter

This is completely a noninvasive technique and is based on the ordinary magnetic resonance imaging (MRI) technique. Many different methods of tissue blood flow measurement and perfusion imaging using nuclear magnetic resonance have been offered and some of them have been applied to clinical diagnosis [5]:

- Continuous Wave Excitation
- Time-of-Flight Method
- Phase Contrast Method

8.4 MEASUREMENT OF BLOOD PRESSURE AND SOUND

Regarding location, blood pressure transducers (sensors) can be classified into two groups:

1. *Extravascular pressure transducers*: This type of transducer (sensor) is the most common transducer in clinical systems and it is known as a disposable blood pressure sensor. It is made up of a catheter connected to a three-way valve and then to the pressure sensor. This sensor area must be flushed with a saline-heparin solution every few minutes to prevent clotting at the tip. The blood flow comes from the patient's artery or vein and then the blood goes inside the sensor chamber on the diaphragm. The deflection of the diaphragm and its strain gauge generates a signal for the signal processing system. Fig. 8.11 shows the schematic of the setup and design of an extravascular transducers or disposable blood pressure sensor.
2. *Intravascular pressure transducers*: This type of transducer (sensor) detects the blood pressure without any liquid coupling or hydraulic connection only by the use of the catheter. This design provides a high frequency response and eliminates the time delay which occurs with an extravascular transducer. Another name of this type of pressure transducer is catheter-tip transducer (sensor). This type of pressure transducer is not common in health care systems due to its design limitation and commercial price in comparison to extravascular transducers. Fig. 8.12 shows an intravascular transducer. It is made of several parts including a diaphragm, sensor, sensor case,

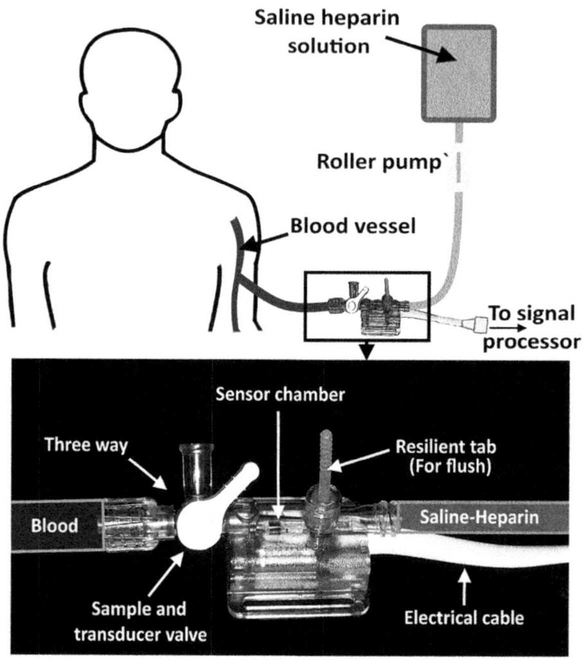

■ **FIGURE 8.11** Blood pressure transducer (Edwards Lifescience). Top: system connections; Bottom: elements of a disposable blood pressure transducer.

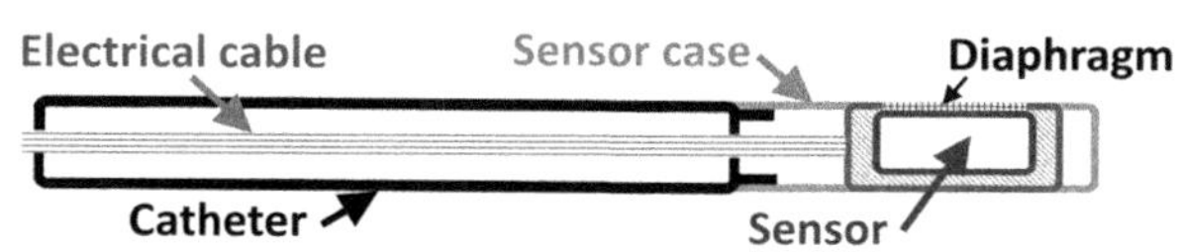

■ **FIGURE 8.12** Model of catheter sensor system.

catheter and electrical cable. An increase in pressure causes a deflection of the diaphragm which is sensed by an electromechanical system (MEMS) sensor. The sensor sends the collected data to the next stage to be analyzed by a signal processor.

3. *Sound measurement*: In medicinal science and especially in the cardiovascular field, heart sound is considered as a symptom of disease. Phonocardiography or PCG is the record of the sounds made by the heart during a cardiac cycle. From a mechanical view, heart sounds are the vibrations caused by the acceleration and deceleration of blood inside the heart chambers. Murmurs are vibration (or sound) due to blood turbulence.

Heart sounds and murmurs have a small amplitude, with frequency from 0.1 to 2000 Hz. There are some limitations to detecting heart sounds and murmurs since a frequency less than 20 Hz is not audible. The noises have to be minimized for low amplitude of heart sounds and murmurs [1].

There are several methods to analyze and measure biosounds:

- *Stethoscope*:

  The stethoscope is probably the first medical device for auscultation. This device transmits sounds from the chest wall to the human ear. Also it can be used by clinicians to listen to the lung, the blood flow in large blood vessels (arteries and veins) and intestines. Two types of stethoscope are used in the health care system: acoustic (mechanical) stethoscope and electronic stethoscope. Advanced technology is used in the electronic stethoscope but a huge number of physicians prefer to use an acoustic (mechanical) stethoscope due to its reliability, convenience and other important considerations.

- *Phonocardiography*:

  This technology is based on the recording of the heart sounds and murmurs. This method evaluates the heart sounds which are thought to result from heart valves operation and vibrations. This method respects to the electric and mechanical functions in cardiac cycles and, possibly, many other evaluations (ECG, arterial and venous pulse tests, apex cardiogram) are measured by clinicians with PCG.

8.4.1 Cardiac Catheterization (Heart-Cath)

This is a comprehensive technique to examine cardiac functions. By this method, clinicians can organize diagnosis or treatment for their patients. Clinicians insert a catheter into a large vessel that leads to the heart. Cardiac catheterization measures cardiac output, blood gases and other vital elements in an inclusive range. Also they can measure blood pressure in all chambers and valves of the heart as well as in large vessels. In addition to these features, clinicians can use balloons tipped to open the wedged or blocked vessels. Angiographic visualization of cardiac catheterization is one of the most important tools in cardiovascular diseases and real-time monitoring. Radiopaque dyes are injected into the bloodstream and clinicians can view or record the blood dynamic characters of the heart or cardiovascular system on an X-ray film or movie. Fig. 8.13 shows the vascular access paths for cardiac catheterization (left heart). As the figure shows, there are three locations to insert the catheter into a large vessel.

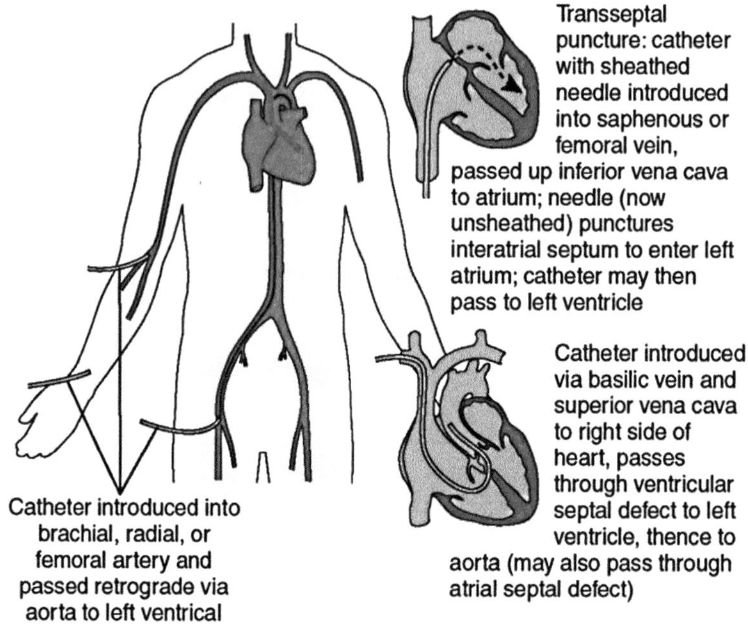

■ **FIGURE 8.13** Schematic of the vascular access paths for the cardiac catheterization (inside left heart). *Adapted from M.J. Kern, Cardiac Catheterization Handbook: Expert Consult - Online and Print, Saunders-Elsevier, Philadelphia, PA, 2011*

8.5 MEASUREMENT OF THE RESPIRATORY SYSTEM

Flow rate and pressure measurements are the best characters to reveal the operational condition of the respiratory system. For pressure measurement, two aspects are very important: measured pressure and pressure differences. For pressure measurement, the best options are regular electronic pressure sensors with a catheter (see Section 8.4). For static or hydrostatic pressure, these sensors can be used, however for pressure difference, the best option is a differential pressure sensor, which is made of two separated chambers with a diaphragm between the chambers and a sensor which is connected to the diaphragm.

Several methods have been used and proposed to measure flow rate in the respiratory system:

8.5.1 Rotary Flowmeter

This flowmeter is made of a rotor, a stator, light source and a photocell. Rotation of the rotor is related to the flow rate of gas and this rotation

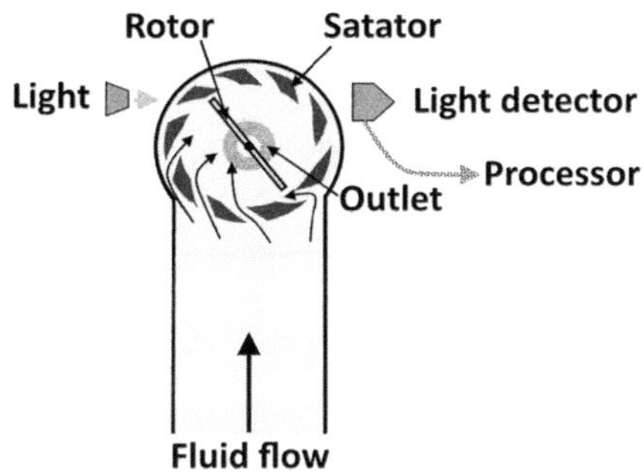

■ **FIGURE 8.14** Schematic of the Wright Respirometer.

interrupts the light beam. This interruption is detected by a sensor and creates a voltage signal to the processor. The advantage and disadvantage of this method are its simple design and high measurement error, respectively.

The most common industrialized devices for this type of flowmeter are the Wright Respirometer (its rotor rotates in one direction), the Spirostat (its rotor rotates in both directions) and Dräger volumeter (its rotor rotates in both directions). Fig. 8.14 illustrates a schematic of the Wright Respirometer, including rotor, stator, light and light detector and flowmeter chamber.

8.5.2 Thermal Flowmeter

This technology uses convection–conduction principles and thermal elements such as metal wires, metal surfaces and thermistors to sense a temperature difference which occurs in a system due to heat transfer between the thermal element and the fluid flow.

A thermal flowmeter uses a heated wire (hot wire anemometer) with a linearizing circuit to provide enough temperature for unidirectional flow measurement. Heat transfer rate in this method demonstrates the flow rate of the fluid. Heat transfer in this system depends on the geometry of the wire, temperature, viscosity, mass flow, specific heat and thermal conductivity coefficient of the fluid.

The heat transfer H can be written as [5]:

$$H = \alpha \pi d l (T_w - T_s) \tag{8.17}$$

where d is diameter of the wire, l is length of the wire, T_w and T_s are the wire and gas temperature, respectively, and α is the heat transfer coefficient.

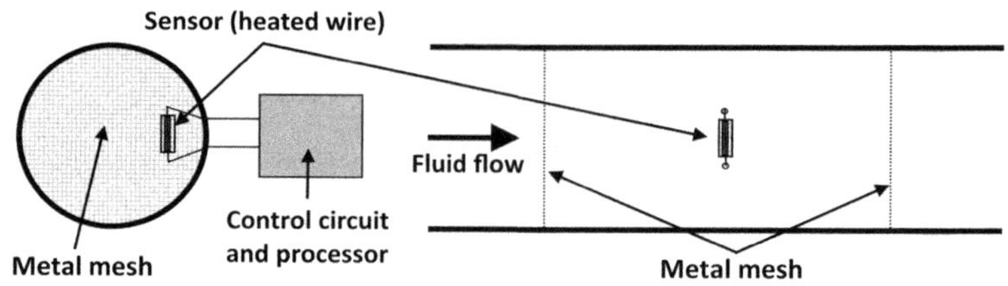

■ **FIGURE 8.15** Schematic of thermal flowmeter, one sensor. Left: diagonal cutaway view including sensor, control circuit and processor. Right: longitudinal cutaway view of tube with the sensor.

Electrically, the heat dissipation H can be calculated by the wire resistance R and the current I as:

$$H = RI^2 \tag{8.18}$$

The important aspect in this method is a constant wire temperature to measure the flow rate in an accurate way. For this purpose, a feedback circuit is added to the design to provide a constant temperature for the heated wire [5]. Fine metal meshes are located at both ends of the duct to provide a definite flow profile and a unique relationship between gas flow rate and flow velocity at the wire [5]. Also, this type of flowmeter can be equipped with several sensors to provide directional sensitivity [1]. Fig. 8.15 shows the thermal flowmeter structure for one sensor (heated wire).

8.5.3 **Ultrasonic Flowmeter**

The ultrasonic technique for respiratory fluid flow measurement is similar to that which is defined in the ultrasonic blood flowmeters (see Section 8.4). The difference between the gas and blood measurement for this method is only in the formed elements or the blood cells which are not in the gas flow. Such as blood flow ultrasonic method, this difference restricts the use of Doppler shift measurement due to the scattered ultrasound.

There are several configurations for respiration ultrasound:

- Axial configuration
- Diagonal configuration
- Cylindrical shell configuration

Fig. 8.16 shows the configurations of an ultrasonic flowmeter in respiratory systems.

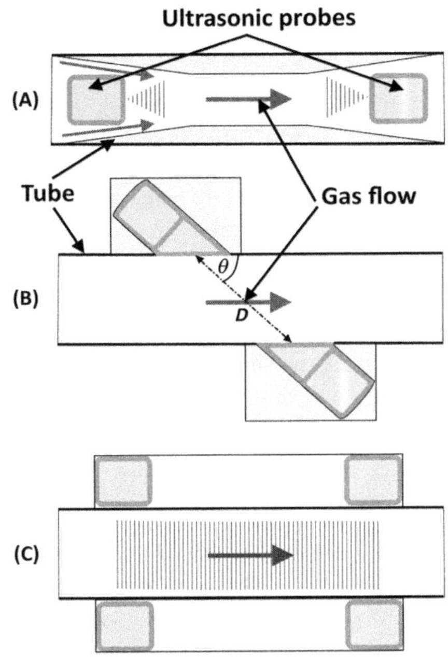

■ **FIGURE 8.16** Schematic of ultrasonic configurations: (A) axial; (B) diagonal and (C) cylindrical shell configuration.

The flow rate is a function of the fluid velocity and the area of the channel. The fluid velocity can be calculated from

$$T_1 = \frac{D}{C+\overline{U}} \tag{8.20}$$

$$T_2 = \frac{D}{C-\overline{U}} \tag{8.21}$$

where T_1 and T_2 are transmit time for downstream and upstream, C and $\overline{U}$ are sound and fluid velocities between probes, respectively.

$$\Delta T = T_1 - T_2 = \frac{2D\overline{U}}{C^2 - \overline{U}^2} \tag{8.22}$$

And the normal fluid velocity, U is;

$$\overline{U} = U\cos\theta \tag{8.23}$$

where θ is the beam angle with respect to the flow channel. In the axial or the cylindrical shell configurations theta is zero, $\theta = 0$.

Usually, U and $\overline{U} \ll C$, and considering Eqs. 8.22 and 8.23, the normal fluid velocity, U, can be represented as:

$$U = \frac{C^2 \Delta T}{2D \cos \theta} \tag{8.24}$$

Fig. 8.16 shows these variables on a diagonal ultrasonic arrangement.

8.5.4 Vortex Flowmeter

This method uses vortex principles to measure the fluid velocity. The quantity of vortices in the flow shows the fluid velocity and an ultrasonic device detects the vortices. Two techniques are employed to measure the flow rate: the Kármán flowmeter and the swirl flowmeter.

8.5.4.1 *Kármán Vortex*

The Kármán vortex flowmeter is based on the development of a Kármán vortex shedding street which are generated behind a rod placed in the stream perpendicularly to the flow and the quantity of vortices produced in unit time is related to the flow velocity (see Fig. 8.17).

The dimensionless Strouhal number, S_t defines the relationship between vortex shedding frequency and fluid velocity:

$$S_t = \frac{fd}{V} \tag{8.26}$$

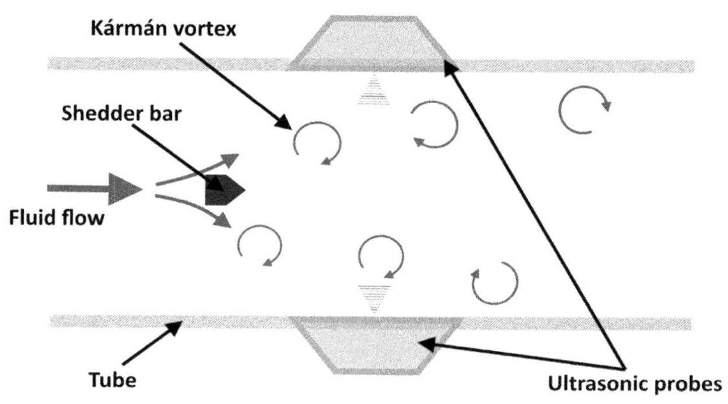

■ **FIGURE 8.17** Schematic of a vortex flowmeter using the Kármán technique. The ultrasonic probes detect vortices and the number of them is related to velocity of fluid and flow rate.

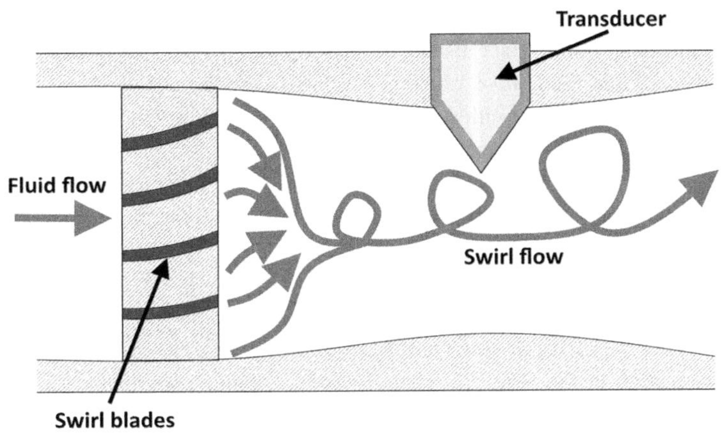

■ **FIGURE 8.18** Cutaway view of the swirl flowmeter and its elements.

or

$$V = \frac{fd}{S_t} \tag{8.27}$$

where f is frequency of vortex shedding, V is fluid velocity and d is diameter or width of shedder bar (obstacle).

8.5.4.2 *Swirl Flowmeter*

This flowmeter consists of swirl blades (like a helical gear) installed inside a tube. Passing fluid (gas) through the swirl blades generates a swirling motion (vortices). A transducer (sensor) detects these motions and the fluid flow rate can be calculated by the number of vortices per unit of time. Fig. 8.18 shows a cross-section of a swirl flowmeter and its elements: swirl blades and transducer.

8.6 MEDICAL IMAGING SYSTEMS

Medical image processing is a technique to examine the internal organs and tissues for clinical analysis, diagnosis and treatment. This technique has a wide range of tools to analyze internal organs, including the cardiovascular system.

- *Echocardiogram (Echo)*:

  Echocardiogram is a type of ultrasound test to analyze the heart function and heart diseases, such as heart valve disease, pericardial disease, myocardial disease, infective endocarditis and congenital heart disease. Furthermore, it is a useful tool for the assessment of medical or surgical treatment, and this technique also provides ultrasound images for the cardiovascular system plus an accurate flow rate estimation. It has several subsystems: transthoracic echocardiography, transesophageal echocardiography, stress echocardiography (Stress Echo) and 3D echocardiography.

- *Nuclear Cardiac Imaging*:

  Nuclear cardiac imaging is a method for producing images of cardiac parts using small amounts of radioactive tracers. After injection of the radiotracer into a vein, a gamma camera and single-photon emission-computed tomography obtains images of the cardiac parts. Clinicians can then analyze the blood flow distribution to the heart muscles, the function of the heart and the blood vessels' conditions; these are technically called myocardial perfusion imaging and multigated acquisition.
- *Cardiac positron emission tomography*:

  Cardiac positron emission tomography (PET) is an imaging test that uses a radioactive tracer to detect the heart diseases or poor blood flow in the heart. The advantage of this technique is accuracy and more information about the function of heart muscles and tissues after a heart attack or cardiac diseases.
- *Cardiac MRI (Heart MRI)*:

  MRI uses a powerful magnetic field, radiofrequency waves and a computer to create high resolution photos of body organs and tissues. This method can be used for cardiac and cardiovascular systems and reveals cardiac problems such as reduction of blood flow in heart muscles, heart muscle diseases, tissue damage due to heart attack, heart valve disorders, control of tissue after surgical repair and congenital cardiac problems.
- *Heart Computed Tomography (CT) and Coronary Computed Tomography Angiography*:

Heart CT is an imaging method that creates high resolution 3D pictures of the heart and cardiovascular system. Two types of CT scan are used by clinicians in the cardiovascular system:

1. Coronary computed tomography angiography, to examine inside the heart chambers, coronary arteries and pulmonary veins by using X-ray contrast.
2. Calcium score screening heart scan, to control arteries for calcium deposition inside the heart vessels or coronary arteries.

CHAPTER SUMMARY

- A sensor is a device to sense physical quantity such as voltage, current, temperature, pressure, mass, strain, humidity, light, flow, etc.
- There are two sensing systems for biomedical purposes: regular transducers and biosensors.
- The transducer transfers generated signal to an analog/digital converter before transferring the signal to a microprocessor which only recognizes digital signals. It is then converted to analog again by a digital/analog converter to show the output.
- Biosensor is a sensor to detect an analyte and is made of a biological element which acts as a sensor and a transducer for detecting and transmitting the signal.
- The biosensor construction has two main parts (layers): (1) transducer or electrode and (2) bioreaction material or biointerface.
- The temperature units can be converted to each other:

$$°C = (°F - 32) \times 1.8$$

$$°C = K - 273.15$$

$$K = (°F - 32) \times 1.8 + 273.15$$

- The average normal body temperature in humans is about 37°C or 98.6°F.
- Several types of measurement are used to collect thermal information from a patient's body, including fluid/glass thermometer, thermocouple, thermistor, radiation and optical systems.
- Several techniques are used to measure flow and blood volume:
  - Indicator-Dilution method by using continuous infusion and concentration and Fick's technique
  - Indicator-Dilution method by using rapid injection:
    - dye-dilution method
    - thermodilution method
  - Electromagnetic flowmeter: This type of flowmeter is based on the electromagnetic principle; when a fluid containing particles with electric charges (such as ions) flows in a magnetic field, an electromagnetic force is generated.
  - Ultrasonic flowmeter: An ultrasonic probe can create (transit) sound waves through the living tissues to analyze tissue biocharacters, such as blood flow rate. The difference between the integrated transit times of upstream and downstream derives a measurement for the blood flow rate and the velocity inside vessels.

- Blood flow analyzing and recording by laser Doppler flowmeter: In this method three approaches have been used: airborne beams, catheters of fiber-optic have been inserted into blood vessels and application of the laser Doppler principle at the microscopic level.
- Implantable blood sensors: These sensors have to be sensitive to the heart rhythm disorders and have long-term stability.
- Plethysmography: This technique measures and records the blood volume change in tissues. This method is known as one of the most accurate techniques to measure changes in blood volume in limbs, organs and tissues.
- Heat transport blood flowmeter: The blood flow can transfer heat to tissues and it can be considered as a mechanism for blood flow estimation.
- Nuclear magnetic resonance tissue blood flowmeter: This is a completely noninvasive technique and is based on the ordinary MRI technique.

- Blood pressure transducers (sensors) can be classified into two groups:
  - Extravascular pressure transducers: This type of transducer (sensor) is the most common transducer in clinical systems and it is known as the disposable blood pressure sensor.
  - Intravascular pressure transducers: This type of transducer (sensor) detects the blood pressure without any liquid coupling or hydraulic connection only by the use of the catheter.
- Sound measurement:
  - Stethoscope: This device transmits sounds from the chest wall to the human ear.
  - PCG: This technology is based on the recording of the heart sounds and murmurs. This method evaluates the heart sounds which are thought to result from heart valves operation and vibrations.
- Cardiac catheterization (Heart-Cath): This a comprehensive technique to examine cardiac functions. Cardiac catheterization measures cardiac output, blood gases and other vital elements in an inclusive range. Moreover, it can measure blood pressure in all chambers and valves of the heart as well as in large vessels.
- Measurement of the Respiratory System:
  - Rotary flowmeter: This flowmeter is made of a rotor, a stator, light source and a photocell. Rotation of the rotor is related to flow rate of gas and this rotation interrupts the light beam. This interruption is detected by a sensor and creates a voltage signal to the processor.

- Thermal flowmeter: This technology uses convection–conduction principles and thermal elements to sense temperature differences which occur in a system due to heat transfer between the thermal element and the fluid flow.
- Ultrasonic flowmeter: The ultrasonic technique for respiratory fluid flow measurement is similar to that which is defined in the ultrasonic blood flowmeters. There are several configurations for respiration ultrasound: (1) axial configuration; (2) diagonal configuration; (3) cylindrical shell configuration.
- Vortex flowmeter: This method uses vortex principles to measure the fluid velocity. The quantity of vortices in the flow shows the fluid velocity and an ultrasonic device detects the vortices. Two techniques are employed to measure flow rate: the Kármán flowmeter and the swirl flowmeter.
- The Kármán vortex flowmeter is based on the development of a Kármán vortex shedding street which is generated behind a rod placed in the stream perpendicularly to the flow, and the quantity of vortices produced in unit time is related to the flow velocity.
- Swirl flowmeter: This flowmeter consists of swirl blades (like a helical gear) installed inside a tube. Passing fluid (gas) through the swirl blades generates a swirling motion (vortices). A transducer (sensor) detects these motions and then the fluid flow rate can be calculated by the number of vortices per unit of time.

- Medical imaging systems: This technique has a wide range of tools to analyze the internal organs including cardiovascular system:
  - Echocardiogram (Echo)
  - Nuclear Cardiac Imaging
  - Cardiac PET
  - Cardiac MRI (Heart MRI)
  - Heart CT and Coronary Computed Tomography Angiography

PROBLEMS

1. What is the difference between a sensor and a transducer?
2. Explain the structure of a biosensor.
3. Explain and compare a thermocouple and a thermistor.
4. Explain the electromagnetic flowmeter method by using Fig. 8.6.
5. Discuss the venous occlusion (VO) technique.
6. Which technique provides a high frequency response and eliminates the time delay in measuring blood pressure?
7. Explain the cardiac catheterization method.

8. Explain the flow rate measurement methods in the respiratory system, which one is more accurate and why?
9. What is the difference between the Kármán flowmeter and the swirl flowmeter?
10. Classify medical imaging systems.

REFERENCES

[1] J.G. Webster, Medical Instrumentation Application and Design, third ed., Wiley, New York, NY, 1998.

[2] A.Y.K. Chan, Biomedical Device Technology: Principles And Design, Charles C Thomas Pub Ltd, Springfield, CT, 2008.

[3] S. I. m. i. solutions, Thermographic images, 2015 [Online]. Available from: <http://spectronir.com/image>.

[4] J.M. Capek, R.J. Roy, Fick techniques, in: J.G. Webster (Ed.), Encyclopedia of Medical Devices and Instrumentation, Wiley, New York, NY, 1988, pp. 1302–1314.

[5] T. Togawa, T. Tamura, P.Å. Oberg, Biomedical Sensors and Instruments, CRC-Taylor & Francis Group, Boca Raton, FL, 2011.

[6] F. Bingger, J. Fiala, A. Seifert, N. Weber, A. Moser, F. Goldschmidtboeing, et al., Implantable multisensor system for in vivo monitoring of cardiovascular parameters, IEEE Conference on Transducers, Denver, 2009.

[7] M.J. Kern, Cardiac Catheterization Handbook: Expert Consult - Online and Print, Saunders-Elsevier, Philadelphia, PA, 2011.

Chapter 9

Tissue Engineering of Cardiovascular System

LEARNING OBJECTIVES

- Explanation of tissue engineering.
- Classification of tissue engineering techniques and their procedures.
- Definition of micro- and nanofabrication techniques; bottom-up or top-down.
- Classification of applied materials in tissue engineering.
- Fabrication techniques in tissue engineering of the blood vessels; heart valves and cardiac tissues, including cell sources, cell culture, scaffolds and implantation.

9.1 INTRODUCTION

Tissue engineering is the application of methods and principles of engineering, medicine and physiology to obtain novel tissues with similar functions and characters as a biological substitute for implantation into the body for the purpose of replacing, repairing or enhancing organ function.

Tissue engineering can generate the new tissues using human tissues, animal tissues, a combination of human/animal-synthetic materials or totally synthetic materials. Generally, tissue engineering uses multiple engineering design approaches to generate the tissue. Original cells are extracted from a tissue source, then the cells are expanded in culture. In the next stage the cells will be located into a scaffold and will be prepared for implantation. The last stage is implantation into the human body.

Fig. 9.1 illustrates the simple schematic of the tissue engineering procedure to produce blood vessels. In this procedure, the biophysical properties of the engineered tissue have to be appropriate.

Biofluid Mechanics. DOI: http://dx.doi.org/10.1016/B978-0-12-802408-9.00009-0

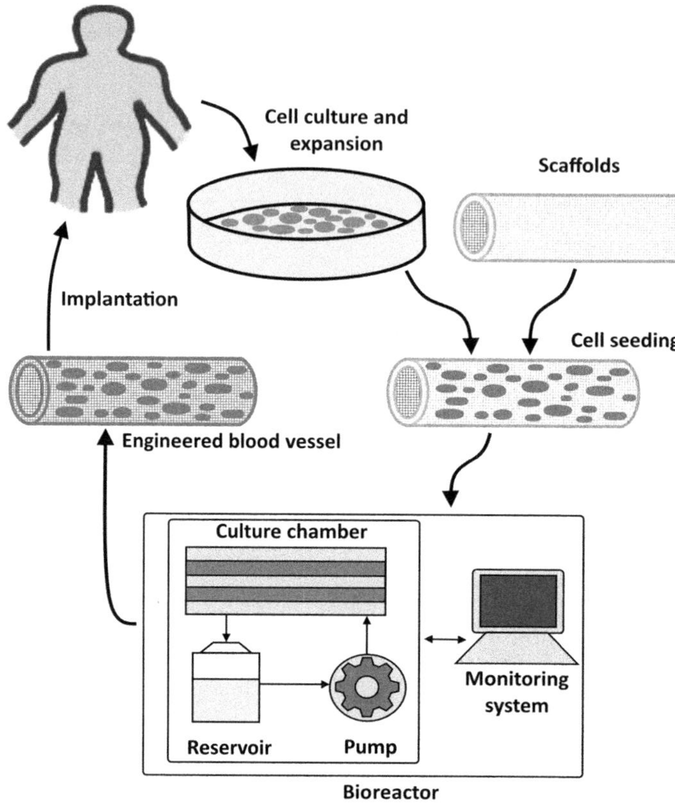

■ **FIGURE 9.1** Simple schematic of the tissue engineering procedures to produce blood vessels.

As an engineering and scientific field, tissue engineering is very young and it needs more time to achieve its goal. To approach the goals, a series of technological developments must be made. Some of its earliest aims have been achieved with clinical success, such as engineered tissue of the human skin [1–3]. In all cases, cells, scaffolds and growth factors/cytokines, or a combination of these elements have been implanted [3].

Understanding cellular physiology, cell dividing, growth and death of cells have key roles in tissue engineering. One of the famous stem cells is the mesenchymal stem cell (MSC). MSCs are multi-application cells and can differentiate into several types of cell, including muscle cells (myocytes), bone cells (osteoblasts), cartilage cells (chondrocytes), etc.

Research introduced and suggested MSCs as stem cell for cardiovascular regeneration. The research suggested that in vitro condition MSCs can differentiate into cardiomyogenic and vasculogenic lineages, presenting an alternative cell source for cardiovascular regeneration, and in vivo condition MSCs may offer regrowth and protection of vasculature and cardiomyocytes [4]. During the last decade, several methods and techniques have been developed for cardiac applications in vitro or in vivo. These techniques have examined several related factors for the function and structure of cardiac tissues, including biophysical principles, cell separation, seeded cells composition, bioreactors, scaffold features and culture medium [5].

Tissue engineering methods depend on micro- and nanotechnology principles. For instance, micro structure porous scaffolds are used in most tissue engineering applications to support the primary cell attachment and tissue formation [6,7]. Furthermore, micro- and nanotechnology can be used to fabricate biomimetic scaffolds. Micro- and nanofabrication techniques can be classified into two areas [7]:

- top-down, or
- bottom-up.

The top-down is a traditional method and involves seeding cells into porous scaffolds to create constructs of tissue. Several limitations are posed by this technique, including diffusion limitations, slow vascularization, nonuniform cell distribution and low cell density. In contrast, the bottom-up method involves cell-laden modules to form larger structures, involves assembling small without any limitation for diffusion, so the bottom-up method eliminates the deficiencies of the traditional (top-down) method [8]. The bottom-up method is progressively becoming a capable tool to study and recreate vascular physiology in tissue engineering [8–10]. Fig. 9.2 shows the structure of both methods.

In tissue engineering, the term of cardiovascular involves the creation of the cardiac system and vascular tissues, for instance, the blood vessels, myocardium and heart valves.

9.2 TISSUE ENGINEERING OF BLOOD VESSELS

Blood vessel bypass is a significant therapeutic choice for cardiovascular diseases, such as coronary artery and renal blood vessel diseases. Tissue engineering introduced and created a solution for these problems—tissue-engineered blood vessels.

The general methodology of tissue engineering for blood cells is to seed cells on scaffolds and the process is followed by an in vitro culture or in vivo implantation [11]. Fig. 9.1 shows the schematic of tissue

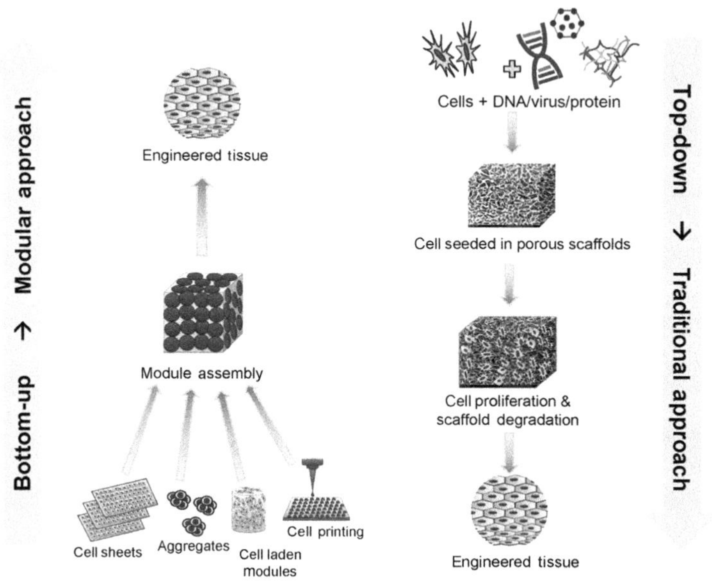

■ **FIGURE 9.2** Schematic of fabrication methods in tissue engineering. Left: bottom-up method, Right: top-down method. *Reprinted from R. Tiruvannamalai-Annamalai, D.R. Armant, H.W.T. Matthew, A glycosaminoglycan based, modular tissue scaffold system for rapid assembly of perfusable, high cell density, engineered tissues, PLoS One 9 (1) (2014) 1—15 with permission.*

engineering procedures to produce the blood vessels. Considering the blood pressure and pulsatile behavior of the blood flow in cardiovascular system, there are numerous challenges to produce tissue-engineered blood vessels, for instance, elasticity, strength, fatigue, compliance and other hemodynamics or physiological aspects. The most important aspect of tissue-engineered blood vessel is their functionality immediately after implantation. The regenerated blood vessels must work and be compatible with other vessels and immune system. After implantation, the engineered blood vessels have to perform similarly to native vessels and keep their functional characteristics for several years. The problems, such as thrombocyte, immune system reaction, fatigue and wall plasticity, of engineered vessels must be minimized to increase durability and reliability.

Ideal properties of a tissue-engineered blood vessel can be classified as [12]:

- Biological characters:
  - Vasoreactive: dilate/constrict to neural and chemical stimuli
  - Nonthrombogenic
  - Biostable: does not weaken in vivo to result in aneurysms and/or rupture

- Biocompatible: not inflammatory, toxic, carcinogenic or immunogenic
- Infection-resistant

- Mechanical:
  - Strength to resist burst pressures
  - Avoids kinking even over joints
  - Hold sutures under circumferential and longitudinal tension
  - Retains axial and radial compliance and pulsatility
- Physical:
  - Leak-proof: avoids hemorrhage through its walls
  - Porosity for healing and angiogenesis
- Commercial:
  - Can be tailored to an individual's requirements, for example, length and diameter
  - Inexpensive to manufacture
  - Short time period from request to implantation

9.2.1 Tissue Engineering Procedures for the Blood Vessels

Considering Fig. 9.3, native blood vessels are made of three main layers: tunica externa; tunica media and tunica intima. To create engineered blood vessels, all these layers and their biophysical aspects must be taken into consideration, since each layer has an important role in the normal function of the vascular system.

As mentioned before, the general procedure of blood vessel tissue engineering is to seed cells on scaffolds first, followed by in vitro culture and in vivo implantation. In this procedure, the operational stages can be

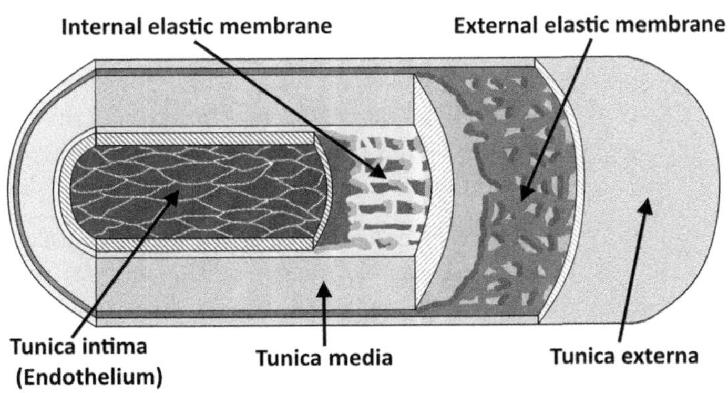

FIGURE 9.3 Cutaway view of an artery and the three main layers: tunica intima; tunica media and tunica externa.

denoted as cell sources, cell culture, scaffolds, vessel bioreactors and implantation.

9.2.1.1 *Cell Sources*

Nonimmunogenic autologous endothelium cells and smooth muscle cells (tunica media area) isolated from patients are the first choice for vessel tissue engineering. Mature vascular cells and embryonic and adult stem cells have been tested in vessel tissue engineering to be alternative cell types [11]. Isolated cells from autologous vessels have been used for engineered vessels [13,14]. Endothelial progenitors cells (type of adult stem cells) have the capability to proliferate, migrate and differentiate into mature endothelium cells, and also studies have revealed that bone marrow-derived MSCs could be differentiated into smooth muscle phenotypic cells by using some certain factors [11].

9.2.1.2 *Scaffolds*

A 3D scaffold is the main structure of engineered tissue and creates a pattern to support cell growth, differentiation, extracellular matrix (ECM) proteins secretion, etc. The scaffolds should be biocompatible, biodegradable and able to absorb inside the bioenvironments.

For fabricating tissue-engineered vascular grafts, the range of scaffolds materials and techniques differs from decellularized tissue skeletons to biodegradable synthetic polymers. However, several issues, such as various mechanical properties, have not yet been studied and they may affect on in vivo function of engineered tissues [15].

Fig. 9.4 shows scanning electron microscope micrographs of a type of scaffold, the electrospinning (ES)—thermally induced phase separation (TIPS)—polyester-urethane urea (PEUU) scaffold with internal diameter of 1.3 mm. To fabricate ES-TIPS PEUU scaffold, first the internal PEUU tubular scaffolds are made by TIPS method, then, nanofibers of PEUU are deposited on the external surface of TIPS tubular scaffolds by ES to obtain ES-TIPS bilayered scaffolds [16]. Fig. 9.4 shows several pictures from different views of ES-TIPS PEUU scaffold.

There are two main types of scaffolds for vascular tissue engineering: (1) biological, such as natural proteins and (2) synthetic, such as synthetic biodegradable polymers. Many different scaffolds can be used for tissue-engineered blood vessels [11]:

- Natural scaffolds
- Permanent synthetic scaffolds

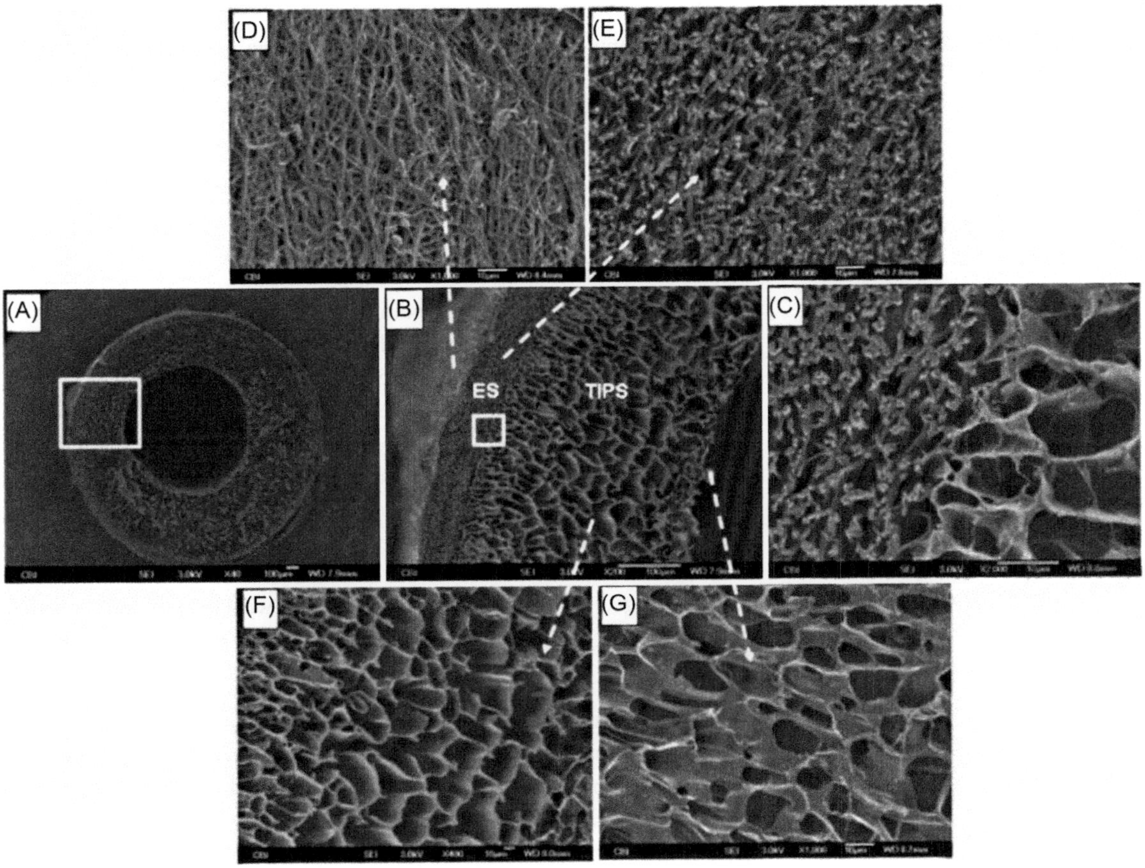

■ **FIGURE 9.4** Scanning electron microscope (SEM) micrographs of ES-TIPS PEUU scaffold. (A) Tubule cross-section, (B) higher magnification of the close-up in (A), (C) higher magnification of the close-up in (B), showing adherence between the ES and TIPS layer, (D) external surface of the ES layer, (E) cross-section of the ES layer, (F) cross-section of the TIPS layer, (G) internal surface of the TIPS layer. *Reprint from W. He, A. Nieponice, L. Soletti, Y. Hong, B. Gharaibeh, M. Crisan, et al., Pericyte-based human tissue engineered vascular grafts, Biomaterials 31 (32) (2010) 8235–8244 with permission.*

- Biodegradable synthetic scaffolds
- Nonscaffold-based tissue-engineered blood vessels.

Advantage and disadvantage of the scaffold materials for engineered blood vessels scaffolds can be defined as [17]:

- Biological scaffolds:
  - Advantages: Naturally occurring, nontoxic, favorable for cell binding, generally biocompatible.
  - Disadvantages: May degrade rapidly, may have weak mechanical properties, inconsistency between batches, chance of disease transmission.

- Synthetic scaffolds:
  - Advantages: Precise control over material properties, easily available and cheap, easy to process, little or no batch-to-batch variations.
  - Disadvantages: Toxic residual monomers or catalysts and degradation by-products may illicit inflammation, poor cellular interaction.

9.2.1.3 *Vessel Bioreactors*

Considering hemodynamics conditions of the cardiovascular system, the engineered blood vessel must be fully functional after implantation. Vessel bioreactors were designed and used to mimic the physiological aspects (stress, strain, pulsatile flow and pressure) that a native blood vessel undergoes in real cardiovascular conditions. Furthermore, some chemical substances may be added to maximize and normalize the physiological aspects of the engineered blood vessel.

9.2.1.4 *Implantation*

In this stage, the engineered blood vessel will be implanted inside a patient's body and it will be observed by image processing methods for any damage or abnormal conditions.

As an alternative method, several novel techniques are employed by investigators in this field to create tissue-engineered blood vessels. For instance, one of these methods used photocross-linkable gelatin hydrogel to create a trilayer biomimetic blood vessel by MEMS and microfluidic platform. The fabrication procedure is simple and rapid, the physical fabrication and formation of the vascular structure and the internal layer of vascular endothelial occurs in minutes and 3–5 days, respectively [18].

9.3 TISSUE ENGINEERING OF HEART AND HEART VALVES

Heart diseases are important cause of illness and death around the world. Several hundred thousand procedures and surgical operations are performed for heart and heart valve replacements annually. Tissue engineering could be an alternative for these problems. The procedures of heart and heart valves tissue engineering are similar to engineered blood vessels with some changes due to their different functions.

9.3.1 Heart Valves

Heart valve diseases are a growing problem around the world and increase the demand for the replacement of heart valves. Several type of prosthetic heart valves including mechanical and biological valves are made and used by clinicians (see chapter: Biofluid Flow in Artificial, Assistive and Implantable Devices) but due to their disadvantages and weaknesses, clinicians and biomedical engineers have introduced tissue-engineered heart valves as an emerging alternative for prosthetic models.

To produce a fully functional tissue-engineered heart valve, an appropriate scaffold is required to be seeded using source cells. There are three common techniques for scaffold formation: bioscaffold; synthetic materials and preseeded composites. The majority of investigators suggest that synthetic materials are the most suitable method for the formation of the heart valve scaffold. The best cell selection is by collecting autologous material directly from the patients due to minimize the immune system reaction. Human and animal sources could possibly be used as well, but the risk of the immune reaction and rejection of the engineered tissues would be increased. Then, the cells will be seeded inside the scaffold, followed by placing the seeded scaffold in a bioreactor as an essential tool in finding the optimal hemodynamics properties [19].

One of the methods to fabricate the scaffold is stereolithographic. In this technique, first, a computed tomography (CT) device scans a human heart valve homograft, then the data will be used to construct a three-dimensional (3D) stereolithographic model. Based on this model, the scaffold will be fabricated [20]. Fig. 9.5 shows (A) 3D recreated stereolithographic model from the inside of an aortic homograft; (B) a tri leaflet heart valve scaffold which is fabricated from the stereolithographic model and (C) a tissue engineered heart valve after 14 days of static and dynamic conditions inside bioreactor.

As it mentioned before, culture conditions and bioreactors have key roles in the tissue engineering of the heart valves. The bioreactor provides similar in vivo conditions for seeded scaffolds; during this procedure the cells in scaffolds have to be qualified and the engineered heart valve will be adapted for in vivo hemodynamics and mechanical properties.

9.3.2 Heart

The heart is one of the largest vascular parts of the human body. Considering this feature, a tissue-engineered heart is still a very significant challenge in biomedical engineering. According to the stage of development, the number of capillaries per unit of area in the human heart is between 2400 and 3300 capillaries/mm^2 [21]. Currently, there is no fully tissue-engineered heart for implant to the human body and all

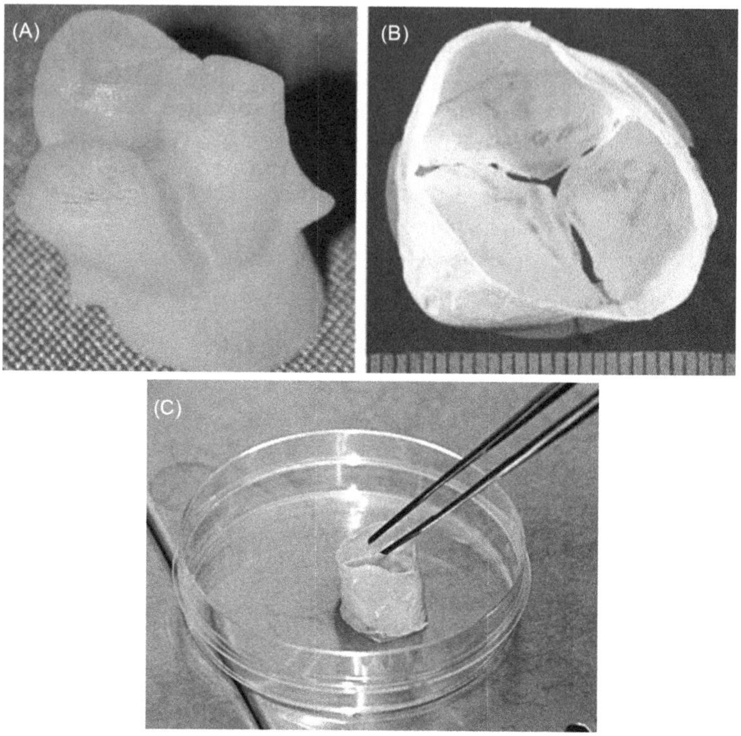

■ **FIGURE 9.5** (A) Three-dimensional reconstructed stereolithographic model from the inside of an aortic homograft, (B) tri leaflet heart valve scaffold fabricated from the stereolithographic model, (C) tissue-engineered heart valve after 7 days of static conditioning and an additional 7 days of dynamic conditioning in a bioreactor. *From R. Sodian, C. Lueders, L. Kraemer, W. Kuebler, M. Shakibaei, B. Reichart, et al., Tissue engineering of autologous human heart valves using cryopreserved vascular umbilical cord cells, Ann. Thorac. Surg. 81 (6) (2006) 2207—2216 with permission.*

works and data are only for use in research in labs. There are still many huge challenges to create a fully functional tissue-engineered heart.

One of the traditional ideas in the tissue engineering of the heart is the transplant or injection of normal and native heart muscle tissues to restore the normal function for damaged hearts which have malfunction due to myocardial infarction. To increase the performance of cell therapy, researchers have injected normal cells in hydrogel solutions or as microspheres into the heart. As a hypothesis, some researchers think that the hydrogel environment has a positive role in cell retention, survival and growth [22,23].

There is no ideal source of cells and most cell types can be characterized by their advantages and disadvantages. Table 9.1 describes the implanted cell types with cardiac regeneration potential and their benefits and detriments.

Table 9.1 Advantages and Disadvantages of Implanted Cells with Cardiac Regeneration Potential

Cell Type	Advantages	Disadvantages
Skeletal myoblasts	Easily isolated High rate of proliferation Hypoxia-resistant Autologous	High incidence of arrhythmias
Bone marrow-derived stem cells Endothelial progenitor cells Hematopoietic stem cells Mesenchymal stem cells	Autologous Easily isolated Multipotent Low immune response	Limited availability Cases of bone or cartilage formation in the myocardium
Adipose tissue-derived stem cells	Easily isolated High availability Multipotent Low immune response	Low survival
Cardiac stem cells	Multipotent Autologous	Limited availability
Embryonic stem cells	Pluripotent Easy to expand	Teratogenic Limited availability Host immune response Ethical problems
Induced pluripotent stem cells (iPSC)	Pluripotent Easy to expand Good availability Autologous	Potentially teratogenic Possible oncogenic potential
Fetal cardiomyocytes	Cardiomyocyte phenotype	Limited availability Low survival Host immune response Ethical problems

From C. Gálvez-Montón, C. Prat-Vidal, S. Roura, C. Soler-Botija, A. Bayes-Genis, C. Gálvez-Montón, Cardiac tissue engineering and the bioartificial heart, Rev. Esp. Cardiol. 66 (5) (2013) 391–399 with permission.

During the last years, different methods have been developed to understand the generation of cardiac tissue. Important parameters for the generation of cardiac tissues in in vitro conditions can be represented as [25]:

1. the presence of biological matrix components
2. mechanical and/or electrical stimulation
3. the presence of all cardiac specific cell types
4. a spontaneously occurring or electrically inducible contractile activity.

Currently, several methods are available to create heart muscles in combination with human embryonic stem cells, human induced pluripotent stem cells or human heart muscles. Among the main tissue engineering approaches are hydrogel techniques, prefabricated matrices, cell sheets and decellularized heart tissue. The hydrogel method was originally designed for in vitro testing and has the widest range of applications in this field due to straightforward measurements of force and other biophysical parameters of heart muscle function. In addition it can simply be miniaturized and automated [22].

9.3.3 Cell–Hydrogel Injection Method

Hydrogels are polymeric chain networks and made of natural or synthetic polymers with water insoluble and superabsorbent features. Due to these characters hydrogels are known as an alternative for the damaged ECM of cardiac system. Hydrogel are delivered by direct injection or by catheter methods through cardiac blood vessels.

Fig. 9.6 shows three methods of cell injections into cardiac muscles, including intramyocardial injection and catheter techniques. Fig. 9.6A illustrates the intracoronary infusion method, in this method a balloon infusion catheter is guided through the aorta into the coronary vessels and the cell–hydrogel is transported through the designated area. Fig. 9.6B shows transendocardial delivery through a catheter from the aortic valve path. The injection catheter injects the cell–hydrogel directly into the infarct and/or borderline areas. Fig. 9.6C shows the intramyocardial injection. This method needs surgical access and the cell–hydrogel is directly injected into ischemic tissue [26].

9.3.4 Decellularization Method

In this method the ECM of tissue is isolated from cells and the end product of this process is an ECM scaffold of the original tissue. This product can be used in tissue engineering or implantable artificial organs.

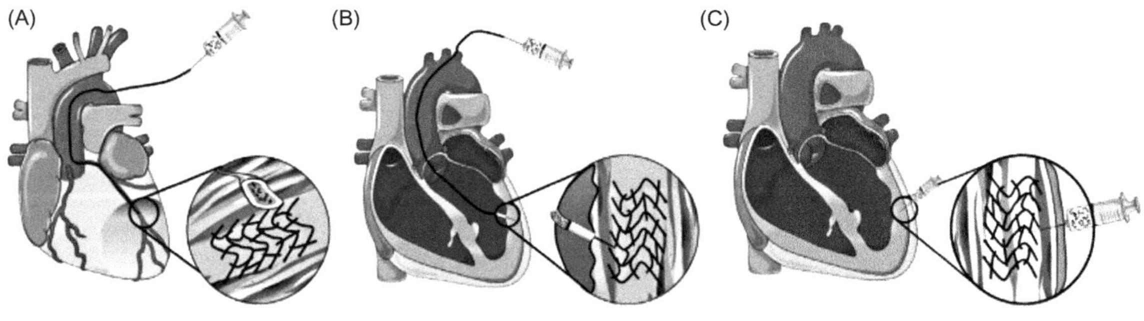

■ **FIGURE 9.6** Schematic of delivery approaches for cell–hydrogel delivery and injection. (A) Intracoronary (IC) infusion method, (B) transendocardial by catheter and (C) intramyocardial injection. *Adapted from E.C. Perin, L.W. Miller, D. Taylor, J.T. Willerson, Stem Cell and Gene Therapy for Cardiovascular Disease, Elsevier Academic Press, 2015 with permission.*

The decellularization technique is a combination of physical, chemical and enzymatic procedures. Generally, the technique begins with lysis of the cell membrane using physical treatments or ionic solutions. Then, an enzymatic solution is used to separate cellular components from the ECM. To solubilize the cytoplasmic and nuclear cellular components, clinicians use detergents. In the last part, cellular debris is removed from the tissue. A mechanical agitation can be coupled to increase these stages. All residual chemicals must be removed to prevent an adverse host tissue reaction to the chemicals [27]. Fig. 9.7 shows the decellularization and recellularization techniques and their procedure. During the recellularization procedure, the scaffold is seeded with cells, and will be ready for the next stage to form a practical organ.

Tissue engineering is one of the new engineering fields in applied sciences and has numerous hidden questions to answer. At present, except for a few applicable clinical and research works undertaken for skin, heart valves and trachea, the other organs and tissues are still under research to produce the implantable/transplantable engineered tissue.

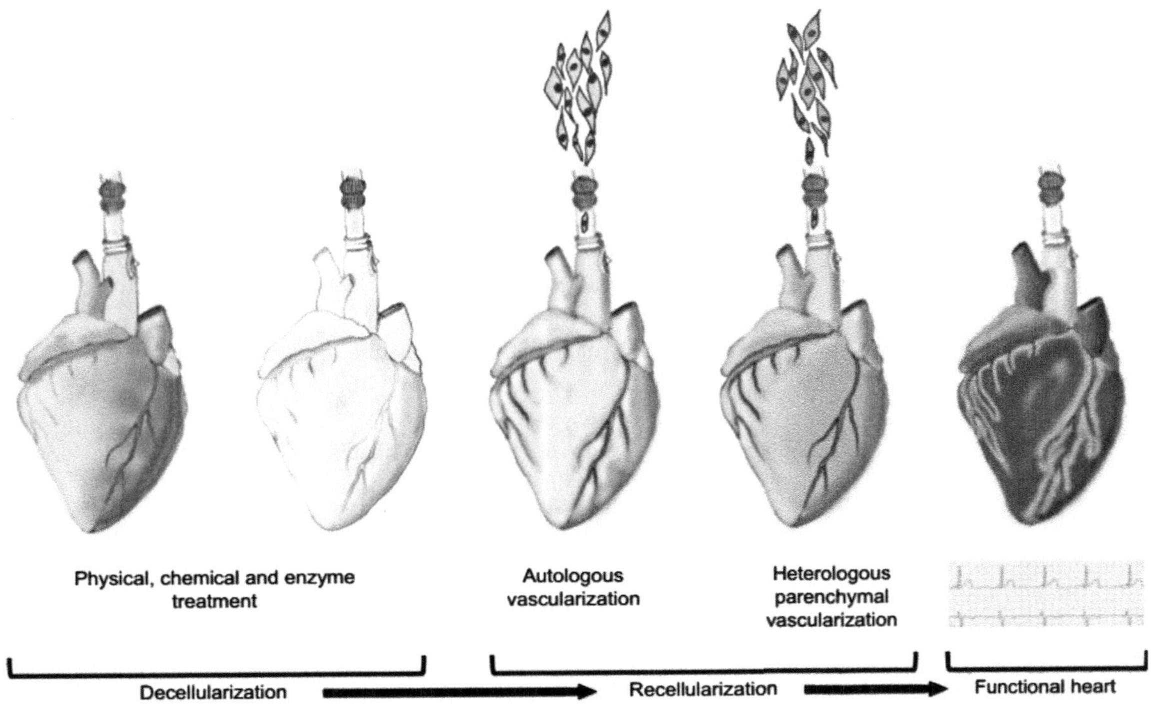

■ **FIGURE 9.7** Schematic representation of the different stages in the decellularization and recellularization of a heart. *Reprinted from C. Gálvez-Montón, C. Prat-Vidal, S. Roura, C. Soler-Botija, A. Bayes-Genis, Cardiac tissue engineering and the bioartificial heart, Rev. Esp. Cardiol. 66 (5) (2013) 391–399 with permission.*

CHAPTER SUMMARY

- Tissue engineering can generate new tissues using human tissues, animal tissues, a combination of human/animal-synthetic materials or totally synthetic materials.
- In tissue engineering, original cells are extracted from a tissue source, then the cells are expanded in culture. In the next stage the cells will be located into scaffold and will be prepared for implantation, and the last stage is implantation into the human body.
- Micro- and nanofabrication techniques can be classified into two areas: bottom-up or top-down.
- Top-down is the traditional method and involves seeding cells into porous scaffolds to create constructs of tissue.
- The bottom-up method involves cell-laden modules forming larger structures, also it involves assembling small without any limitation for diffusion, so the bottom-up method eliminates the deficiencies of the traditional (top-down) method.
- The most important aspect of tissue-engineered blood vessel is its functionality immediately after implantation. After implantation, the engineered blood vessels have to perform similarly to native vessels and keep their functional characteristics for several years.
- Ideal properties of a tissue-engineered blood vessel can be classified as: biological; biophysical; biomechanical and commercial.
- In the procedure of blood vessel tissue engineering, the operational stages are cell sources, cell culture, scaffolds, vessel bioreactors and implantation.
- Nonimmunogenic autologous endothelium cells and smooth muscle cells (tunica media area) isolated from patients are the first choice for vessel tissue engineering.
- A 3D scaffold is the main structure of engineered tissue and creates a pattern to support cell growing, differentiation, ECM proteins secretion, etc.
- The scaffolds should be biocompatible, biodegradable and able to absorb inside the bioenvironments.
- Many different scaffolds can be used for tissue-engineered blood vessels
  - Natural scaffolds
  - Permanent synthetic scaffolds
  - Biodegradable synthetic scaffolds
  - Nonscaffold-based tissue engineered blood vessels
- Vessel bioreactors were designed and used to mimic the physiological aspects (stress, strain, pulsatile flow and pressure) for engineered

blood vessels would be undergone by a native blood vessel in real cardiovascular conditions.

- In the implantation stage, the engineered blood vessel will be implanted inside a patient's body and it will be observed by image processing methods for any damage or abnormal conditions.
- In tissue engineering of heart valves, there are three common techniques for scaffold formation: bioscaffold; synthetic materials and preseeded composites.
- The majority of investigators suggest that synthetic materials are the most suitable method for the formation of a heart valve scaffold.
- One of the methods to fabricate the scaffold is stereolithographic. In this technique, first, a CT device scans a human heart valve homograft, then the data will be used to construct a 3D stereolithographic model. Based on this model, the scaffold will be fabricated.
- One of traditional ideas in the tissue engineering of the heart is the transplant or injection of normal and native heart muscle tissues (cell–hydrogel) to restore the normal function for hearts damaged which have malfunction due to myocardial infarction.
- Important parameters for the generation of cardiac tissues in in vitro conditions can be represented as:
  - the presence of biological matrix components
  - mechanical and/or electrical stimulation
  - the presence of all cardiac specific cell types
  - a spontaneously occurring or electrically inducible contractile activity.
- The decellularization technique begins with lysis of the cell membrane. Then, an enzymatic solution is used to separate cellular components from the ECM. To solubilize the cytoplasmic and nuclear cellular components, clinicians use detergents. In the last part, cellular debris is removed from the tissue.

PROBLEMS

1. Explain the general procedures of tissue engineering.
2. Discuss and compare two fabrication techniques; bottom-up or top-down.
3. Which cell sources may be used in tissue engineering of the blood vessels?
4. Which aspects of scaffolds are the most important in tissue engineering?
5. Discuss the advantages and disadvantages of scaffold types.

6. Explain the fabrication process of tissue engineered heart valve.
7. Explain cell–hydrogel injection and decellularization methods in tissue engineering, which method do you recommend and why? Discuss their application.

REFERENCES

[1] E. Bell, H.P. Ehrlich, D.J. Buttle, T. Nakatsuji, Living tissue formed in vitro and accepted as skin-equivalent tissue of full thickness, Science 211 (4486) (1981) 1052–1054.

[2] S.T. Boyce, Chap. 4, Cultured skin for wound closure, Skin Substitute Production by Tissue Engineering: Clinical and Fundamental Applications, Landes Bioscience, Austin, TX, 1997.

[3] A.L. Caplan, Fundamentals of stem cells tissue engineering, The Biomedical Engineering Handbook: Tissue Engineering and Artificial Organs, third ed., Taylor & Francis Group, LLC, Boca Raton, FL, 2006, pp. 62–1.

[4] D. Kuraitis, M. Ruel, E.J. Suuronen, Mesenchymal stem cells for cardiovascular regeneration, Cardiovasc. Drugs Ther. 25 (4) (2011) 349–362.

[5] N. Bursac, Cardiac tissue engineering: matching native architecture and function to develop safe and efficient therapy, The Biomedical Engineering Handbook, Tissue Engineering and Artificial Organs, Taylor & Francis Group CRC press, Boca Raton, FL, 2006, pp. 56–1, 56–24.

[6] L.S. Nair, C.T. Laurencin, Biodegradable polymers as biomaterials, Prog. Polym. Sci. 32 (8–9) (2007) 762–798.

[7] D. Coutinho, P. Costa, N. Neves, M.E. Gomes, R.L. Reis, Micro- and nanotechnology in tissue engineering, Tissue Engineering, From Lab to Clinic, Springer-Verlag Berlin Heidelberg, 2011, pp. 3–30.

[8] R. Tiruvannamalai-Annamalai, D.R. Armant, H.W. Matthew, A glycosaminoglycan based, modular tissue scaffold system for rapid assembly of perfusable, high cell density, engineered tissues, PLoS One 9 (1) (2014) 1–15.

[9] M. Lovett, K. Lee, A. Edwards, D.L. Kaplan, Vascularization strategies for tissue engineering, Tissue Eng. Part B Rev. 15 (2009) 353–370.

[10] A.D. van der Meer, A.A. Poot, M.H.G. Duits, J. Feijen, I. Vermes, Microfluidic technology in vascular research, J. Biomed. Biotechnol. 2009 (2009) 823148. Available from: http://dx.doi.org/10.1155/2009/823148.

[11] S. Sorrentino, H. Haller, Tissue engineering of blood vessels: how to make a graft, in: Tissue Engineering From Lab to Clinic, Springer-Verlag Berlin Heidelberg, 2011, pp. 263–278.

[12] S.T. Rashid, H.J. Salacinski, B.J. Fuller, G. Hamilton, M. Seifalian, Engineering of bypass conduits to improve patency, Cell Prolif. 37 (5) (2004) 351–366.

[13] N. L'Heureux, S. Pâquet, R. Labbé, L. Germain, F.A. Auger, A completely biological tissue-engineered human blood vessel, FASEB J. 12 (1) (1998) 47–56.

[14] L.E. Niklason, J. Gao, W.M. Abbott, K.K. Hirschi, S. Houser, R. Marini, et al., Functional arteries grown in vitro, Science 284 (5413) (1999) 489–493.

[15] A. Hasan, A. Memic, N. Annabi, M. Hossain, A. Paul, M.R. Dokmeci, et al., Electrospun scaffolds for tissue engineering of vascular grafts, Acta Biomater. 10 (1) (2014) 11–25.

[16] W. He, A. Nieponice, L. Soletti, Y. Hong, B. Gharaibeh, M. Crisan, et al., Pericyte-based human tissue engineered vascular grafts, Biomaterials 31 (32) (2010) 8235–8244.

[17] A. Patel, B. Fine, M. Sandig, K. Mequanint, Elastin biosynthesis: the missing link in tissue-engineered blood vessels, Cardiovasc. Res. 71 (1) (2006) 40–49.

[18] A. Hasan, A. Paul, A. Memic, A. Khademhosseini, A multilayered microfluidic blood vessel-like structure, Biomed. Microdevices 17 (5) (2015) 88. Available from: http://dx.doi.org/10.1007/s10544-015-9993-2.

[19] R.A. Rippel, H. Ghanbari, A.M. Seifalian, Tissue-engineered heart valve: future of cardiac surgery, World J. Surg. 36 (7) (2012) 1581–1591.

[20] R. Sodian, C. Lueders, L. Kraemer, W. Kuebler, M. Shakibaei, B. Reichart, et al., Tissue engineering of autologous human heart valves using cryopreserved vascular umbilical cord cells, Ann. Thorac. Surg. 81 (6) (2006) 2207–2216.

[21] K. Rakusan, M.F. Flanagan, T. Geva, J. Southern, R. Van Praagh, Morphometry of human coronary capillaries during normal growth and the effect of age in left ventricular pressure-overload hypertrophy, Circulation 86 (1) (1992) 38–46.

[22] M.N. Hirt, A. Hansen, T. Eschenhagen, Cardiac tissue engineering: state of the art, Circ. Res. 114 (2) (2014) 354–367.

[23] W. Bian, B. Liau, N. Badie, N. Bursac, Mesoscopic hydrogel molding to control the 3D geometry of bioartificial muscle tissues, Nat. Protoc. 4 (10) (2009) 1522–1534.

[24] C. Gálvez-Montón, C. Prat-Vidal, S. Roura, C. Soler-Botija, A. Bayes-Genis, Cardiac tissue engineering and the bioartificial heart, Rev. Esp. Cardiol. 66 (5) (2013) 391–399.

[25] A. Haverich, M. Wilhelmi, Heart and cardiovascular engineering, Tissue Engineering From Lab to Clinic, Springer, Berlin Heidelberg, 2011, pp. 317–333.

[26] E.C. Perin, L.W. Miller, D. Taylor, J.T. Willerson, Stem Cell and Gene Therapy for Cardiovascular Disease, Elsevier Academic Press, Waltham, MA, 2015.

[27] T.W. Gilbert, T.L. Sellaro, S.F. Badylak, Decellularization of tissues and organs, Biomaterials 27 (19) (2006) 3675–3683.

Appendix

A CONVERSION FACTORS AND UNIT

A.1 Main Units of Measurement Systems

SI or Metric	Imperial and United States
Length: meter (m)	Length: foot (ft)
Mass: kilogram (kg)	Mass: pound (lb)
Time: seconds (s)	Time: seconds (s)

A.2 Special Constants

$\pi = 3.14159\ldots$

$e = 2.71828\ldots$

$1° = \pi/180$ radians $= 0.01745\ldots$ radians

$1° = 60' = 3600''$

1 radian $= 180°/\text{p} = 57.29577\ldots°$

g (Gravity acceleration) $= 9.80655$ m/s^2 $= 32.174$ ft/s^2

Avogadro constant: $N_a = 6.02214 \times 10^{23}$ per mol

A.3 Conversion Factors

Length	Mass
1 m = 100 cm	1 kg = 1000 g
1 cm = 10 mm	1 kg = 2.205 lb
1 mm = 10^{-3} m	1 kg = 35.274 oz
1 μm = 10^{-6} m	1 lb = 454 g
1 nm = 10^{-9} m	1 slug = 32.17 lb
1 m ≈ 3.280 ft	
1 ft = 12 in	
1 in ≈ 2.54 cm	

Continued

Continued

Volume	Pressure
1 liter (L) = 0.001 m^3	1 atm = 1.01325 bar
1 L = 1000 cm^3	1 atm = 14.696 psia
1 L = 1,000,000 mm^3	1 psia = 6.89476 kPa
1 L = 1000 mL	1 atm = 1.01325 × 10^5 Pa
1 L = 1,000,000 μL	1 bar = 10^5 Pa
1 L = 1,000,000 mm^3	1 dyn/cm^2 = 1 μbar
1 L ≈ 0.035314 ft^3	1 μbar = 0.1 Pa
1 L ≈ 61.02374 in^3	1 μbar = 2.089 lbf/ft^2
	1 mmHg = 133.3 Pa
	1 Torr = 1 mmHg
	1 in H_2O = 249.1 Pa
	1 in Hg = 70.73 lbf/ft^2
Force	**Energy**
1 N = 10^5 dyn	1 W (J/s) ≈ 0.23884 cal/s
1 lbf = 4.448 N	1 cal/s = 4.1868 W
1 dyn = 2.248 × 10^{-6} lbf	1 Btu = 252 cal
	1 Btu = 778 ft lbf
	1 kW h = 3413 Btu
	1 horse power (hp) ≈ 178.107 cal/s
Kinematic Viscosity	**Dynamic Viscosity**
1 Stoke (St) = 10^{-4} m^2/s = 1 cm^2/s	1 Poise (P) = 10^{-2} centi P = 1 dyn-s/cm^2
1 S = 0.36 m^2/h = 100 mm^2/s	1 P = 0.1 N-s/m^2 = 0.1 kg/m-s
1 St = 100 centi St	1 P = 0.0278 kg/m-h
1 St = 3.88 ft^2/h = 1.0764 × 10^{-3} ft^2/s	1 P = 98.1 kg_f-s/m^2
1 St = 558 in^2/h = 0.155 in^2/s	1 P = 6.89 × 10^4 lb_f-s/in^2
	1 P = 479 lb_f-s/ft^2
	1 P = 14.9 lb/ft-s
Flow Rate	**Temperature**
1 L/s = 0.264172 gal/s (United States)	Fahrenheit, °F
1 gal/s = 230.999 in^3/s	Celsius, °C
1 in^3/s = 0.01638 L/s	Kelvin, K
1 cubic foot/min (cfm) = 28.3168 L/min	Rankine, °R
1 cfm = 0.028 m^3/min	
	°F = °C × 1.8 + 32
	K = °C + 273.15
	°R = °F + 460

B REVIEW OF CALCULUS AND DIFFERENTIAL EQUATIONS

B.1 Special Products and Factors

$(x+y)^2 = x^2 + 2xy + y^2$ (see Fig. B1 for the calculation technique)

$$(x-y)^2 = x^2 - 2xy + y^2$$

$$(x+y)^3 = x^3 + 3x^2y + 3xy^2 + y^3$$

$$(x-y)^3 = x^3 - 3x^2y + 3xy^2 - y^3$$

$$(x+y)^4 = x^4 + 4x^3y + 6x^2y^2 + 4xy^3 + y^4$$

$$(x-y)^4 = x^4 - 4x^3y + 6x^2y^2 - 4xy^3 + y^4$$

$$(x+y)^5 = x^5 + 5x^4y + 10x^3y^2 + 10x^2y^3 + 5xy^4 + y^5$$

$$(x-y)^5 = x^5 - 5x^4y + 10x^3y^2 - 10x^2y^3 + 5xy^4 - y^5$$

$$(x^2 - y^2) = (x + y)(x - y)$$

$$(x^3 - y^3) = (x - y)(x^2 + xy + y^2)$$

$$(x^3 + y^3) = (x + y)(x^2 - xy + y^2)$$

$$(x^4 - y^4) = (x - y)(x + y)(x^2 + y^2)$$

$$(x^5 - y^5) = (x - y)(x^4 + x^3y + x^2y^2 + xy^3 + y^4)$$

$$(x^5 + y^5) = (x + y)(x^4 - x^3y + x^2y^2 - xy^3 + y^4)$$

Factorial:

$$n! = n(n-1)(n-2)\cdots 3 \times 2 \times 1$$

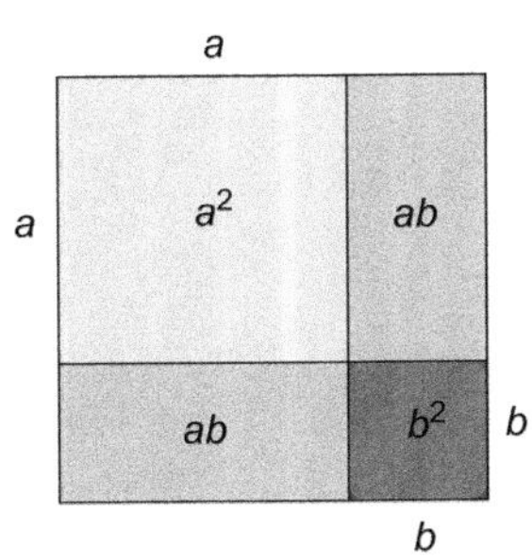

FIGURE B1 Visual illustration to show the calculation technique for $(x+y)^2 = x^2 + 2xy + y^2$.

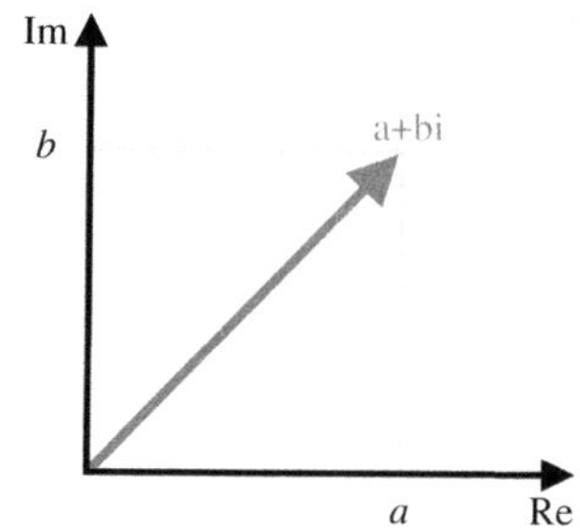

FIGURE B2 Schematic of a complex number vector with real and imaginary parts.

where $n = 1, 2, 3, \ldots,$

$$0! = 1$$

Binomial coefficients:

$$\binom{n}{k} = \frac{n!}{k!(n-k)!}$$

Complex numbers:

Generally, a complex number is written in the form of

$$z = a + bi$$

where a and b are real numbers and i is an imaginary unit and equal to $\sqrt{-1}$. In the equation, bi is called an imaginary part. Fig. B2 shows a vector diagram of a complex number and a real (Re) and imaginary (Im) axis.

B.2 Geometry

Areas:

Square

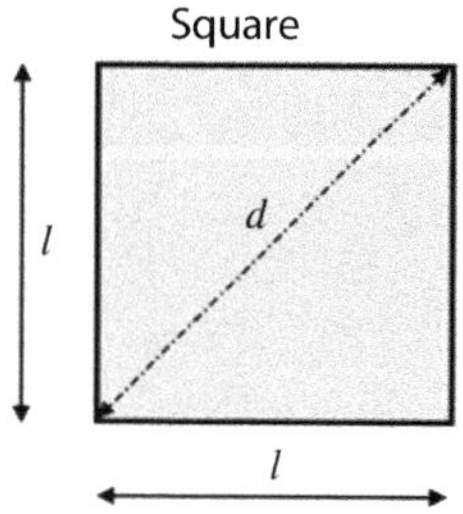

A: area, *d*: diagonal length, *l*: length of side

$$A = l^2$$
$$d = \sqrt{2}l$$

Continued

Continued

Shape	Definitions and formulas
Rectangle 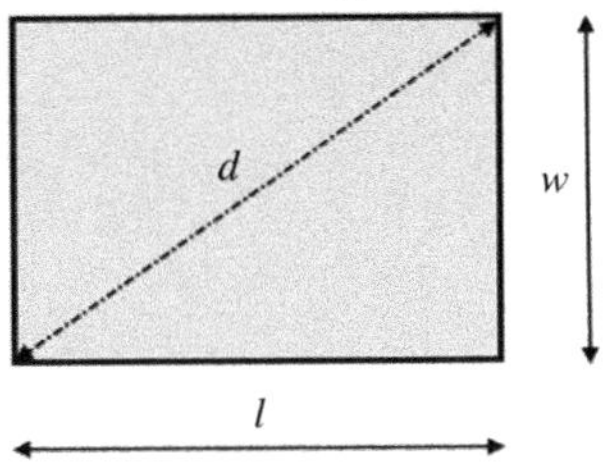	*A*: area, *w*: width, *l*: length of side, *d*: diagonal length $A = l \cdot w$ $d = \sqrt{l^2 + w^2}$
Rhombus 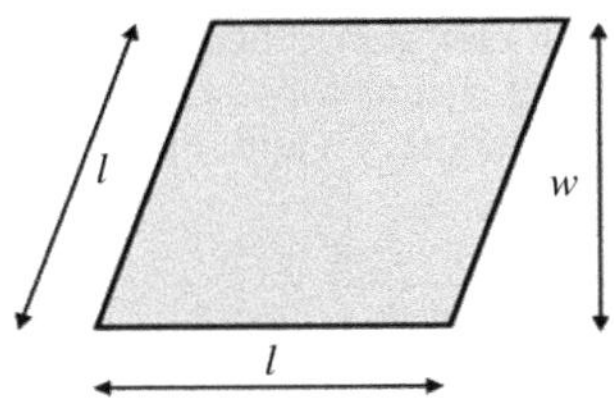	*A*: area, *w*: width, *l*: length of side $A = l \cdot w$
Rhomboid 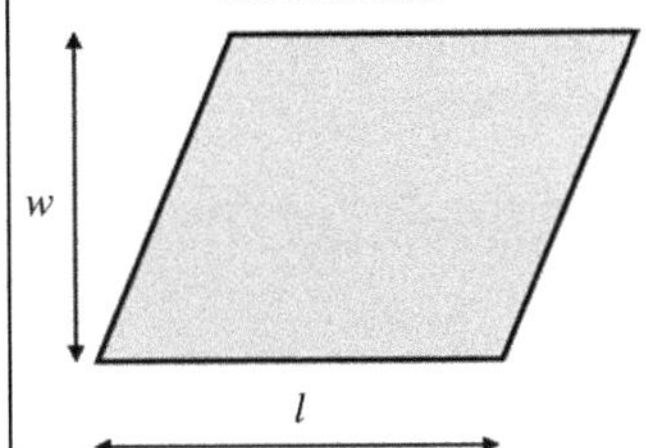	*A*: area, *w*: width, *l*: length of side $A = l \cdot w$
Trapezoid 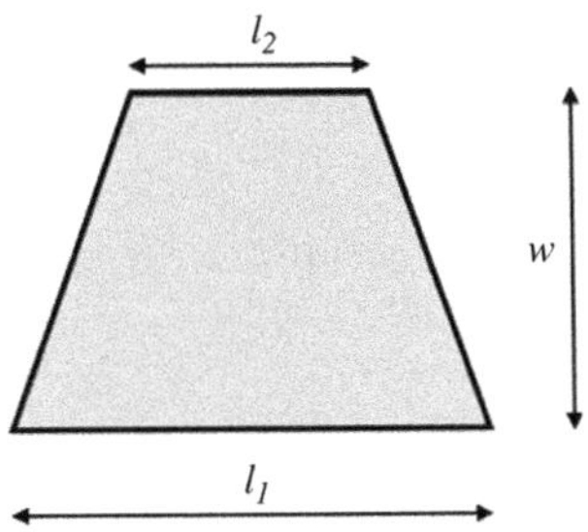	*A*: area, *w*: width, l_1: longer length, l_2: shorter length $A = \frac{l_1 + l_2}{2} \cdot w$
Triangle 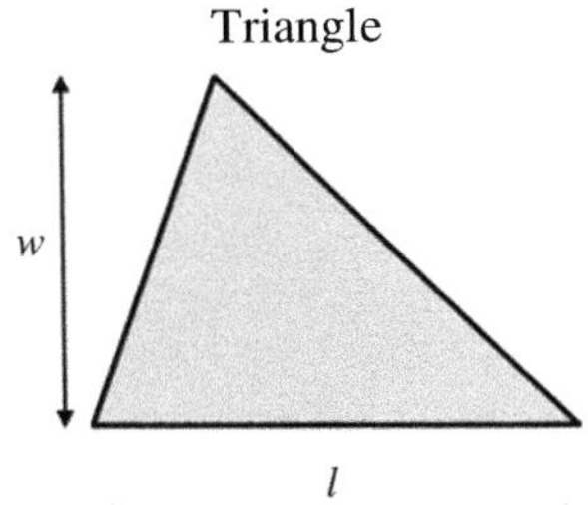	*A*: area, *w*: width, *l*: length of side $A = \frac{l \cdot w}{2}$

Continued

Continued

Equilateral triangle

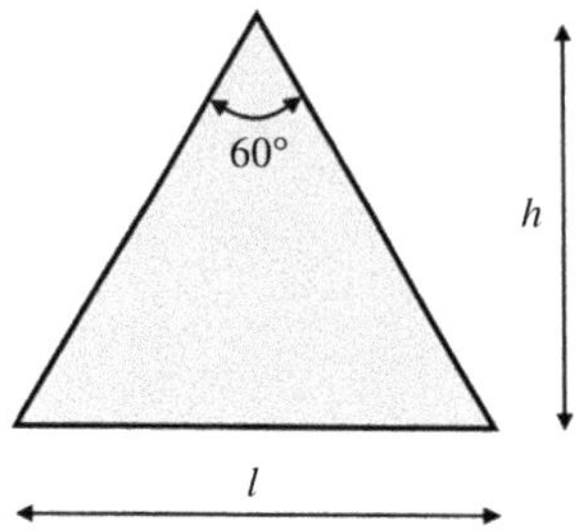

A: area, *h*: height, *l*: length of side

$$A = \frac{\sqrt{3}}{4} \cdot l^2$$

$$h = \frac{\sqrt{3}}{2} \cdot l$$

Regular polygons

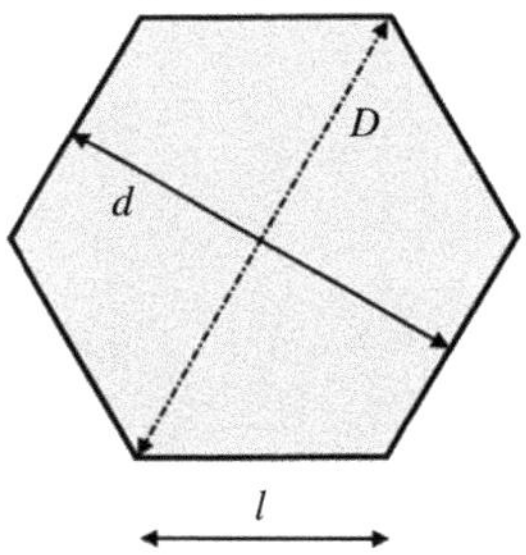

A: area, *d*: diameter of inscribed circle, *D*: diameter of circumscribed circle, *n*: number of vertices, *l*: length of side

$$A = \frac{n \cdot l \cdot d}{4}$$

$$D^2 = d^2 + l^2$$

$$l = D \cdot \sin\left(\frac{180°}{n}\right)$$

Circle

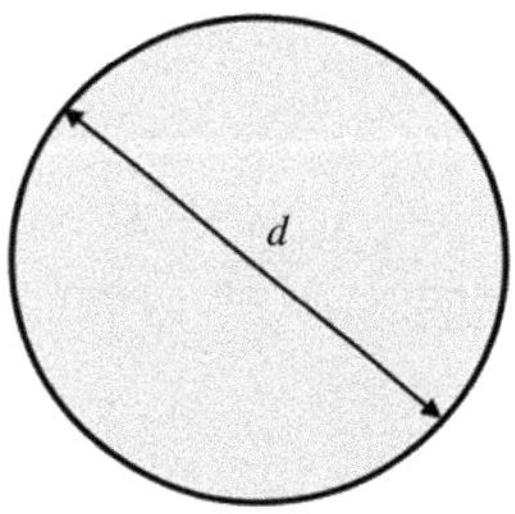

A: area, *d*: diameter, *C*: circumference

$$A = \frac{\pi \cdot d^2}{4}$$

$$C = \pi \cdot d$$

Circular sector

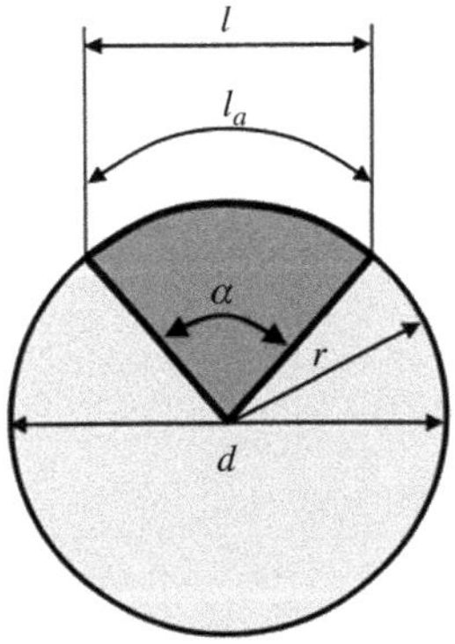

A: area, *d*: diameter, *r*: radius, l_a: arc length, *l*: cord length, α: angle

$$A = \frac{\pi \cdot d^2}{4} \cdot \frac{\alpha}{360°}$$

$$A = \frac{l_a \cdot r}{2}$$

$$l = 2 \cdot r \cdot \sin\frac{\alpha}{2}$$

$$l_a = \frac{\pi \cdot r \cdot \alpha}{180°}$$

Continued

Continued

Circular segment with $\alpha \leq 180°$

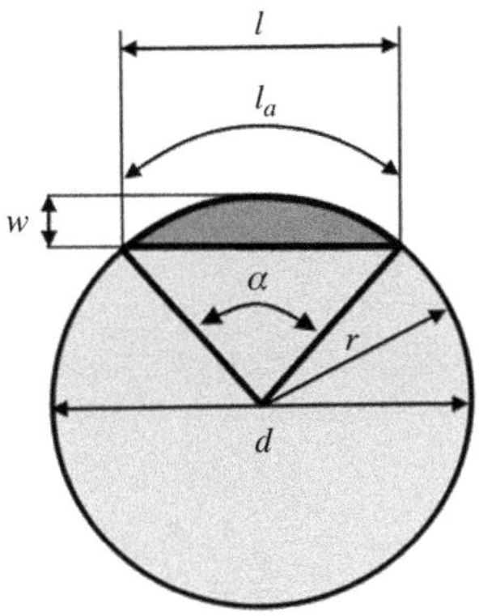

A: area, *d*: diameter, *r*: radius, l_a: arc length, *l*: cord length, α: angle, *w*: width of segment

$$A = \frac{\pi \cdot d^2}{4} \cdot \frac{\alpha}{360°} - \frac{l \cdot (r - w)}{2}$$

$$A = \frac{l_a \cdot r - l \cdot (r - w)}{2}$$

$$l = 2 \cdot r \cdot \sin\frac{\alpha}{2}$$

$$l = 2 \cdot \sqrt{w \cdot (2 \cdot r - w)}$$

$$w = \frac{l}{2} \cdot \tan\frac{\alpha}{4}$$

$$w = r - \sqrt{r^2 - \frac{l^2}{4}}$$

$$l_a = \frac{\pi \cdot r \cdot \alpha}{180°}$$

Circular ring

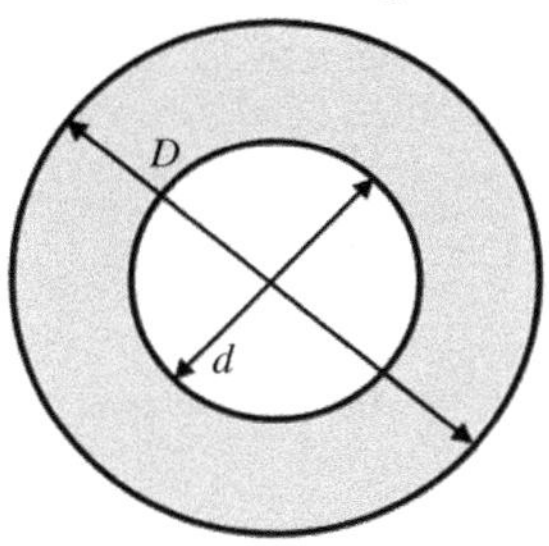

A: area, *d*: inside diameter, *D*: outside diameter

$$A = \frac{\pi \cdot (D^2 - d^2)}{4}$$

Volume:

Cube

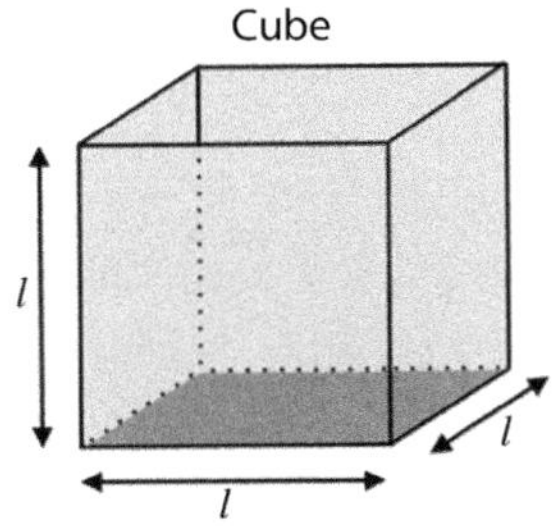

V: volume, *l*: length of side, A_s: surface area

$$V = l^3$$

$$A_s = 6 \cdot l^2$$

Continued

Continued

Rectangular cuboid

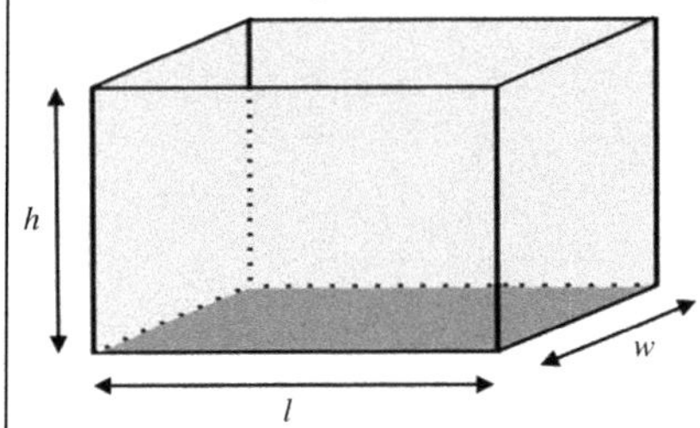

V: volume, l: length of side, w: width, h: height, A_s: surface area

$$V = l \cdot h \cdot w$$
$$V = 2 \cdot (l \cdot w + h \cdot l + h \cdot w)$$

Cylinder

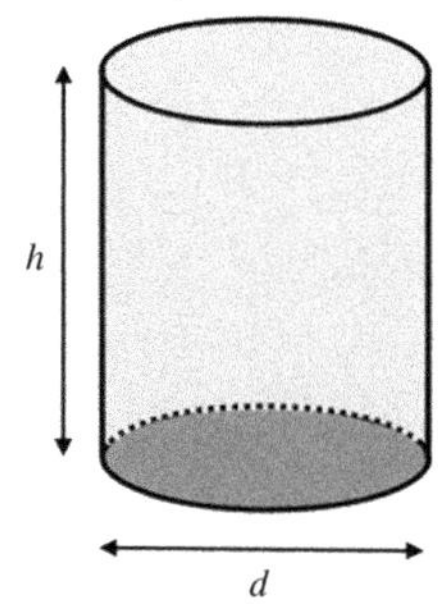

V: volume, d: diameter, h: height, A_s: total surface area, A_l: lateral surface area

$$V = \frac{\pi \cdot d^2}{4} \cdot h$$
$$A_s = \pi \cdot d \cdot h + 2 \cdot \frac{\pi \cdot d^2}{4}$$
$$A_l = \pi \cdot d \cdot h$$

Hollow cylinder

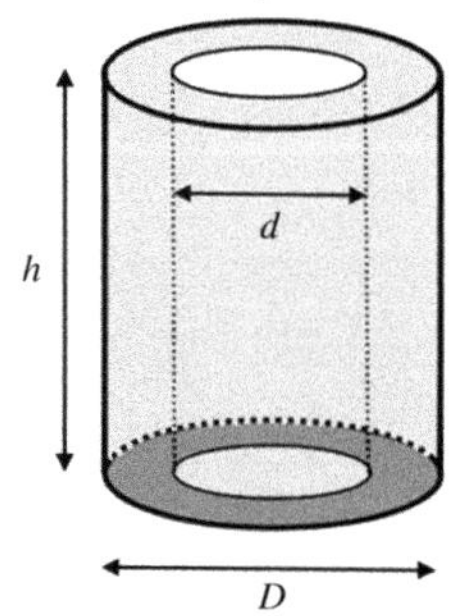

V: volume, d: inside diameter, D: outside diameter, h: height, A_s: surface area

$$V = \frac{\pi \cdot (D^2 - d^2)}{4} \cdot h$$
$$A_s = \pi \cdot (D + d) \cdot \left[\frac{1}{2} \cdot (D - d) + h\right]$$

Pyramid

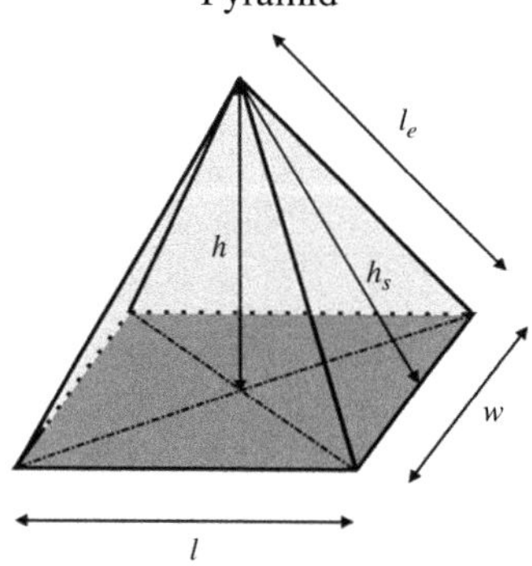

V: volume, h: height, h_s: slant height, l: length of base, w: width of base, l_e: edge length

$$V = \frac{l \cdot w \cdot h}{3}$$
$$l_e = \sqrt{\left(h_s^2 + \frac{w^2}{4}\right)}$$
$$h_s = \sqrt{\left(h^2 + \frac{l^2}{4}\right)}$$

Continued

Continued

Truncated pyramid

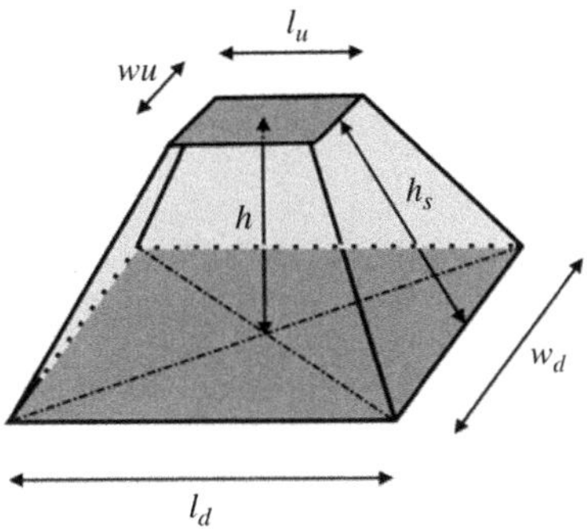

V: volume, h: height, h_s: slant height, l_d: length of base (down), w_d: width of base, l_u: length top, w_u: width of top

$$V = \frac{h}{3} \cdot \left(l_d \cdot w_d + l_u \cdot w_u + \sqrt{l_d \cdot w_d \cdot l_u \cdot w_u} \right)$$

$$h_s = \sqrt{h^2 + \frac{(l_d - l_u)^2}{4}}$$

Cone

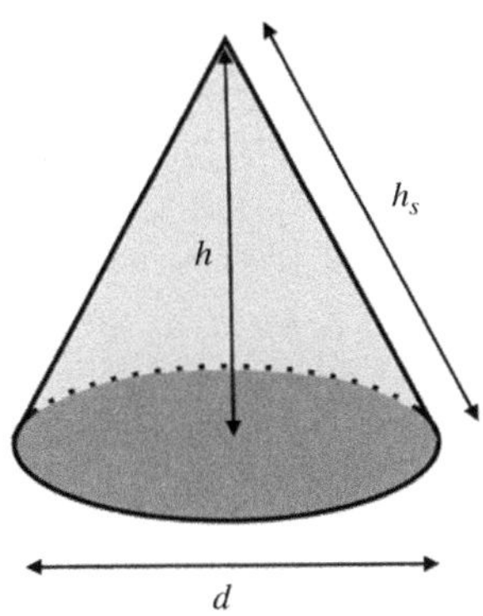

V: volume, h: height, h_s: slant height, d: diameter, A_{cs}: conical surface area

$$V = \frac{\pi \cdot h \cdot d^2}{12}$$

$$A_{cs} = \frac{\pi \cdot d \cdot h_s}{2}$$

$$h_s = \sqrt{\frac{d^2}{4} + h^2}$$

Truncated cone

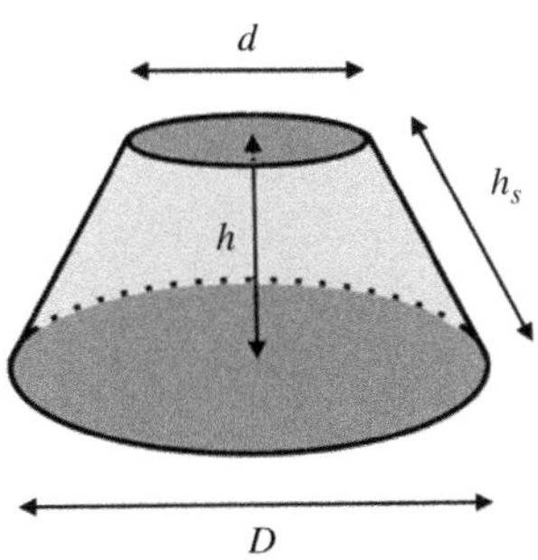

V: volume, h: height, h_s: slant height, d: diameter of top, D: diameter of base, A_{cs}: conical surface area

$$V = \frac{\pi \cdot h \cdot (d^2 + D^2 + d \cdot D)}{12}$$

$$A_{cs} = \frac{\pi \cdot h_s \cdot (D + d)}{2}$$

$$h_s = \sqrt{\frac{(D - d)^2}{4} + h^2}$$

Continued

Continued

Sphere

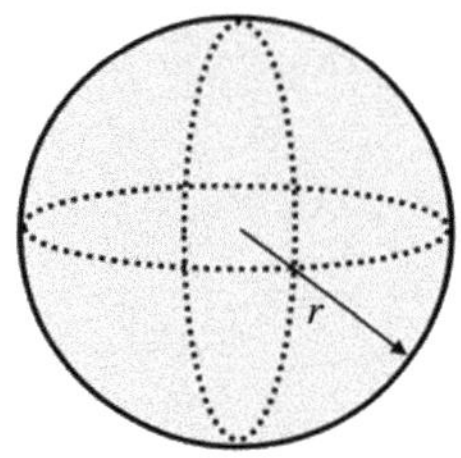

V: volume, r: radius, A_s: surface area

$$V = \frac{4 \cdot \pi \cdot r^3}{3}$$

$$A_s = 4 \cdot \pi \cdot r^2$$

Spherical segment

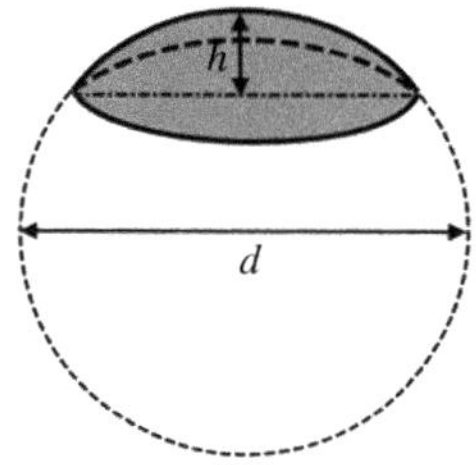

V: volume, d: diameter, h: height, A_s: surface area, A_l: lateral surface area

$$V = \pi \cdot h^2 \cdot \left(\frac{d}{2} - \frac{h}{3}\right)$$

$$A_s = \pi \cdot h \cdot (2 \cdot d - h)$$

$$A_l = \pi \cdot h \cdot d$$

Torus of inner radius a and outer radius b

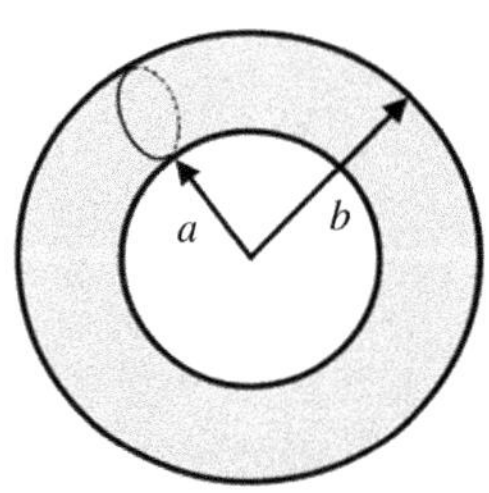

V: volume, A_s: surface area

$$V = \frac{1}{4} \cdot \pi^2 \cdot (a + b) \cdot (b - a)^2$$

$$A_s = \pi^2 \cdot (b^2 - a^2)$$

Ellipsoid of semi-axes a, b, c

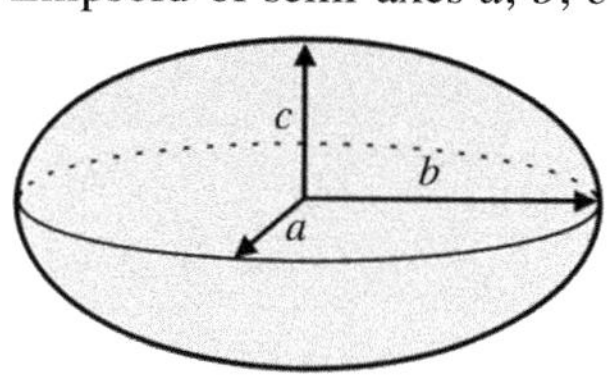

V: volume

$$V = \frac{4}{3} \cdot \pi \cdot a \cdot b \cdot c$$

Paraboloid of revolution

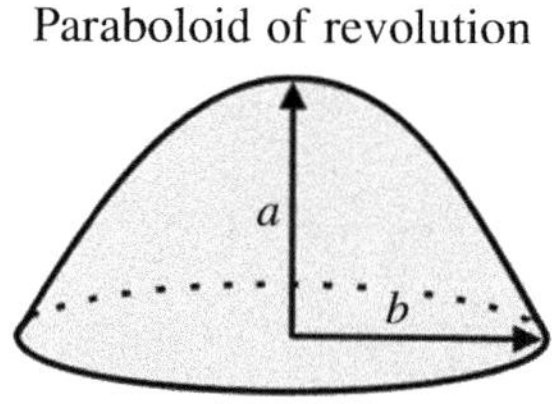

V: volume

$$V = \frac{1}{2} \cdot \pi \cdot a \cdot b^2$$

B.3 Limits

Limit means that real valued function, $f(x)$, can be made as close to Z as desired by making x sufficiently close to b:

$$\lim_{x\to b} f(x) = Z$$

B.4 Differentiation

Differentiation is about rates of change of one element compared to another. The most common formulas for differentiation are:

$$\frac{d}{dx}(cx^n) = ncx^{n-1}$$

$$\frac{d}{dx}(c) = 0$$

$$\frac{d}{dx}(cx) = x$$

$$\frac{d}{dx}(cu) = c\frac{du}{dx}$$

$$\frac{d}{dx}(u \pm v \pm w \pm z \pm \cdots) = \frac{du}{dx} \pm \frac{dv}{dx} \pm \frac{dw}{dx} \pm \frac{dz}{dx} \pm \cdots$$

$$\frac{d}{dx}(uvw) = uv\frac{dw}{dx} + vw\frac{du}{dx} + uw\frac{dv}{dx}$$

$$\frac{d}{dx}\left(\frac{u}{v}\right) = \frac{v(du/dx) - u(dv/dx)}{v^2}$$

$$\frac{d}{dx}(u^n) = nu^{n-1}\frac{du}{dx}$$

$$\frac{dy}{dx} = \frac{dy}{du}\frac{du}{dx}$$

$$\frac{d}{dx}\sin u = \cos u\frac{du}{dx}$$

$$\frac{d}{dx}\cos u = -\sin u\frac{du}{dx}$$

$$\frac{d}{dx}\tan u = \sec^2 u\frac{du}{dx}$$

$$\frac{d}{dx}\cot u = -\operatorname{cosec}^2 u\frac{du}{dx}$$

B.5 Integral

The term of integral may refer to the related idea of the antiderivation. Mathematically, $dy/dx = f(x)$, so y is the function whose derivative is $f(x)$ and can be denoted as the integral of $f(x)$ or $\int f(x)dx$.

In the following common integral formulas, u, v, and w are functions of x and a, b,... are constant:

$$\int x^n dx = \frac{x^{n+1}}{n+1}, \quad \text{if } n \neq -1$$

$$\int x^n dx = \ln|u|, \quad \text{if } n = -1$$

$$\int a\, dx = ax$$

$$\int af(x)dx = a\int f(x)dx$$

$$\int f(ax)dx = \frac{1}{a}\int f(u)du$$

$$\int F\{f(x)\}dx = \int F(u)\frac{dx}{du}du, \quad \text{where } u = f(x)$$

$$\int u\, dv = uv - \int vdu$$

$$\int (u \pm v \pm w \pm \cdots)dx = \int udx \pm \int vdx \pm \int wdx \pm \cdots$$

$$\int e^u du = e^u$$

$$\int \sin u\, du = -\cos u$$

$$\int \cos u\, du = \sin u$$

$$\int \tan u\, du = -\ln \cos u$$

$$\int \cot u\, du = \ln \sin u$$

Definite integral in an interval $a \leq x \leq b$ can be denoted for $f(x) = (d/dx)g(x)$ as:

$$\int_a^b f(x)dx = g(x)\big|_a^b = g(b) - g(a)$$

$$\int_a^a f(x)dx = 0$$

$$\int_a^b f(x)dx = -\int_b^a f(x)dx$$

$$\int_a^b f(x)dx = \int_a^c f(x)dx + \int_c^b f(x)dx$$

B.6 Linear Ordinary Differential Equations

Mathematically, a differential equation is an equation involving a function ($f(x) = y$) and its derivatives (y', y'', y''',...). Derivation can be written as dy/dx for y′, d^2y/dx^2 for y″,....

Considering the first order differential equation as:

$$\frac{dy}{dx} + P(x)y = Q(x)$$

The general solution for this equation is

$$ye^{\int Pdx} = \int Qe^{\int Pdx}dx + c$$

There are two types of equations, homogeneous and nonhomogeneous equations. The nonhomogeneous equations are not considered here.

For linear, homogeneous nth order equations:

$$y^n + a_{n-1}y^{n-1} + \cdots + a_1 y' + a_0 y = 0$$

Considering $y = e^{\lambda x}$, the characteristic equation is:

$$\lambda^n + a_{n-1}\lambda^{n-1} + \cdots + a_1\lambda + a_0 = 0$$

For real numbers of λ, the general solution of the linear, homogeneous nth order equation is:

$$y = c_1 e^{\lambda_1 x} + c_2 e^{\lambda_2 x} + \cdots + c_n e^{\lambda_n x}$$

For nonreal and complex numbers of λ, the term of the complex exponential will be considered, for instance, for a second order equation:

$$\frac{d^2y}{dx^2} + a\frac{dy}{dx} + by = 0$$

Both a and b are real constants. The λ's can be written as

$$\lambda_1 = p + qi, \quad \lambda_2 = p - qi$$

And the general solution is

$$y = e^{px}(c_1 \cos qx + c_2 \sin qx)$$

where $p = -a/2$, $q = \sqrt{b - (a^2/4)}$.

B.7 Bessel Function

Bessel functions are kind of differential equations and have a key role as a part of the solutions for many problems of wave propagation, such as in the blood circulation system.

Solutions of the following equation are known as the Bessel function of order n:

$$x^2 y'' + xy' + (x^2 - n^2)y = 0, \quad n \geq 0$$

There are two kinds of Bessel function: the first kind and the second kind.

- Bessel functions of the first kind of order n:

$$J_n(x) = \sum_{k=0}^{\infty} \frac{(-1)^k (x/2)^{n+2k}}{k!\Gamma(n+k+1)},$$
$$J_{-n}(x) = (-1)^n J_n(x), \quad n = 0, 1, 2, 3, \ldots$$

  Example; for $n = 0$:

$$J_0(x) = 1 - \frac{x^2}{2^2} + \frac{x^4}{2^2 \cdot 4^2} - \frac{x^6}{2^2 \cdot 4^2 \cdot 6^2} + \cdots$$

- Bessel functions of the second kind of order n:

$$Y_n(x) = \frac{J_n(x) \cos n\pi - J_{-n}(x)}{\sin n\pi}, \quad n \neq 0, 1, 2, 3, \ldots$$

and

$$Y_n(x) = \lim_{p \to n} \frac{J_p(x) \cos p\pi - J_{-p}(x)}{\sin p\pi}, \quad n = 0, 1, 2, 3, \ldots$$

C BASICS OF VECTORS AND TENSORS

C.1 Vectors

Vectors associate a magnitude with a direction to show the properties of physical quantities such as; force, pressure, speed, etc. Vectors follow algebra laws, including commutative, associative for adding, associative for multiplying and distributive laws.

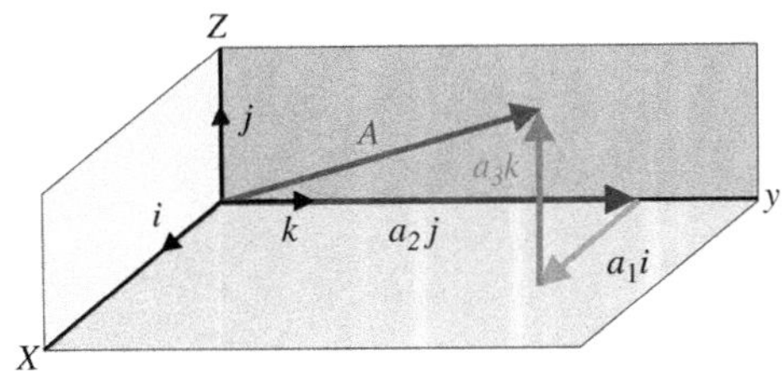

■ **FIGURE C1** Schematic of vector "**A**" and its unit vectors on *xyz* axes in a 3D space.

- commutative: $\mathbf{A} + \mathbf{B} = \mathbf{B} + \mathbf{A}$,
- associative for adding: $\mathbf{A} + (\mathbf{B} + \mathbf{C}) = (\mathbf{A} + \mathbf{B}) + \mathbf{C}$,
- associative for multiplying: $x(y\mathbf{A}) = (xy)\mathbf{A} = y(x\mathbf{A})$ and
- distributive: $(x + y)\mathbf{A} = x\mathbf{A} + y\mathbf{A}$.

A vector in 3D space can be represented as:

$$\mathbf{V} = a_1 i + a_2 j + a_3 k$$

where a_1, a_2, a_3 are the components of $\mathbf{V}$, also i, j, k are the standard unit vectors in the direction of x, y and z axes (Fig. C1).

Vector calculus:

- Scalar (dot) product:

$$A \cdot B = AB \cos\theta$$

where θ is the angle between A and B and its range is $0 \le \theta \le 180°$

$$A \cdot B = B \cdot A$$

$$A \cdot (B + C) = A \cdot B + A \cdot C$$

$$A \cdot B = a_1 b_1 + a_2 b_2 + a_3 b_3$$

- Cross product:

$$A \times B = AB \sin\theta\, n$$

where θ is the angle between A and B and its range is $0 \le \theta \le 180°$, also n is a unit vector which is perpendicular to the AB plane (Fig. C2).

$$A \times B = \begin{vmatrix} i & j & k \\ a_1 & a_2 & a_3 \\ b_1 & b_2 & b_3 \end{vmatrix} = (a_2 b_3 - a_3 b_2)i + (a_3 b_1 - a_1 b_3)j + (a_1 b_2 - a_2 b_1)k$$

$$A \times B = -(B \times A)$$

$$A \times (B + C) = (A \times B) + (A \times C)$$

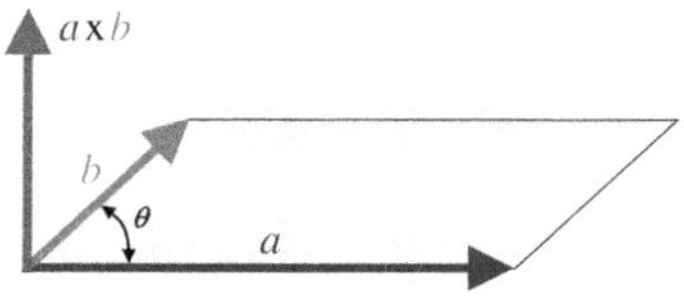

■ **FIGURE C2** Schematic of cross product or $A \times B$.

- *The Del* (∇) *operator*:

  The Del is a vector differential operator in vector calculus. It has a main role in mechanical engineering fields, especially in fluid mechanics and Navier–Stokes equations.

$$\nabla = i\frac{\partial}{\partial x} + j\frac{\partial}{\partial y} + k\frac{\partial}{\partial z}$$

- *The gradient of vector* (∇V):

  The gradient is the change rate of a function. Considering vector **V**, the gradient of the vector, grad V or ∇V, is:

$$\nabla V = \frac{\partial V}{\partial x}i + \frac{\partial V}{\partial y}j + \frac{\partial V}{\partial z}K$$

- *The divergence of vector* (div V):

  To explain in simple words, the divergence is the quantity of flux entering or leaving a point in 3D space. The divergence of vector or div V is:

$$\text{div } V = \nabla \cdot V = \left(i\frac{\partial}{\partial x} + j\frac{\partial}{\partial y} + k\frac{\partial}{\partial z}\right) \cdot (a_1 i + a_2 j + a_3 k) = \frac{\partial a_1}{\partial x} + j\frac{\partial a_2}{\partial y} + k\frac{\partial a_3}{\partial z}$$

- *The curl of vector* (Curl V):

  Curl means circulation per unit of area. For a vector field, the curl is a perpendicular vector to the surface of the field. Mathematically, the curl is:

$$\text{Curl } V = \nabla \times V = \left(i\frac{\partial}{\partial x} + j\frac{\partial}{\partial y} + k\frac{\partial}{\partial z}\right) \times (a_1 i + a_2 j + a_3 k)$$

$$= \begin{vmatrix} i & j & k \\ \dfrac{\partial}{\partial x} & \dfrac{\partial}{\partial y} & \dfrac{\partial}{\partial z} \\ a_1 & a_2 & a_3 \end{vmatrix} = \left(\frac{\partial a_3}{\partial y} - \frac{\partial a_2}{\partial z}\right)i + \left(\frac{\partial a_1}{\partial z} - \frac{\partial a_3}{\partial x}\right)j + \left(\frac{\partial a_2}{\partial x} - \frac{\partial a_1}{\partial y}\right)k$$

- *Laplace operator* ($\nabla^2 V$):

Considering divergence and gradient, the Laplace operator is divergence of gradient.

$$\nabla^2 V = \nabla \cdot (\nabla V) = \frac{\partial^2 V}{\partial x^2} + \frac{\partial^2 V}{\partial y^2} + \frac{\partial^2 V}{\partial z^2}$$

C.2 Tensors

Generally, tensors are defined as a geometric object to describe the relations between vectors and scalar quantities. The tensors can create an organized framework to analyze the physical aspects such as stress, pressure, temperature, etc. The tensors can represent the numerical values as 2D (square) or 3D (cube) objects. One of the most famous tensors in fluid mechanics is the stress tensor (see chapter: Fluid Mechanics and Biofluids Principles).

A tensor rank n in m dimensional space has n indices and m^n components. For instance:

- Scalar: tensor of rank 0 with 1 component;
- Vector: tensor of rank 1 with 3 components;
- Dyad: tensor of rank 2 with 9 components;
- Triad: tensor of rank 3 with 27 components.

D STRESSES IN THIN WALL CYLINDRICAL TUBE

Thin wall cylindrical tubes provide an important application of the plain stress analysis which is useful to analyze the stresses inside blood vessel walls. We used a small element of the thin wall with sides parallel and perpendicular to the axis of the cylindrical tube. Two normal stresses σ_l and σ_h are defined as principle stresses on the element of the tube. The σ_l is known as the longitudinal stress and σ_h is called the hoop stress. Fig. D1 shows schematic of a cylindrical tube, analyzed element and the internal forces (the forces are equal to normal stresses multiply related area).

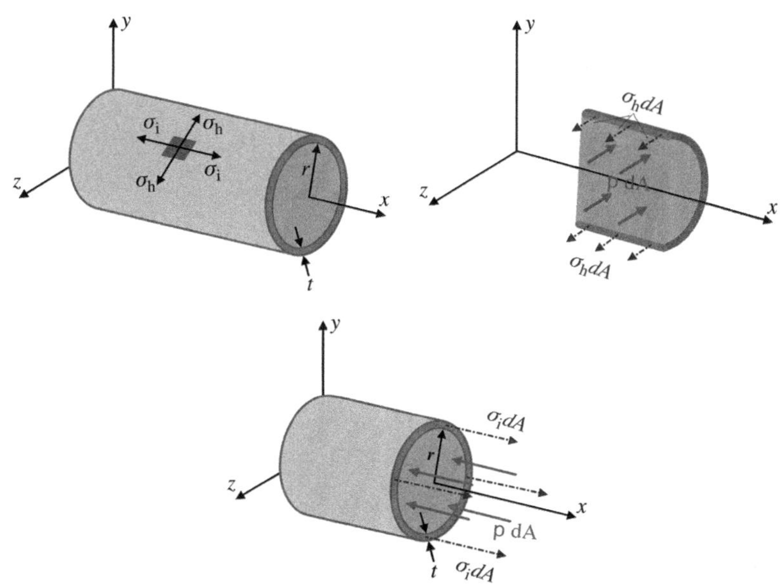

■ **FIGURE D1** Schematic of stresses and internal forces in thin wall cylindrical tube under fluid pressure of p.

Mathematically, the stresses can be written as:

$$\sigma_h = \frac{pr}{t}$$

$$\sigma_l = \frac{pr}{2t}$$

$$\tau_{max} = \sigma_l = \frac{pr}{2t}$$

where p is the fluid pressure through the tube, r is the tube radius, t is the wall thickness of the tube and τ_{max} is the maximum shearing stress.

Index

Note: Page numbers followed by "*f*" and "*t*" refer to figures and tables, respectively.

C

I

9780128024089